SO-FKS-653

Biotechnology of Crucifers

Surinder Kumar Gupta

Editor

Biotechnology of Crucifers

Springer

Editor
Surinder Kumar Gupta
Division of Plant Breeding and Genetics
S.K. University of Agricultural Sciences and Technology
Chatha, India

QK
495
.C9
B56
2013

ISBN 978-1-4614-7794-5   ISBN 978-1-4614-7795-2 (eBook)
DOI 10.1007/978-1-4614-7795-2
Springer New York Heidelberg Dordrecht London

Library of Congress Control Number: 2013941733

© Springer Science+Business Media, LLC 2013
This work is subject to copyright. All rights are reserved by the Publisher, whether the whole or part of
the material is concerned, specifically the rights of translation, reprinting, reuse of illustrations, recitation,
broadcasting, reproduction on microfilms or in any other physical way, and transmission or information
storage and retrieval, electronic adaptation, computer software, or by similar or dissimilar methodology
now known or hereafter developed. Exempted from this legal reservation are brief excerpts in connection
with reviews or scholarly analysis or material supplied specifically for the purpose of being entered and
executed on a computer system, for exclusive use by the purchaser of the work. Duplication of this
publication or parts thereof is permitted only under the provisions of the Copyright Law of the Publisher's
location, in its current version, and permission for use must always be obtained from Springer.
Permissions for use may be obtained through RightsLink at the Copyright Clearance Center. Violations
are liable to prosecution under the respective Copyright Law.
The use of general descriptive names, registered names, trademarks, service marks, etc. in this publication
does not imply, even in the absence of a specific statement, that such names are exempt from the relevant
protective laws and regulations and therefore free for general use.
While the advice and information in this book are believed to be true and accurate at the date of
publication, neither the authors nor the editors nor the publisher can accept any legal responsibility for
any errors or omissions that may be made. The publisher makes no warranty, express or implied, with
respect to the material contained herein.

Printed on acid-free paper

Springer is part of Springer Science+Business Media (www.springer.com)

B3395260

Preface

Despite the recent advances made in the improvement of crucifer crops using conventional breeding techniques, the yield levels and the oil and meal quality that were expected could not be achieved. The understanding of genetic material (DNA/RNA) and its manipulation by scientists have provided the opportunity to improve crucifers by increasing their diversity beyond conventional genetic limitations. The application of biotechnological techniques will have two major benefits: first, it allows to choose from a number of techniques/methods for appropriate selection of favorable variants, and second, it gives an opportunity to utilize alien variation available in the crucifers to develop high-yielding varieties with good nutritional quality and resistance to insects, pests, and diseases.

Realizing the importance of biotechnology, there is an urgent need to update current techniques for enhancing crucifer crop production at the global level. The editor approached the leading scientists of the world for write-ups on the advances made in the area of crucifer biotechnology to be packaged into one volume for the benefit of students, nutritionists, and biotechnologists as well as researchers engaged in the improvement of Brassicas. The book consists of 12 chapters. Chapter 1 deals with the importance, origin, and evolution of Brassicas, while Chaps. 2 and 3 describe the major advances made in cytogenetics at the molecular level and the introgression of genes from wild species. Chapter 4 deals with microspore culture and double haploid technology, while Chap. 5 describes phytoremediation in crucifers. These are followed by chapters on genome analysis (Chap. 6) and genetic engineering of lipid biosynthesis in seeds (Chap. 7). Metabolism and detoxification of crucifer phytoalexins, the molecular basis of cytoplasmic male sterility, and self-incompatibility have been discussed in detail in Chaps. 8, 9, and 10. Chapters 11 and 12 provide brief accounts of the molecular basis of hybrid technology and genetic modifications for pest resistance.

I am highly indebted to Prof. D. K. Arora, Honorable Vice-Chancellor, Sher-e-Kashmir University of Agricultural Sciences and Technology of Jammu, India, for encouraging me to carry out oilseed research work with all required facilities for the same.

I am indeed grateful to Prof. W. J. Zhou, Institute of Crop Science, Hangzhou, China, for providing crucial inputs and critically reviewing some of the chapters. Help rendered by Prof. Y. Takahata, Iwate University, Japan; Prof. Graham King, Southern Cross University, Australia; Prof. Randall Weselake, University of Alberta, Canada; Prof. M. S. C. Pedras, University of Saskatchewan, Canada; Prof. Takeshi Nishio, Tohoku University, Japan; and Prof. M. H. Fulekar, Central University of Gujarat, India, is gratefully acknowledged.

Joni Fraser, Developmental Editor for Springer Science + Business Media, deserves special thanks for her hard work in bringing this book to life. I owe a lot to my wife, Dr. Neena Gupta, for her support and patience during the preparation of this manuscript.

Chatha, India   Surinder Kumar Gupta

Contents

Contributors

Liping Chen, Ph.D. Department of Horticulture, Zhejiang University, Hangzhou, Zhejiang Province, China

Peng Cui, M.S. Institute of Crop Science and Zhejiang Key Laboratory of Crop Germplasm, Zhejiang University, Hangzhou, China

Jyoti Fulekar, M.Phil. School of Environment and Sustainable Development, Central University of Gujarat, Gandhinagar, Gujarat, India

M.H. Fulekar, Ph.D. School of Environment and Sustainable Development, Central University of Gujarat, Gandhinagar, Gujarat, India

Xinxin Geng, B.S. Institute of Crop Science and Zhejiang Key Laboratory of Crop Germplasm, Zhejiang University, Hangzhou, China

Muhammad Awais Ghani, M.S., M.D. Department of Horticulture, Zhejiang University, Hangzhou, Zhejiang Province, China

Michael S. Greer, M.Sc. Agricultural, Food Nutritional Sciences, University of Alberta, Edmonton, Alberta, Canada

Surinder Kumar Gupta, Ph.D. Division of Plant Breeding and Genetics, S.K. University of Agricultural Science and Technology, Chatha, India

Changrong Huang, M.S. Institute of Crop Science and Zhejiang Key Laboratory of Crop Germplasm, Zhejiang University, Hangzhou, China

Razia Khan, M.Phil. School of Environment and Sustainable Development, Central University of Gujarat, Gandhinagar, Gujarat, India

Graham J. King, B.Sc., Ph.D. Southern Cross Plant Science, Southern Cross University, Lismore, Australia

Hiroyasu Kitasiba, Ph.D. Graduate School of Agricultural Science, Tohoku University, Sendai, Miyagi, Japan

Junxing Li, M.D. Department of Horticulture, Zhejiang University, Hangzhou, Zhejiang Province, China

Bin Liu, B.S. Department of Horticulture, Zhejiang University, Hangzhou, Zhejiang Province, China

Dan Liu, Ph.D. Institute of Crop Science and Zhejiang Key Laboratory of Crop Germplasm, Zhejiang University, Hangzhou, China

Institute of Tobacco Research, Chinese Academy of Agricultural Sciences, Qingdao, Shandong, China

Hongbo Liu, Ph.D. Institute of Crop Science and Zhejiang Key Laboratory of Crop Germplasm, Zhejiang University, Hangzhou, China

College of Agricultural and Food Sciences, Zhejiang A&F University, Lin'an, Zhejiang, China

Bizeng Mao, Ph.D. Institute of Biotechnology, Zhejiang University, Hangzhou, China

Annaliese S. Mason, Ph.D., B.Sc. (Hons) School of Agriculture and Food Sciences, The University of Queensland, Brisbane, Australia

Elzbieta Mietkiewska, Ph.D. Agricultural, Food Nutritional Science, University of Alberta, Edmonton, Alberta, Canada

Takeshi Nishio, Ph.D. Graduate School of Agricultural Science, Tohoku University, Sendai, Miyagi, Japan

Xue Pan, M.Sc. Agricultural, Food Nutritional Science, University of Alberta, Edmonton, Alberta, Canada

Bhawana Pathak, Ph.D. School of Environment and Sustainable Development, Central University of Gujarat, Gandhinagar, Gujarat, India

M. Soledade C. Pedras, Lic, Ph.D., D.Sc. University of Saskatchewan, Saskatoon, Saskatchewan, Canada

Stacy D. Singer, Ph.D. Agricultural, Food Nutritional Science, University of Alberta, Edmonton, Alberta, Canada

Yoshihito Takahata, Ph.D. Iwate University, Ueda, Morioka, Iwate, Japan

Yu Takahashi, Ph.D. Iwate University, Ueda, Morioka, Iwate, Japan

Tian Tian, B.S. Institute of Crop Science and Zhejiang Key Laboratory of Crop Germplasm, Zhejiang University, Hangzhou, China

Ryo Tsuwamoto, Ph.D. Misato Agricultural Extension Centre, Miyagi prefecture, Misato-machi, Toda-gun, Miyagi, Japan

Bing Wang, M.S. Institute of Crop Science and Zhejiang Key Laboratory of Crop Germplasm, Zhejiang University, Hangzhou, China

Randall J. Weselake, B.Sc., M.Sc., Ph.D. Agricultural, Food Nutritional Science, University of Alberta, Edmonton, Alberta, Canada

Ling Xu, Ph.D. Institute of Crop Science and Zhejiang Key Laboratory of Crop Germplasm, Zhejiang University, Hangzhou, China

Xi Xu, B.S. Institute of Crop Science and Zhejiang Key Laboratory of Crop Germplasm, Zhejiang University, Hangzhou, China

Jinghua Yang, Ph.D. Institute of Vegetable Science, Hangzhou, Zhejiang Province, China

Langlang Zhang, B.S. Department of Horticulture, Zhejiang University, Hangzhou, Zhejiang Province, China

Mingfang Zhang, Ph.D. Institute of Vegetable Science, Hangzhou, Zhejiang Province, China

Zhenchao Zhang, Ph.D. Institute of Crop Science and Zhejiang Key Laboratory of Crop Germplasm, Zhejiang University, Hangzhou, China

Zhenjiang Agricultural Research Institute, Jurong; Jiangsu, China

Weijun Zhou, Ph.D. Institute of Crop Science and Zhejiang Key Laboratory of Crop Germplasm, Zhejiang University, Hangzhou, China

Agricultural Experiment Station, Zhejiang University, Hangzhou, China

Yuanfei Zhou, Ph.D. Agricultural Experiment Station, Zhejiang University, Hangzhou, China

Chapter 1
The Importance, Origin, and Evolution

Surinder Kumar Gupta

Abstract The family Brassicaceae constitutes one of the world's most economically group of plants which includes important vegetable oilseeds and condiment crops. Amongst the crucifer crops, rapeseed is the main source of fats and oil and shown an upward trend during the past 25years (Kalia and Gupta 1997). Besides improvement in the nutritional profile of the Brassica oil and its meal, the conventional breeding as well modern biotechnological tools have led to the improvement of various agronomically important quantitative and qualitative characters. The nuclear restriction fragment length polymorphism technology has greatly aided in determining the degree of genetic variability among various Brassicas as well in studying their evolution pattern. The oldest references regarding origin and cultivation of rapeseed come from Asia, though the evolution of this crop took place in many countries throughout the globe. Lack of consistency in names, inclusion of too many forms in one species, and the entirely different forms of present day Brassicas from their ancestors make this genus a complex member of Brassicaceae and poses several taxonomic and classification problems. Still many attempts have been made to establish the origin of various Brassica species and their interrelationships through cytogenetic, chemotaxonomic, and molecular studies. The present chapter focuses on the importance origin and evolutionary developments in crucifers.

Keywords *Brassicaceae* • Rapeseed • *B. rapa* • *B. juncea* • *B. carinata* • Origin • Evolution • RFLP • Oilcrops

S.K. Gupta, Ph.D. (✉)
Division of Plant Breeding and Genetics, S.K. University of Agricultural
Science and Technology, Chatha 180009, India
e-mail: guptaskpbg@rediffmail.com

S.K. Gupta (ed.), *Biotechnology of Crucifers*, DOI 10.1007/978-1-4614-7795-2_1,
© Springer Science+Business Media, LLC 2013

1.1   Introduction

Brassicas are the world's third important source of vegetable oils after palm and soya bean (Beckman 2005) and contribute 14 % to the world's total vegetable oil pool. The production has shown a steady upward trend during the past 25 years. Brassica oilseed crops grow at relatively low temperature. In temperate regions of the world, oilseed rape *(B. napus)* and toria /turnip rape/Indian mustard are grown in subtropical parts of the Asia and is the main source of oil. The mode of reproduction varies from species to species. *B. napus, B. juncea and B. carinata* are predominately self- pollinated, although they show some degree of cross- pollination ranging from 5 to 30 %, whereas, *B. rapa, B. oleracea and B. nigra* show cross- pollination due to sporophytic self- incompatibility. All the cultivated Brassica species are highly polymorphic including oilseed crops, root crops, and vegetables such as Chinese cabbage, Broccoli, and Brussels sprouts. These Brassica vegetables are dietary staple food in various parts of the world. However, our discussion in this chapter shall concentrate on the importance and origin of major species of Brassicas.

B. napus and *B. campestris* with both spring and winter type are grown in Canada and Europe. However, in countries like India and China, the production is also shared by other species, viz, *B. oleracea* and *B. juncea*. Rapeseed oil has gradually become important domestic and industrial oil in the western nations as a result of breeding for improved oil and meal quality and better processing techniques.

1.2   History

The family *Brassicaceae* contains over 338 genera and 3,709 species (Al-Shehbaz et al. 2006). The crop Brassicas have been very important as food crops in the form of vegetables, oilseeds, feed and fodder, green manure, and condiments and have played a great role in the human history by contributing a good share of food in one form or another. The Greek, Roman, and Chinese writings of 500–200 BC refer to *rapiferous* forms of *B. rapa* and also described their medicinal values (Downey and Robellen 1989).

Species grown as oilseed crops are *B. napus, B. junca, B. rapa* and *B. carinata*. The vegetable Brassica include *B. napus, B. rapa* (Chinese Cabbage, pak-choi, Chinese mustard, broccoli and kale); *B. oleracea* (cabbage, broccoli, cauliflower, Brussels sprouts, kale, etc.,) *Raphanus sativus* and *Lepidum sativum*, *B. nigra* (black mustard), *B. juncea*, (brown mustard) and *Sinapis alba* are the main condiment of crops.

Early records indicate that Brassicas cultivated for several years in Asia. Seeds *of B. juncea* have been excavated from Chanhundaro, a site of Indus Valley civilization that existed in the plains of Punjab along the river of Indus ca 2300–1750 (Piggot 1950). Species from the genus Brassica were in use and also in Gallia (Fussel 1955) and the seeds of the species had also been found in old German graves and Swiss constructions from the Bronze Age (Neuweiller 1905; Schiemann 1932;

Witmack 1904). In Dodoneus's "Herbalist" (1578), a mention has been made regarding the growing of *B. rapa var. rapifera* in 1470 as a winter crop. In his "Herball," Gerarde (1597) had very clearly differentiated between turnips *(B. rapa)* and navews *(B. napus)*. Rape has been recorded as an oilseed crop in Europe at least since the Middle Ages., but it is still uncertain which species was cultivated (Appelquist and Ohlson 1972).

Domestication of rapeseed in Europe appears to have started in the early Middle Ages, although the true turnip was probably introduced by Romans since many other oil-yielding plants, particularly olive tree, were available in Southern Europe, *B. rapa* initially spread mainly as turnip rape crop within Europe. *B. rapa* had a wide distribution before the recorded history. Indian Sanskrit literature first mentions the plants about 1599 BC as Siddharth (Prakash 1961). Seed of both *B. rapa* and *juncea* were found in the archaeological excavation of ancient village Banpo, China, that existed in Neolithic times 6,000–7,000 years ago (Liu 1985). *B. nigra* (black mustard) is mentioned in Greek literature for its medicinal value. Ancient records indicated the cultivation rape seed was predominant during the thirteenth century. The rapeseed oil is used as major source of lamp oil and it was replaced by petroleum by the end of nineteenth century

A high quality of rapeseed named as Canola developed through genetic modification following the conventional plant breeding. Canola emerged in the 1970s as a viable oilseed, with high quality oil and meal for both human as well as livestock consumption (Shahidi 1990). Today, the fatty acid profile of Canola is considered as the most desirable, of all vegetable oil profiles by nutritionists (Stringam et al. 2003). The occurrence of two important components, glucosinolates and erucic acid were considered antinutritional for animal and humans, respectively. The high amount of glucosinolates in the meal still remained a major concern in the expansion of market of the vegetable oil derived from rapeseed. Prior to 1960, the erucic acid (a long chain fatty acid) content of rapeseed oil was not of particular interest while evaluating the oil use for edible purposes. The concern was felt by the European Economic Community (EEC) in 1960 with France. West Germany, Italy, the Netherland, Belgium, and Luxemburg as the founder members, for the development of low erucic acid varieties (less than 5 %). As a result, the traditional rapeseed oil started being considered as unsafe for human health. This led to the concentration of rapeseed breeding efforts toward the development of such varieties in late 1960s and early 1970. The application of gas liquid chromatography (Craig and Murphy 1959) led to the identification of low erucic acid plants in *B. napus* and *B. campestris* with the first low erucic acid plants in them identified in 1968 and the first *B. campestris* variety in 1971. In 1977, the cultivation of such varieties was made mandatory.

The oilseed Brassica has another important byproduct known as meal/cake. It is an excellent source of protein with a favourable balance of amino acids. However, its use was limited by its high glucosinolate content, which is a constituent of most of the plants of Brassicas. Traditional rapeseed varieties contained high levels of glucosinolates in the meal which when fed to livestock in sufficient quantities led to the problems related with nutrition, digestion, and thyroid. The development of fast

Brassica species
B. nigra
BB
n=8

B. juncea
AABB, n=18

B. carinata
BBCC, n=17

B. oleracea
CC, n=9 ——————→ B. napus ←—————— AA, n=10
AACC, n=19

B. campestris

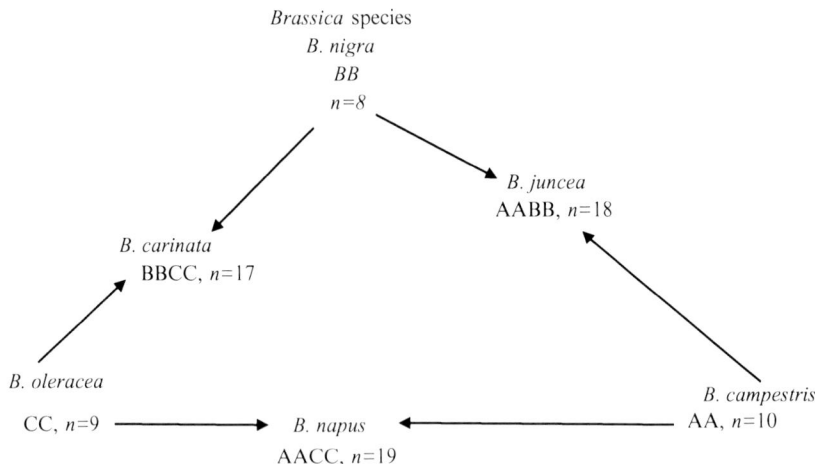

Fig. 1.1 Evolution of cultivated Brassica species and its relatives (Nagaharu 1935)

and accurate chemical methods led to the identification of plants of the B. napus cultivar Bronowski from Poland, which was essentially free of the harmful gluco-sinolates normally found in rapeseed. The low glucosinolate genes were then incorporated in the well adapted and high yielding cultivars of B. napus and subsequently transferred to B. campestris.

The Brassicaceae family comprise of 25 tribes (Al-Shehbaz et al. 2006). The tribe Brassiceae contains genus Brassica and its wild relatives. It comprised of 48 genera and approximate 240 species (Warwick and Hall 2009). Schulz (1919, 1936) established the basic taxonomic classification and he recognized ten sub-tribes whereas Gómez-Campo (1999) recommended 9 subtribes. The subtribes Brassicinae, Moricandiinae and Raphaninae are of the great relevance to the scientists who are working on the Brassica species. The relationship among the species viz., *Brassica, Sinapis, Diplotaxis, Erucastrum, Herschfeldia, Eruca* and *Raphanus*. These sub tribes have been studied by Prakash and Hinata (1980) and Takahata and Hinata (1983, 1986).

Further more during the domestication, man has modified the entire plant and the present day Brassicas are entirely different from their ancestors. Also the occurrence of similar plant forms in more than one Brassica species resulted in considerable confusion and misclassification by early botanists (Downey and Robellen 1989). The cytogenetic relationships between the Brassica species as well as their closest allies were first explained systematically by Nagaharu (1935) about 70 years ago (Fig. 1.1) These relationships show that B. campestris (2n=20, AA), B. nigra (2n=16,BB) and B. oleracea (2n=18,CC) are the primary species and B. napus (2n=38, AACC), B. carinata (2n=34,BBCC) and B. juncea (2n=36, AABB) are the amphidiploids resulting from paired crossings between the primary species. Morinaga (1928, 1929a, b, 1934a, b) discussed that crop Brassicas include six cytodemes, three elementary ones with 16, 18 and 20 chromosomes as diploid and three with higher chromosomes number of 34, 36 and 38 as tetraploid, the latter having evolved through interspecific hybridization in nature between any two of the elementary taxa. Herberd (1972, 1976) defined coeno species and on the basis of their chromosome number, they have been classified into 43 diploid and 13 tetra ploid cytodemes (Warwick and Black 1993).

Morinaga and his associates carried extensive cytogenetic studies in oilseed Brassicas and clarified the relationships between them (Prakash and Hinata 1980). According to the hypothesis of Morinaga and his student Nagaharu (1935), the three species with the higher chromosome number, *B. napus* L, *B juncea* L, Czern and Coss, and *B. carinata* A. Braun, are amphidiploids combining in pairs the chromosome sets of the low chromosome number species *B. nigra*, *B. oleracea* and *B. rapa*. Nagaharu (1935) verified the hypothesis with successful resynthesis of *B. napus*. Resynthesis of *B. juncea* and *B. carinata* was accomplished by Frandsen (1943, 1947). Further verification of these species relationships were obtained from the studies on phenolic compound (Dass and Nybom 1967), protein pattern (Vaughan 1977), isozymes (Coulthart and Denford 1982; Chen et al. 1989), nuclear DNA and RFLP (Song et al. 1988a, b), molecular analysis of nuclear and chloroplast DNA and fluorescence in situ hybridization (Snowdon 2007; Warwick and Sauder 2005; Lysak et al. 2005). Robellen (1960) suggested that the low chromosome number species might have developed from the ancestral species, which could have even lower chromosome numbers. Also the chromosome analysis of the monogenomic species revealed that only six chromosomes were distinctly different, the remaining being homologous with one or another of the basic set of six.

Olsson (1954) suggested that all the 20 chromosome forms of leafy, *oleiferous*, and *rapiferous* Brassicas should be grouped into one species *B. campestris*. This was in support of the views of Howard (1940) that the name *B. campestris* should be reserved for the forms with 2n = 20 and *napus* for the forms with 2n = 38. He proposed that the name sarson and toria should be *B. campestris* L. var. sarson and B. *campestris* L. var. toria, respectively. Singh (1958) considered yellow and brown sarson as varieties whereas Prakash (1973) considered them as the form of subspp. *oleifera*. Toxopens et al. (1984) suggested a classification and nomenclature of *B. campestris* should be changed to *B. rapa*.

1.2.1 **B. rapa**

The name *B. rapa* was mentioned as annual weed by Linnaeus (1973) in "Species Planatarum". It was described as a plant with rough, stiff hairs when young, and just like *B. rapa* by DeCandolle et al. (1824). However, when it was realized that *B. campestris* and the turnip rape *B. rapa* have been classified as same species, a confusion was created in nomenclature and the wild type was often subordinated under *B. rapa* (Reiner et al. 1995). *B. rapa* subspp *campestris* (formerly subspp. *sylvestris*), the wild form of *oleifera* is morphologically indistinguishable from the cultivated spring oilseed rapa. *B rapa* subspp. *dichotoma* commonly referred as toria, is an oilseed crop grown in Indian subcontinent. The yellow sarson and brown sarson (*B. rapa* subspp. *trilocularis*) are also grown in this continent.

B. *rapa* is thought to have originated in the mountainous areas near the Mediterranen sea (Tsunoda 1980). The orginal progenitor of the Indian and European forms was the same and that the Indian brown sarson evolved in the northwest of the Indian subcontinent from the original stock as suggested by the Russian workers, (Sinskaia 1928; Vavilov and Bukinich 1929), who regarded India as one of the independent centers of origin. The species appears to have attained a wide distribution throughout the Europe, parts of Africa, Asia and the Indian subcontinent before the recorded history. As B. rapa was most intensively grown at that time, it can be concluded that this crop was the major source of producing large quantities of vegetable oils. Seeds of B. rapa were first recorded in Europe in 1620 by the Swiss botanist Casper Banhin. However, Boswell (1949) was of the view that these existed much earlier than this. As per some anonymous authors, rapeseed was grown in Europe as early as in the thirteenth century. Prakash and Hinata (1980) also suggested that oleiferous *B. campestris* subspecies developed in two places giving rise to two different races, one European and other Asian. There is a lot of evidence that European oilseed type *B. rapa* must be very close to the turnip type *B. rapa* genetically because it was produced out of it only some 100 years ago. On other hand in China Lintao Caizi very well known to the world as *B. chinensis* (leafy type B. rapa n=10) is used as oilseed crop. This can interpreted as parallel to the evolution of the oilseed type of the turnip type *B. rapa* in Europe (Sun et al. 1991). Alam (1945) concluded that sarson and toria types of *B. rapa* grown as oilseed crops in India and Pakistan evolved in Afghan that Persian area and migrated South India and further East. Song et al. (1988a, b) studied the phylogenetic analysis of 17 cultivated and 5 wild population of *B. rapa*. All the 17 cultivated forms were designated into two distinct groups as European and East Asia group. The phylogenetic grouping seems to correspond with the respective geographic distribution of the cultivated and wild forms of Brassica.

1.2.2   **B. napus**

B. *napus* is an amphidiploid resulting from the cross between the plants of B. *oleracea* and B. *rapa* and is comparatively of recent origin (Olsson 1960). It is uncertain to maintain if *B. napus* is found wild or not, since wild forms of this crop are difficult to find (Hinata and Prakash 1984). However, if wild *napus* exists, it must be a European- Mediterranean species that originated in the area of overlap between *B. oleracea* and *B. campestris*. Olsson (1960) suggested that *B. napus* could have arisen several times by spontaneous hybridization between the different forms of *B. rapa* and *B. oleracea*. Song and Osborn (1992) on the basis of chloroplast and mitochondrial DNA analysis suggested that *B. montana* (n=9) might be closely related ancestral species of *B. rapa* and *B. montana* was the maternal donor. The parental origins of *B. napus* were also investigated using six microsatellite markers located in the chloroplast genome by Allender et al. (2005). Allender and king (2010) using chloroplast and nuclear markers concluded that it is highly unlikely that *B. oleracea* or any of the C- genome species are closely related to the maternal progenitor of most *B napus* accession. They also suggested that either of *B. rapa* and *montana* or a common ancestor could have been the maternal parent of *B. napus*. Though, they suggested that *B. oleracea* was not the parent of most of *B. napus* accessions, a small number of accessions shared *B. oleracea* haplotype. Similarly, the phylogenetic analysis based on nuclear RFLP data also suggests that *B. napus* has multiple origins (Song et al. 1990, 1993). The various cytoplasm types found in *B. napus* accessions correspond to the progenitor diploid species which provide a strong evidence for the multiple origins of this crop (Song et al. 1997).

In *B. napus* as well as *B. campestris*, a range of morphological forms are found both having annual and biennial types. Keeping this in view, Olsson (1960) suggested that B. *napus* could have arisen several times by spontaneous hybridization of different forms of *B. campestris* and *B. oleracea*. The majority of the cultivated *B. napus* accessions appear to have arisen by an interspecific cross in which a wild nine or ten chromosome species having the B. montana cytoplasm type.

Mizushima and Tsunoda (1967) inferred that *B. napus* was found in the coast of northern Europe because *B. oleracea* extended its territory up to northern Europe from the Irano Turanean regions with its high adaptability to low temperature. Sinskaia (1928) and Schiemann (1932) were also of the view that it might have originated in the Mediterranean region or in the western or northern Europe. In Europe, production of oleiferous, *B. napus* might have started during the middle Ages. In Asia, it was introduced during the nineteenth century. The Chinese and Japanese germplasm was developed by crossing European *B. napus* cultivars with indigenous *B. rapa* cultivars (Shiga 1970). Today most of the oilseed rape produced in China, Korea and Japan is harvested from *B. napus* cultivars. It is less adapted to the Indian sub continent due to the short days and warm growing conditions.

1.2.3 **B. juncea**

B. juncea is an amphidiploid and results from an interspecific cross between the plants of *B. rapa* and *B. nigra* and it has longer history than *B napus*. A number of workers have suggested that China as the centre of origin where the maximum diversity is found (Prain 1898; Sinskaia 1928; Vavilov 1949). It came to India from China through a North Eastern route and its immigration to India has been independent of an Aryan incursion (Prain 1898). According to Sun (1970), *B. juncea* originated in Middle East from where it spread to Asia. Afghanistan is thought to be as secondary centre of origin (Olsson 1960; Mizushima and Tsunoda 1967; Tsunoda and Nishi 1968) from where it spread to secondary centre on the Indian sub continent as an major oilseed crops (Hemingway 1995; Prakash and Hinata 1980). The analysis of Fraction I protein data (Uchimiya and Wildman 1978) and chloroplast DNA established the fact that *B. rapa* served as female parent in the formation of the species (Erickson et al. 1983; Palmer et al. 1983; Palmer 1988; Song et al. 1988a, b; Warwick and Black 1991; Yang et al. 2002). Qi et al. (2007) reported that some phenotypes may have evolved with *B. nigr*a as maternal parent as evidenced from the investigation on nuclear Internal Transcribe Spacer (ITS) regions of ribo-somal DNA from 15 different Chinese vegetables and one oilseed form. Wu et al. (2009) studied the relationship among 95 *B. juncea* accessions originated from China, India, Pakistan and Japan following the sequenced Related Amplified Polymorphisms (SRAPs). Although winter sown accessions exhibited more genetic diversity than the spring sown accessions yet, SRAP markers did not provide clear cut separation between Indian/Pak and China winter sown mustard. Data supporting the polyphyletic origin are parallel variation observed a nuclear RFLP pattern of *B. campestris* and *B. juncea* (Song et al. 1988a, b). Wu et al. (2009) and Qi et al. (2007) also supported the idea that vegetables and oilseed forms have polyphyletic origin and evolved separately during the course of evolution.

1.2.4   **B. carinata**

B. carinata is commonly known as Abyssinian or Ethiopian mustard and it is indige-nous Brassica oilseed and vegetable crop in Ethiopia. It is also an amphidiploid species derived from two parental species *B. nigra* as a female and *B. oleracea* a male parent (Uchimiya and Wildman 1978; Palmer et al. 1983; Song et al. 1988b; Erickson et al. 1983). Quiros et al. (1988) suggested on the basis DNA analysis that *B carinata* is an amphidiploid of the recent origin and may have the multiple origin. Song et al. (1988a, b) also confirmed on the basis of RFLP study that *B. carinata* came from *B. nigra* and *B. oleracea*. This species is a new introduction to India however it is being bred for potential commercial production in Spain, Canada, India and Australia.

1.2.5   **B. nigra**

B. nigra is an ancient crop which finds mention in the Sanskrit Upnisdas as a Sarshap (Prakash 1961). Hemingway (1995) placed it in Irano-Turanian, Saharo-Sindian region This species became wide spread in old world probably having its origin in Asia minor. The distribution in Europe, Mediterranan and Ethopian plateau (Bailey 1930; Schulz 1919; Mizushima and Tsunoda 1967) suggest that *B. nigra* originated in central and Southern Europe.

1.3   Conclusion

Brassica have a range of morphotypes and accordingly vary in their origin, cultivation, use, and history. The evolution of each species of Brassica has witnessed a shift in their morphophysiological traits from their original form to present day cultivars. Canola is one of the examples in rapeseed. In *B. oleracea* present day cultivars have resulted from mutation followed by adaptation and selection.

References

Alam Z (1945) Nomenclature of oleiferous Brassicas cultivated in Punjab. Indian J Agric Sci 15:173–181

Allender CJ, King GJ (2010) Origins of the amphiploid species *Brassica napus* L. investigated by chloroplast and nuclear molecular markers. BMC Plant Biol 10:54

Allender C, Evered C, Lynn J, Graham K (2005) Tracing the origins of *Brassica napus* using chloroplast microsatellites. In: Proceedings of plant and animal genomes XIII conference, 15–19 Jan 2005. Town & country convention center, San Diego, p 411

Al-Shehbaz LA, Beilstein MA, Kellog EA (2006) Systematics and phylogeny of the Brassicaceae (Cruciferae): an overview. Plant Systemat Evol 259:89–130

Appelquist LA, Ohlson R (1972) Rapeseed: cultivation, composition, processing and utilization. Elsevier, Amsterdam/London/New York

Bailey LH (1930) The cultivated Brassicas: second paper. Gentes Herb 2:211–267

Beckman C (2005) Vegetable oils: competition in a changing market. Bi-weekly Bulletin. Agriculture and Agri-Food Canada 18(11). Available at http://www.agr.gc.ca/mad-dam/e/bulletine/v18e/v18n11_e.htm

Boswell VR (1949) Our vegetable travelers. Natl Geogr Mag 96:145–217

Chen BY, Heneen WK, Simonsen V (1989) Comparative and genetic studies of isozymes in resynthesized and cultivated *Brassica napus* L., *B. campestris* L. and *B. alboglabra* Bailey. Theor Appl Genet 77:673–679

Coulthart MB, Denford KE (1982) Isozyme studies in *Brassica*. I lectrophoretic techniques for leaf enzymes and Comparison of *B. napus*, *B. campristris* and *B. oleracea* using phosphoglucomutase. Can J Plant Sci 62:621–630

Craig BM, Murphy NL (1959) Quantitative fatty acid analysis of vegetable oil by gas liquid chromatography. J Am Oil Chem Soc 36:549–552

Dass H, Nybom N (1967) The relationships between *Brassica nigra*, *B. campestris*, *B. oleracea*, and their amphidiploid hybrids studied by means of numerical chemotaxonomy. Can J Genet Cytol 9:880–890

DeCandolle AP (1824) Translated into German by Berg, C.W.1824. Die verschiedenenArten. Unterarten und Spielarten des Kohls und der Rettige, Welche in Europa gebauet warden, Leipzig

Dodoneus R (1578) A nievve Herball. Antwerp. Translated by H. Lyte, London

Downey RK, Robellen G (1989) Brassica species. In: Robellen G, Downey RK, Ashri A (eds) Oil crops of the world. McGraw Hill, New York, pp 339–362

Erickson LR, Straus NA, Beversdorf WD (1983) Restriction patterns reveal origins of chloroplast genomes in *Brassica amphidiploids*. Theor Appl Genet 65:201–206

Frandsen KJ (1943) The experimental formation of B. juncea. Dansk Bot. Archiv11:1–17

Frandsen KJ (1947) The experimental formation of Brassica *napus* L. va. *Oleifera* DC. and Brassica *carinata* Braun. Dansk Bot. Arkiv 12:1–16

Fussel GE (1955) History of cole (Brassica sp). Nature 176:48–51

Gerarde J (1597) Herball or generall historie of plantes. Norton, J, London

Gómez-Campo C (1999) Taxonomy. In: Gómez-Campo C (ed) Biology of Brassica coenospecies. Elsevier, Amsterdam, pp 3–32

Herberd DJ (1972) A contribution to the cytotaxonomy of Brassica (cruciferae) and its allies. Bot J Linn Soc 65:1–23

Herberd DJ (1976) Cytotaxonomic studies of *Brassica* and related genera. In: Vaughan JG et al (eds) The biology and chemistry of the cruciferae. Academic, London, pp 47–68

Hedge IC (1976) A systematic and geographical survey of the world cruciferae. In: Vaughan JG, McLeodand AJ, Jones BMG (eds) The biology and chemistry of cruciferae. Academic, New York, pp 1–45

Hemingway JS (1995) Mustards: *Brassica spp.* and *Sinapis alba* (Cruciferae). In: Smartt J, Simmons NW (eds) Evolution of crop plants. Longman, London, pp 82–86

Hinata K, Prakash S (1984) Ethnobotany and evolutionary origin of Indian oleiferous Brassicae. Indian J Genet 44:102–112

Holzner W (1981) Acker-Unkra¨uter-Bestimmung, Verbreitung, Biologie und O kologie. Leopold Stocker Verlag, Graz/Stuttgart

Howard HW (1940) Nomenclature of Brassica species. Curr Sci 9:494–495

Kalia HR, Gupta SK (1997) Importance, nomenclature and origin. In: Kalia HR, Gupta SK (eds) Recent advances in oilseed Brassicas. Kalyani, New Delhi, pp 1–11

Linnaeus C (1973) Species planatarum. Holmiae (Stockholm) (Reprint London, 1957)

Liu H (1985) Genetics and breeding of rapeseed. Shanghai Sci. and Tech., Shanghai, p 592

Lysak MA, Koch MA, Pecinka A, Schubert I (2005) Chromosome triplication found across the tribe Brassiceae. Genome Res 15:516–525

Mizushima U, Tsunoda S (1967) A plant exploration in Brassica and Allied Genera. Tohoku J Agr Res 17:249–276

Morinaga T (1928) Preliminary note on interspecific hybridization in Brassica. Proc Imper Acad Tokyo 4:620–622

Morinaga T (1929a) Interspecific hybridization in Brassica I. The cytology of F1 hybrids of B. nepella and various other species with 10 chromosomes. Cytologia 1:16–27

Morinaga T (1929b) Interspecific hybridization in Brassica II. The cytology of F1 hybrids B. cerna and various other species with 10 chromosomes. Jpn J Bot 4:277–280

Morinaga T (1934a) Interspecific hybridization in Brassica VI. The cytology of F1 hybrids of B. juncea and. B. nigra. Cytologia 6:62–67

Morinaga T (1934b) On the chromosome number of Brassica juncea and Brassica napus on the hybrid between these two and on cv spring of the hybrid. Jpn J Genet 9:161–163

Nagaharu U (1935) Genome analysis in Brassica with special reference to the experimental formation of Brassica napus and peculiar mode of fertilization. Jpn J Bot 7:389–452

Neuweiller E (1905) Die prahistorishe pflanzenreste mitteleuropas. Albert Raustein, Zurich

Olsson G (1954) Crosses within the campestris group of the genus Brassica. Hereditas 40:398–418

Olsson G (1960) Species crosses within the genus Brassica II. Artificial Brassica napus L. Hereditas 46:351–386

Palmer JD (1988) Intraspecfic variation and multicircularity in Brassica mitochondrial DNAs. Genetics 118:341–351

Palmer JD, Shields CR, Cohen DB, Orton TJ (1983) Chloroplast DNA evolution and the origin of amphiploid Brassica species. Theor Appl Genet 65:181–189

Piggot S (1950) Prehistoric India to 1000 BC. Penguin Books, Harmondsworth

Prain D (1898) The mustards cultivated in Bengal. Agr Ledger 5:1–80

Prakash S (1961) Food and drinks in ancient India. Mitra R, Delhi, pp 265–266

Prakash S (1973) Artificial Brassica juncea Coss. Genetica 44:249–263

Prakash S, Hinata K (1980) Taxonomy, cytogenetics and origin of crop Brassica, a review. Opera Botanica 55:11–57

Qi X, Zhang MF, Yang JH (2007) Molecular phylogeny of Chinese vegetable mustard (Brassica juncea) based on the Internal Transcribed Spacers (ITS) of nuclear ribosomal DNA. Genet Res Crop Evol 54:1709–1716

Quiros CF, Ochoa O, Douches DS (1988) Exploring the role of x = 7 species in Brassica evolution: hybridization with B. nigra and B. oleracea. J Hered 79:351–358

Reiner H, Holzner W, Ebermann R (1995) The development of turnip type and oilseed type Brassica rapa crops from the wild type in Europe-an overview of the botanical, historical and linguistic facts: rapeseed today and tomorrow. In: Proceedings of ninth international rapeseed congress, vol 4. Cambridge, 4–7 July 1995, pp 1066–1069

Robellen G (1960) Beitrage zur analyse des Brassica-genoms. Chromosoma 11:205–228

Schiemann E (1932) Entstehung der kulturpflan zen Handlab. Vererbwis Lfg 15

Schulz OE (1919) IV. 105 Cruciferae-Brassiceae. Part 1. Subtribes I. Brassicinae and II. Raphaninae. In: Engler A (ed) Das Pflanzenreich, Heft 68–70. Wilhelm Engelmann, Leipzig, pp 1–290

Schulz OE (1936) Cruciferae. In: Engler A, Harms H (eds) Die Natürlichen Pflanzenfamilien, 2nd edn. 17B, Verlag von Wilhelm Engelmann, Leipzig, pp 227–658

Shahidi F (1990) Rapeseed and canola: global production and distribution. In: Shahidi F (ed) Canola and rapeseed: production, chemistry, nutrition and processing technology. Van Norstrand Reinhold, New York, p 13

Shiga T (1970) Rape breeding by interspecific crossing between *Brassica napus* and *Brassica campestris* in Japan. Jpn Agr Res Q 5:5–10

Singh D (1958) Rape and mustard. The Indian Central Oilseeds Committee, Bombay

Sinskaia EN (1928) The oleiferous plants and roots of the family cruciferae. Bull Appl Bot Genet Plant Breeding 10:1–648

Snowdon RJ (2007) Cytogenetics and genome analysis in *Brassica* crops. Chromosome Res 15:85–95

Song K, Osborn TC (1992) Polyphyletic origins of *Brassica napus*: new evidence based on organelle and nuclear RFLP analyses. Genome 35:992–1001

Song KM, Osborn TC, William PH (1988a) Brassica taxonomy based on nuclear restriction fragment length polymorphisms(RFLPs). 1. Genome evolution of diploid and amphidiploid species. Theor Appl Genet 75:784–794

Song KM, Osborn TC, Williams PH (1988b) Brassica taxonomy based on nuclear restriction length polymorphisms (RFLPs) 2. Preliminary analysis of subspecies with *B. rapa* (syn. campestris) and *B. oleracea*. Theor Appl Genet 76:593–600

Song KM, Osborn TC, Williams PH (1990) Brassica taxonomy based on nuclear restriction length polymorphisms (RFLPs) 3. Genome relationships in Brassica and related genera and the origin of *B. oleracea* and *B. rapa*. Theor Appl Genet 79:497–506

Song KM, Tang KL, Osborn TC (1993) Development of synthetic Brassica amphidiploids by reciprocal hybridization and comparison to natural amphidiploids. Theor Appl Genet 86:811–821

Song KM, Osborn TC, Williams PH (1997) Taxonomy based on nuclear RFLP analysis. In: Kalia HR, Gupta SK (eds) Recent advances in oilseed brassicas. Kalyani, New Delhi, pp 12–24

Stringam GR, Ripley VL, Love HK, Mitchell A (2003) Transgenic herbicide tolerant canola. The Canadian experience. Crop Sci 43:1590–1593

Sun VG (1970) Breeding plants of Brassica. J Agr Assoc China 71:41–52

Sun WC, Pan QY, An XH, Yang YP (1991) Brassica and Brassica related oilseed crops in Gansu, China. In: McGregor DI (ed) Proceedings GCIRC, eighth international rapeseed congress, vol 4. Saskatoon, pp 1130–1135

Takahata Y, Hinata K (1983) Studies on cytodemes in subtribe Brassicinae (Cruciferae). Tohoku J Agri Res 33:111–124

Takahata Y, Hinata K (1986) Consideration of the species relationships in subtribe Brassicinae (Cruficerae) in view of cluster analysis of morphological characters. Plant Species Biol 1:79–88

Toxopens H, Oost EH, Reuling G (1984) Current aspects of the taxonomy of cultivated Brassica species. The use of *B. rapa* L. versus *B. campestris* L. and a proposal for a new intraspecific classification of B. rapa L. Cruciferae Newsl 9:55–58

Tsunoda S (1980) Eco-physiology of wild and cultivated forms in *Brassica* and allied genera. In: Tsunoda S et al (eds) Brassica crops and wild allies. Japan Scientific Societies Press, Tokyo, pp 109–120

Tsunoda S, Nishi S (1968) Origin, differentiation and breeding of cultivated *Brassica*. Proc XII Int Congr Genet 2:77–88

Uchimiya H, Wildman SG (1978) Evolution of fraction I protein in relation to origin of amphidiploid Brassica Species and other members of the Cruciferae. J Hered 69:299–303

Vaughan JG (1977) A multidisciplinary study of the taxonomy and origin of Brassica crops. Bioscience 27:35–40

Vavilov NI (1949) The origin, variation, immunity and breeding of cultivated plants. Chron Bot 13:1–364

Vavilov NI, Bukinich DD (1929) Agriculture in Afghanistan. Bull Appl Bot Genet Plant Breeding 33:378–382

Warwick SI, Black LD (1993) Molecular relationships in the subtribe Brassicinae (Cruciferae, Tribe, Brassiceae). Can J Bot 71:906–918

Warwick SI, Black LD (1991) Molecular systematics of *Brassica* and allied Genera (subtribe Brassicinae, Brassiceae) – chloroplast genome and cytodeme Congruence, Theor Appl Genet 82:839–850

Warwick SI, Hall JC (2009) Phylogeny of *Brassica* and wild relatives. In: Gupta SK (ed) Biology and breeding of crucifers. CRC, Boca Raton, pp 19–36

Warwick SI, Sauder CA (2005) Phylogeny of tribe *Brassiceae* (Brassicaceae) based on chloroplast restriction site polymorphisms and nuclear ribosomal internal transcribed spacer and chloroplast *trn*L intron sequences. Cand J Bot 83:467–483

Warwick SI, Black LD, Aguinagalde I (1992) Molecular systematics of B ASS anCdA allied genera (subtribe Brassicinae, Brassiceae) – chloroplast DNA variation in the genus diplotaxis. Theor Appl Genet 83:839–850

Witmack L (1904) Über die in Pompej gefundenen Pflanzenreste. Englers Bot. Jahrb. Bd 33

Wu X, Chen B, Lu G, Wang H, Xu K, Guizhan G, Song Y (2009) Genetic diversity in oil and vegetable mustard (*Brassica juncea*) landraces by SRAP markers. Genet Resour Crop Evol 56:1011–1022

Yang YW, Tai PY, Chen Y, Li WH (2002) A study of the phylogeny of *Brassica rapa, B. nigra, Raphanus sativa* and their related genera using noncoding regions of chloroplast DNA. Mol Phylogenet Evol 23:268–275

Chapter 2
Molecular Cytogenetics

Annaliese S. Mason

Abstract Cytogenetics has played a key role in the history of scientific research in the Brassicaceae since the start of the last century. The discovery of the *Brassica* "U's Triangle" species, elucidation of phylogenetic relationships and investigations of chromosome evolution all contributed to building up the basic genomic understanding of the Brassicaceae we have today. The advent of molecular cytogenetics in this family in the last 20 years has led to a progressively greater understanding of the factors underlying chromosome dynamics and organisation, meiotic and mitotic mechanisms and cell division processes. In addition, linking molecular cytogenetics with other molecular techniques, such as marker studies, DNA sequencing and protein expression analysis, has bridged the gap between chromosomes and linkage groups, resulting in a wealth of new information in this family. Future prospects for molecular cytogenetics in the Brassicaceae are bright. The recent and imminent release of additional Brassicaceae genomes will greatly facilitate development of probes for fluorescent in situ hybridisation as well as a comprehensive understanding of gene expression and protein interactions during cell division.

Keywords Cytogenetics • Fluorescent in situ hybridisation • Chromosomes • *Brassica* • *Arabidopsis*

2.1 Introduction

Cytogenetics, literally "cell genetics", traditionally refers to the study of chromosomes. Cytogenetics conventionally encompasses studies of chromosome number, structure and organisation, chromosomal aberrations and chromosome behaviour

A.S. Mason, Ph.D., B.Sc. (Hons) (✉)
School of Agriculture and Food Sciences, The University of Queensland 4072,
Brisbane, Australia
e-mail: Annaliese.mason@uq.edu.au

S.K. Gupta (ed.), *Biotechnology of Crucifers*, DOI 10.1007/978-1-4614-7795-2_2,
© Springer Science+Business Media, LLC 2013

during mitosis and meiosis. Early cytogenetics studies were crucial in defining species and genera in the Brassicaceae, and for elucidating complex species relationships in the *Brassica* crop species and wild allies (Morinaga 1934; Nagaharu 1935; Mizushima 1980). However, the advent of molecular cytogenetics has opened up whole new avenues of exploration in this family. Molecular cytogenetics describes techniques made possible by breakthroughs in molecular genetics over the last 30 years or so, starting with radioactive labelling and genomic in situ hybridisation of DNA to chromosome spreads, and moving through to chromosome and protein labelling using fluorescently tagged antibodies and other molecules (Speicher and Carter 2005). Modern molecular cytogenetics offers a suite of tools suitable for integrating physical and linkage maps, observing protein expression and co-localisation during mitosis and meiosis, investigating chromosomal architecture such as nucleolar organisational regions (NORs), centromere and telomeres, identifying locations of genes and repetitive elements and molecular karyotyping in hybrid and mutant studies.

A great deal of molecular cytogenetic functionality is currently in use or in the process of development in the Brassicaceae. The Brassicaceae is informally known as the crucifer or cabbage family, and comprises approximately 3,700 species in about 338 genera (Warwick and Al-Shehbaz 2006). The bulk of this review will focus on the well-studied and agriculturally important *Brassica* genus and on model plant species *Arabidopsis thaliana*, as these comprise the species upon which the majority of molecular cytogenetics studies in the Brassicaceae have been pioneered. Molecular cytogenetics has been used extensively in the Brassicaceae to characterise and track chromosome behaviour in interspecific hybrids, synthetic canola lines, chromosome addition and genomic introgression lines and in experimental genotypes. Outcomes of molecular cytogenetics in this family include elucidation of nucleolar organisational regions (NORs) and characterization of other meiotic mechanisms, tracking and identifying genomic introgressions between *Brassica* species for breeding purposes and the integration of physical and genetic linkage maps.

The advent of next generation and third generation sequencing technologies facilitating the cheap and rapid sequencing of many *Brassica* and wider Brassicaceae species genomes looks to bring forth a new spectrum of uses for molecular cytogenetics in this family. Future prospects for molecular cytogenetics in the Brassicaceae will be discussed.

2.2 Classical Cytogenetics in the Brassicaceae

Classical cytogenetics, or the use of light microscopy to interrogate chromosome number, meiotic chromosome behaviour in natural species, haploids and interspecific hybrids, has taught us a lot about the Brassicaceae family. Classical cytogenetics provides a base upon which molecular techniques can build and expand on, and is still informative and relevant today in many poorly-examined clades. The unfortunately small size and relative lack of differentiation of most Brassicaceae chromosomes relative to some other plant genera has provided somewhat of a handicap, particularly in the use of karyotyping through "C-banding", or observation of heterochromatin patterns in chromosomes, but nevertheless many highly informative studies have been carried out using only classical cytogenetics. The most widely known use of cytogenetics in the Brassicaceae is the discovery of the relationship between the six agriculturally significant species *Brassica rapa* (buk choy, turnip), *B. nigra* (black mustard), *B. oleracea* (cabbage, cauliflower, broccoli), *B. juncea* (Indian mustard), *B. napus* (rapeseed, canola) and *B. carinata* (Ethiopian mustard). Although this relationship was first elucidated almost incidentally by T. Morinaga in a paper entitled "Interspecific hybridization in *Brassica* VI. The cytology of F_1 hybrids of *B. juncea* and *B. nigra*" in table form (Morinaga 1934), it is usually attributed to Nagaharu U, who produced the figure commonly known as the *Brassica* "U's Triangle" or "Triangle of U" (U Nagaharu 1935). The six "U's Triangle" species were found to comprise three diploid species, with genome complements $2n=2x=AA$ (*B. rapa*), $2n=2x=BB$ (*B. nigra*) and $2n=2x=CC$ (*B. oleracea*), and three allotetraploid species formed from additive genome combinations of the diploid species: $2n=4x=AABB$ (*B. juncea*), $2n=4x=AACC$ (*B. napus*), and $2n=4x=BBCC$ (*B. carinata*). Interestingly, Nagaharu U's original figure also contained *Brassica napocampestris* (Frandsen and Winge 1932), $2n=6x=AAAACC$, an artificial hybrid between *B. napus* and *B. rapa* (previously referred to as *B. campestris*). These early cytogenetics studies involved very basic techniques: production of interspecific hybrids through cross-pollination and observations of chromosome number and chromosome behaviour in these hybrids at mitosis and meiosis. The ease in which many different species in the Brassicaceae hybridise, either through handpollination or via embryo rescue, protoplast fusion or other methods (FitzJohn et al. 2007), has provided a wealth of cytogenetic information. Pairing between chromosomes from different *Brassica* genomes was observed in haploids with genome complements A, B, C, AC, BC and AB, as well as in interspecific Brassicaceae hybrids of various types (Harberd and McArthur 1980) (Fig. 2.1). The propensity of chromosomes to associate at meiosis through regions of ancestral homoeology in the absence of meiotic control provides a crude but effective means of determining phylogenetic relationships (Mizushima 1980). Classical cytogenetics in *Brassica* and *Arabidopsis* has also been used in the identification of nuclear organisational regions (NORs) and provision of a basic karyotype using C-banding (Olin-Fatih and Heneen 1992; Koornneef et al. 2003).

The advent of molecular marker technologies initiated a decline in interest in cytogenetics, but modern molecular cytogenetics often provides a means of investigating regions of repetitive DNA which are otherwise recalcitrant to pure molecular studies, as demonstrated in studies of *Arabidopsis* centromere sequences (Murata et al. 1994; Heslop-Harrison et al. 1999). Molecular cytogenetics techniques have enabled further resolution of chromosomal relationships and genetic control of meiosis in the Brassicaceae, with numerous studies utilising genomic in situ hybridisation and more refined fluorescent in situ hybridisation techniques to confirm and expand upon these early findings.

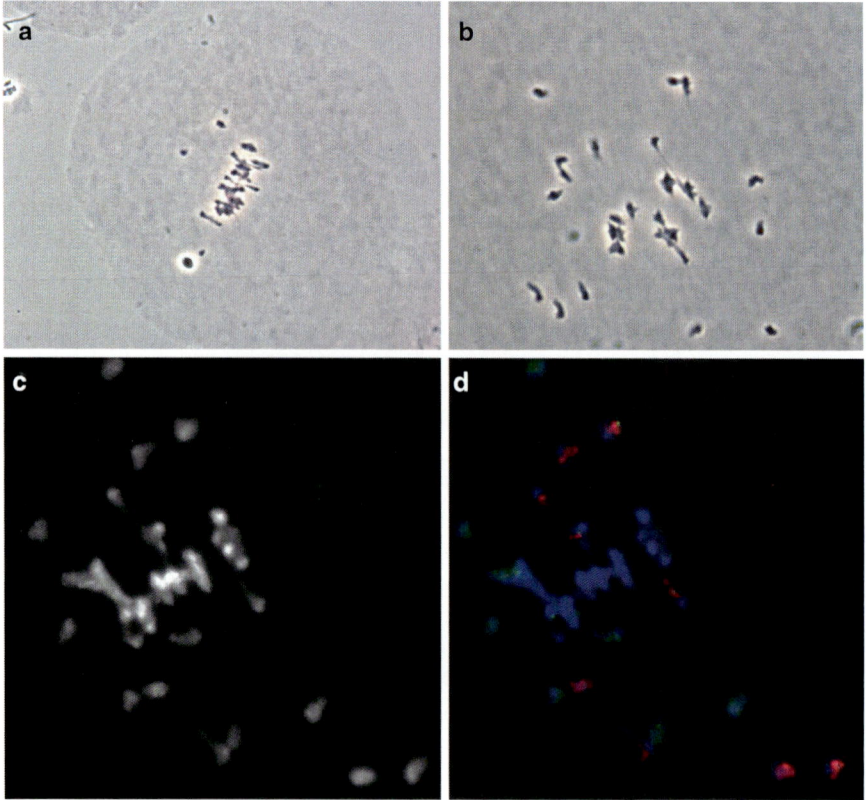

Fig. 2.1 Observations of meiosis in *Brassica* interspecific hybrids using classical cytogenetics and fluorescent in-situ hybridisation. (**a**) Metaphase I in a *B. juncea×B. carinata* hybrid (2n=BBAC=35), taken using 1 % acetic acid carmine stain with a phase contrast microscope and 1,000×magnification. (**b**) Metaphase I in a *B. juncea×B. napus* hybrid (2n=AABC=37), taken using 1 % acetic acid carmine stain with a phase contrast microscope and 1000×magnification. (**c**) Metaphase I in a *B. juncea×B. napus* hybrid (2n=AABC=37), taken using DAPI stain with a fluorescent microscope and 1,000×magnification. (**d**) Metaphase I in a *B. juncea×B. napus* hybrid (2n=AABC=37), taken using a fluorescent microscope and 1,000×magnification, after fluorescent in-situ hybridisation labelling of the B genome (*green*, fluorescein isthiocyanate), C genome (*red, Texas Red*) and background stain using DAPI (A genome, *blue*)

2.3 Fluorescent In Situ Hybridisation

Fluorescent in situ hybridisation (FISH) refers to the use of fluorescently-labelled DNA probes to bind (hybridise) to specific regions on chromosome spreads on microscope slides. Initially, the FISH protocol was based on the methods used previously for RFLPs, given the presumed similarity in the hybridisation protocol, but over the years many research groups have developed in-house modifications and optimisations. A particularly useful approach was taken in recent years: removing each protocol step one by one and assessing if the protocol still worked, the end result of which was a greatly reduced FISH protocol (Kato et al. 2004). However, all FISH protocols involve basically the same process: firstly, the fluorescent labelling of the probe (usually by nick translation or random priming techniques, which insert fluorescently labelled nucleotides into the probe DNA sequence), secondly, the hybridisation of the probe to a denatured genomic chromosome spread on a microscope slide, and thirdly, visualisation of the probe label. Additional steps are mainly concerned with optimisation of the probe signal to background noise ratio, which involves steps such as removal of RNA using RNAse enzyme treatment and washes with salt solutions at various temperatures. Denaturing the genomic DNA to allow probe binding may also be done through a number of different mechanisms, of which heat and treatment with formamide are the most common.

Genomic in situ hybridisation (GISH) is a special case of FISH whereby whole genomic DNA is used as a probe instead of specific DNA sequences. Early FISH studies used radioactive labels, but this has given way to use of fluorescent molecules (such as fluorescein isthiocyanate (FITC)) bound to antibodies such as avidin, biotin or digoxigenin as DNA probes, with a background chromosome stain of 4',6-diamidino-2-phenylindole (DAPI) or propidium iodide (PI). In recent years the use of directly labelled fluorescent dNTPs, which can be integrated into the probe DNA directly using techniques such as nick translation, has also become popular in *Brassica* cytogenetics labs (Xiong and Pires 2011). The addition of fluorescent labels to chromosome spreads can result in a wealth of additional information (Fig. 2.1).

2.4 Chromosome Painting: Early Genomic and Fluorescent In Situ Hybridisation in the Brassicaceae

The earliest use of FISH in the Brassicaceae involved the mapping of repetitive DNA sequences (Iwabuchi et al. 1991) and rDNA loci (Maluszynska and Heslop-Harrison 1993). These and subsequent studies were useful in elucidating the number of rDNA loci (Snowdon et al. 1997a) and in presenting some information about the highly repetitive centromeric regions and conserved telomeric regions in *Brassica* (Hasterok et al. 2005). Molecular cytogenetics studies in model plant *Arabidopsis thaliana* have been highly informative in investigations of meiosis, repetitive DNA sequences and chromosome structure (Koornneef et al. 2003). In the agriculturally important *Brassica* genus, a common application of molecular cytogenetics involves the use of GISH to track chromosome introgressions through generations of backcrosses after interspecific hybridisation (Chen et al. 2011; Snowdon et al. 1997b; Navabi et al. 2011), and more recently the use of FISH probes to differentiate between more closely related genomes for the same purpose (Schelfhout et al. 2006). The ready hybridisation between species in the Brassicaceae has been utilised to effect transfer of useful genetic diversity or genes of interest into crops in breeding programs (Saal et al. 2004; Rygulla et al. 2007), and GISH allows these genomic introgressions to be tracked and analysed.

Another common use of GISH is to investigate meiotic pairing behaviour in interspecific hybrids, in order to assess genomic relationships and predict the success of genomic introgressions for crop improvement. Pairing between each of the genomes in the allotetraploid *Brassica* species *B. juncea* (2n = AABB), *B. napus* (AACC) and *B. carinata* (BBCC) has been observed using FISH and GISH (Mason et al. 2010), as has extensive pairing between *B. napus*, *B. oleracea* and *B. rapa* of different genotypes and ploidy levels (Leflon et al. 2006; Nicolas et al. 2008; Nicolas et al. 2009; Leflon et al. 2010). More distant relatives have also been noted using GISH to form chiasmatic associations at meiosis and recombinant chromosomes with *B. napus* or *B. rapa*, such as *Orychrophragmus violaceus* (Ma et al. 2006), *Isatis indigotica* (Tu et al. 2008), *Lesquerella fendleri* (Du et al. 2008) and *Capsella bursa-pastoris* (Chen et al. 2007).

FISH has also been used for the establishment of monosomic addition lines in the Brassicaceae, whereby a single chromosome from an alien genome is added to a known genomic complement (e.g. chromosome C1 of *B. oleracea* added to the ten chromosomes of the *B. rapa* A genome). This technique allows for the expression of genes belonging to a single linkage group to be analysed in the absence of other genomic loci. The production of C-genome addition lines in a *B. rapa* background allowed the complex trait of seed colour to be investigated (Heneen et al. 2012), as well as the pairing behaviour of individual homoeologous chromosomes. Trisomics and monosomics have also been greatly beneficial in *Arabidopsis* research (reviewed in (Koornneef et al. 2003)). However, the establishment of chromosome addition lines for the small, poorly-differentiated *Brassica* and *Arabidopsis* chromosomes necessitates the use of fluorescent in situ hybridisation. The process of backcrossing to eliminate all but one alien chromosome is not selective for a particular chromosome, and conclusively identifying which chromosome is present and confirming lack of chromosome fusions with the host genome can most readily be achieved by the use of FISH. GISH may also be used to label the alien genome if the alien chromosomes are not readily distinguishable from the host genome, or again to confirm that the alien addition chromosomes are not recombinants (Wang et al. 2006).

FISH can also be used to identify homoeologous non-reciprocal translocations and recombinant chromosomes in other systems, such as synthetic *B. napus* (Szadkowski et al. 2010). In the *Brassica* genus, resynthesis of allotetraploid *B. napus* (2n = AACC), *B. juncea* (2n = AABB) and *B. carinata* (2n = BBCC) from diploid progenitors *B. rapa* (2n = AA), *B. nigra* (2n = BB) *and B. oleracea* (2n = CC) is possible. However, synthetic *B. napus* is generally unstable due to the high degree of homoeology and hence meiotic interaction between the A and C genomes (Szadkowski et al. 2010), which are far more closely related to each other than either is to the B genome (Lagercrantz and Lydiate 1996; Parkin et al. 2003). Although the A and C genomes are known to associate at meiosis due to conservation of the ancestral relationships between the chromosomes, the degree of association is also mediated by genetic factors (Mason et al. 2010; Leflon et al. 2006; Nicolas et al. 2009). Fluorescent in situ hybridisation has been used successfully in the Brassicaceae to determine why particular meiotic behaviours occur, and what kinds of genetic and genomic factors underlie the meiotic process.

Notable successes in using fluorescent in situ hybridisation to investigate meiosis in the Brassicaceae include the discovery of a major gene affecting the number of crossovers during meiosis in *B. napus* (Jenczewski et al. 2003), the elucidation of mechanisms behind stable polyploid formation in *Arabidopsis* (Comai et al. 2003) and the development of an *Arabidopsis thaliana* karyotype (Armstrong and Jones 2003).

2.5   BAC-FISH and Fluorescently Labelled Antibodies

The field of molecular cytogenetics is still expanding. New techniques are constantly being developed in humans and model species (Figueroa and Bass 2010), and many of these have potential applications in the Brassicaceae family. In particular, fluorescently-labelled antibodies and BACs have been used in timing and observing protein expression and co-localization during meiosis and mitosis (Zhang et al. 2008) and molecular karyotyping (Koornneef et al. 2003; Xiong and Pires 2011) in the Brassicaceae. In 2011, a molecular karyotyping set consisting of a chromosome "paintbox" of BAC probes was released for *Brassica napus*, in the hopes of providing a generally applicable means for cytogenetic identification of each of the 19 B. napus chromosomes (Xiong and Pires 2011). This consisted of a set of three BAC probes and four DNA markers (45S, 5S, CentBrI and CentBrII). The BACs were selected from a pool of BACs generated from *B. rapa* sequence: two produced a set of signals across the *B. napus* chromosomes (Xiong and Pires 2011), and the third acted as a GISH-like marker for the C genome, hybridizing to a C-genome specific repetitive sequence in a similar fashion to the BAC BoB014O06 used previously for the same purpose (Alix et al. 2008). The 45S and 5S markers for ribosomal nucleolar organisational regions are well-characterized in *Brassica* and *Arabidopsis*, and added additional chromosome information. Lastly, the CentBrI and CentBrII probes bind to *Brassica* centromeric repeats: the two probes represent different repeats, and chromosomes in *Brassica* have CentBrI, CentBrII or both types of repeats. However, this toolkit has yet to be proven to be of wide use in *Brassica napus*, and may be susceptible to genotypic differences in hybridisation of BAC sequence probes.

FISH can also be used to integrate genetic maps (such as those generated by linkage mapping with molecular markers) with physical maps (such as those produced by classical karyotype observations). This can be done through the hybridisation of fluorescently labelled BACs containing known molecular marker sequence to chromosome spreads, and was successfully used to integrate the *B. oleracea* physical and genetic maps in 2002 (Howell et al. 2002) and the *B. rapa* physical and genetic maps in 2009 (Kim et al. 2009). Co-localization and ordering of FISH markers along single chromosomes can also be used to determine the relative position of difficult-to-map genomic regions and for fine-scale integration of physical and genetic maps, as has been demonstrated in chromosome A7 of *B. rapa* (Xiong et al. 2010).

2.6   Future Prospects

Molecular cytogenetics advances in other fields, particularly in the medical domain, offer a glimpse of a promising future for the Brassicaceae. In future, molecular cytogenetics may aid in genetic engineering, such as development of mini-chromosome platforms for transgene expression (Dhar et al. 2011), and full elucidation of the link between physical behaviour of chromosomes at mitosis and meiosis, DNA sequences and expression of proteins during cell division in the Brassicaceae. The increasingly wide-spread use of fibre-FISH, whereby hybridisation takes place on extended DNA fibres, now allows fluorescently-labelled clones to be separated at a resolution of only a few kB, rather than the few Mb allowed by conventional FISH (Walling and Jiang 2012). In addition, the *B. rapa* genome was released last year (Wang et al. 2011), and the imminent release of the *B. oleracea* and *B. napus* genomes will in upcoming years be joined by a larger set of wild Brassicaceae as well as the genomes of mustard species *B. juncea*, *B. carinata* and *B. nigra*. This unprecedented availability of sequence information will both facilitate the development of FISH probes and complement the ability of molecular cytogenetics to interrogate meiosis and meiotic behaviour. In addition, molecular cytogenetics offers a toolkit to aid in integration of physical and genetic maps, and in particular in addressing chromosome rearrangement, duplication and deletion events that are difficult to identify solely from sequencing or mapping information and are known to be prevalent in the complex ancestral and recent polyploid genomes of the Brassicaceae. From the birth of molecular cytogenetics in the Brassicaceae 20 years ago, technological developments and pioneering research studies have taken molecular cytogenetics in this family to exciting new heights, and the next 20 years of cytogenetics in the Brassicaceae promise to be equally eventful.

References

Alix K, Joets J, Ryder CD, Moore G, Barker G, Bailey JP et al (2008) The CACTA transposon *Bot1* played a major role in *Brassica* genome divergence and gene proliferation. Plant J 56:1030–44

Armstrong SJ, Jones GH (2003) Meiotic cytology and chromosome behaviour in wild-type *Arabidopsis thaliana*. J Exp Bot 54(380):1–10

Chen HF, Wang H, Li ZY (2007) Production and genetic analysis of partial hybrids in intertribal crosses between *Brassica* species (*B. rapa*, *B. napus*) and *Capsella bursa-pastoris*. Plant Cell Rep 26(10):1791–800

Chen S, Nelson MN, Chèvre AM, Jenczewski E, Li Z, Mason AS et al (2011) Trigenomic bridges for *Brassica* improvement. Critical Reviews in Plant Science 30(6):524–47

Comai L, Tyagi AP, Lysak MA (2003) FISH analysis of meiosis in *Arabidopsis* allopolyploids. Chromosome Res 11(3):217–26

Dhar MK, Kaul S, Kour J (2011) Towards the development of better crops by genetic transformation using engineered plant chromosomes. Plant Cell Rep 30(5):799–806

Du XZ, Ge XH, Zhao ZG, Li ZY (2008) Chromosome elimination and fragment introgression and recombination producing intertribal partial hybrids from *Brassica napus* × *Lesquerella fendleri* crosses. Plant Cell Rep 27(2):261–71

Figueroa DM, Bass HW (2010) A historical and modern perspective on plant cytogenetics. Brief Funct Genomics 9(2):95–102

FitzJohn RG, Armstrong TT, Newstrom-Lloyd LE, Wilton AD, Cochrane M (2007) Hybridisation within *Brassica* and allied genera: evaluation of potential for transgene escape. Euphytica 158(1–2):209–30

Frandsen HN, Winge O (1932) *Brassica napocampestris*, a new constant amphidiploid species hybrid. Hereditas 16:212–8

Harberd DJ, McArthur ED (1980) Meiotic analysis of some species and genus hybrids in the Brassiceae. In: Tsunoda S, Hinata K, Gomez-Campo C (eds) Brassica crops and wild allies: biology and breeding. Japan Scientific Societies, Tokyo, pp 65–87

Hasterok R, Ksiazczyk T, Wolny E, Maluszynska J (2005) FISH and GISH analysis of *Brassica* genomes. Acta Biol Cracov Bot 47(1):185–92

Heneen WK, Geleta M, Brismar K, Xiong ZY, Pires JC, Hasterok R et al (2012) Seed colour loci, homoeology and linkage groups of the C genome chromosomes revealed in *Brassica rapa*-B. *oleracea* monosomic alien addition lines. Ann Bot 109(7):1227–42

Heslop-Harrison JS, Murata M, Ogura Y, Schwarzacher T, Motoyoshi F (1999) Polymorphisms and genomic organization of repetitive DNA from centromeric regions of *Arabidopsis* chromosomes. Plant Cell 11(1):31–42

Howell EC, Barker GC, Jones GH, Kearsey MJ, King GJ, Kop EP et al (2002) Integration of the cytogenetic and genetic linkage maps of *Brassica oleracea*. Genetics 161(3):1225–34

Iwabuchi M, Itoh K, Shimamoto K (1991) Molecular and cytological characterization of repetitive DNA sequences in *Brassica*. Theor Appl Genet 81(3):349–55

Jenczewski E, Eber F, Grimaud A, Huet S, Lucas MO, Monod H et al (2003) *PrBn*, a major gene controlling homeologous pairing in oilseed rape (*Brassica napus*) haploids. Genetics 164(2):645–53

Kato A, Lamb JC, Birchler JA (2004) Chromosome painting using repetitive DNA sequences as probes for somatic chromosome identification in maize. Proc Natl Acad Sci U S A 101(37):13554–9

Kim H, Choi SR, Bae J, Hong CP, Lee SY, Hossain MJ et al (2009) Sequenced BAC anchored reference genetic map that reconciles the ten individual chromosomes of *Brassica rapa*. BMC Genomics 10:432

Koornneef M, Fransz P, de Jong H (2003) Cytogenetic tools for *Arabidopsis thaliana*. Chromosome Res 11(3):183–94

Lagercrantz U, Lydiate DJ (1996) Comparative genome mapping in *Brassica*. Genetics 144:1903–10

Leflon M, Eber F, Letanneur JC, Chelysheva L, Coriton O, Huteau V et al (2006) Pairing and recombination at meiosis of *Brassica rapa* (AA)×*Brassica napus* (AACC) hybrids. Theor Appl Genet 113:1467–80

Leflon M, Grandont L, Eber F, Huteau V, Coriton O, Chelysheva L et al (2010) Crossovers get a boost in *Brassica* allotriploid and allotetraploid hybrids. Plant Cell 22:2253–64

Ma N, Li ZY, Cartagena JA, Fukui K (2006) GISH and AFLP analyses of novel *Brassica napus* lines derived from one hybrid between *B. napus* and *Orychophragmus violaceus*. Plant Cell Rep 25(10):1089–93

Maluszynska J, Heslop-Harrison JS (1993) Physical mapping of rDNA loci in *Brassica* species. Genome 36(4):774–81

Mason AS, Huteau V, Eber F, Coriton O, Yan G, Nelson MN et al (2010) Genome structure affects the rate of autosyndesis and allosyndesis in AABC, BBAC and CCAB *Brassica* interspecific hybrids. Chromosome Res 18(6):655–66, Epub 22/6/2010

Mizushima U (1980) Genome analysis in *Brassica* and allied genera. In: Tsunoda S, Hinata K, Gomez-Campo C (eds) Brassica crops and wild allies: biology and breeding. Japan Scientific Societies, Tokyo, pp 89–106

Morinaga T (1934) Interspecific hybridisation in *Brassica* VI. The cytology of F_1 hybrids of *B. juncea* and *B. nigra*. Cytologia 6:62–7

Murata M, Ogura Y, Motoyoshi F (1994) Centromeric repetitive sequences in *Arabidopsis thaliana*. Jpn J Genet 69(4):361–70

Navabi ZK, Stead KE, Pires JC, Xiong Z, Sharpe AG, Parkin IAP et al (2011) Analysis of B-genome chromosome introgression in interspecific hybrids of *Brassica napus*×*B. carinata*. Genetics 187:659–73

Nicolas SD, Leflon M, Liu Z, Eber F, Chelysheva L, Coriton O et al (2008) Chromosome 'speed dating' during meiosis of polyploid *Brassica* hybrids and haploids. Cytogenet Genome Res 120:331–8

Nicolas SD, Leflon M, Monod H, Eber F, Coriton O, Huteau V et al (2009) Genetic regulation of meiotic cross-overs between related genomes in *Brassica napus* haploids and hybrids. Plant Cell 21:373–85

Olin-Fatih M, Heneen WK (1992) C-banded karyotypes of *Brassica campestris*, *Brassica oleracea*, and *Brassica napus*. Genome 35(4):583–9

Parkin IAP, Sharpe AG, Lydiate DJ (2003) Patterns of genome duplication within the *Brassica napus* genome. Genome 46(2):291–303

Rygulla W, Snowdon RJ, Eynck C, Koopmann B, von Tiedemann A, Lühs W et al (2007) Broadening the genetic basis of *Verticillium longisporum* resistance in *Brassica napus* by inter-specific hybridization. Phytopathology 97:1391–6

Saal B, Brun H, Glais I, Struss D (2004) Identification of a *Brassica juncea*-derived recessive gene conferring resistance to *Leptosphaeria maculans* in oilseed rape. Plant Breeding 123(6):505–11

Schelfhout CJ, Snowdon R, Cowling WA, Wroth JM (2006) Tracing B-genome chromatin in *Brassica napus* × *B. juncea* interspecific progeny. Genome 49:1490–7

Snowdon R, Köhler W, Köhler A (1997a) Chromosomal localization and characterization of rDNA loci in the *Brassica* A and C genomes. Genome 40:582–7

Snowdon RJ, Kohler W, Friedt W, Kohler A (1997b) Genomic in situ hybridization in *Brassica* amphidiploids and interspecific hybrids. Theor Appl Genet 95(8):1320–4

Speicher MR, Carter NP (2005) The new cytogenetics: blurring the boundaries with molecular biology. Nat Rev Genet 6(10):782–92

Szadkowski E, Eber F, Huteau V, Lodé M, Huneau C, Belcram H et al (2010) The first meiosis of resynthesized *Brassica napus*, a genome blender. New Phytol 186:102–12

Tu YQ, Sun J, Liu Y, Ge XH, Zhao ZG, Yao XC et al (2008) Production and characterization of intertribal somatic hybrids of *Raphanus sativus* and *Brassica rapa* with dye and medicinal plant *Isatis indigotica*. Plant Cell Rep 27(5):873–83

U Nagaharu (1935) Genome-analysis in *Brassica* with special reference to the experimental formation of *B. napus* and peculiar mode of fertilization. Jpn J Bot 7:389–452

Walling JG, Jiang JM (2012) DNA and chromatin fiber-based plant cytogenetics. Plant Genet Genomics 4:121–30

Wang YP, Sonntag K, Rudloff E, Wehling P, Snowdon RJ (2006) GISH analysis of disomic *Brassica napus-Crambe abyssinica* chromosome addition lines produced by microspore culture from monosomic addition lines. Plant Cell Rep 25(1):35–40

Wang XW, Wang HZ, Wang J, Sun RF, Wu J, Liu SY et al (2011) The genome of the mesopolyploid crop species *Brassica rapa*. Nat Genet 43(10):1035–U157

Warwick SI, Al-Shehbaz IA (2006) Brassicaceae: chromosome number index and database on CD-Rom. Plant Syst Evol 259(2–4):237–48

Xiong Z, Pires JC (2011) Karyotype and identification of all homoeologous chromosomes of allopolyploid *Brassica napus* and its diploid progenitors. Genetics 187:37–49

Xiong Z, Kim JS, Pires JC (2010) Integration of genetic, physical, and cytogenetic maps for *Brassica rapa* chromosome A7. Cytogenet Genome Res 129(1–3):190–8

Zhang WL, Lee HR, Koo DH, Jiang JM (2008) Epigenetic modification of centromeric chromatin: Hypomethylation of DNA sequences in the CENH3-associated chromatin in *Arabidopsis thaliana* and maize. Plant Cell 20(1):25–34

Chapter 3
Distant Hybridization Involving Different In Vitro Techniques

Dan Liu, Ling Xu, Xinxin Geng, Yuanfei Zhou, Zhenchao Zhang, Bing Wang, and Weijun Zhou

Abstract The Brassicaceae family has extensive genetic type and variation, which makes distant hybridization to be a potential strategy for integrating important traits such as quality and resistance to diseases and/or pests into the cultivated species from wild species. However, wide cross incompatibility, often leading to fail to obtain hybrids, limits the application of distant hybridization to crucifers' breeding. Nowadays, with the development of in vitro techniques, it is possible to use these genetic resources by distant hybridization. This chapter reviews the in vitro techniques including somatic hybridization, embryo rescue, microspore embryogenesis, and chromosome doubling that are involved in distant hybridization.

D. Liu, Ph.D.
Institute of Crop Science and Zhejiang Key Laboratory of Crop Germplasm,
Zhejiang University, Hangzhou 310058, China

Institute of Tobacco Research, Chinese Academy of Agricultural Sciences,
Qingdao, Shandong 266101, China

L. Xu, Ph.D. • X. Geng, B.S. • B. Wang, M.S.
Institute of Crop Science and Zhejiang Key Laboratory of Crop Germplasm,
Zhejiang University, Hangzhou 310058, China

Y. Zhou, Ph.D.
Agricultural Experiment Station, Zhejiang University, Hangzhou 310058, China

Z. Zhang, Ph.D.
Institute of Crop Science and Zhejiang Key Laboratory of Crop Germplasm,
Zhejiang University, Hangzhou 310058, China

Zhenjiang Agricultural Research Institute, Jurong; Jiangsu 212400, China

W. Zhou, Ph.D. (✉)
Institute of Crop Science and Zhejiang Key Laboratory of Crop Germplasm,
Zhejiang University, Hangzhou 310058, China

Agricultural Experiment Station, Zhejiang University, Hangzhou 310058, China
e-mail: wjzhou@zju.edu.cn

S.K. Gupta (ed.), *Biotechnology of Crucifers*, DOI 10.1007/978-1-4614-7795-2_3,
© Springer Science+Business Media, LLC 2013

Keywords Cruciferae • Distant hybridization • Embryo rescue • Microspore embryogenesis • Somatic hybridization

3.1 Introduction

Distant hybridization, also known as wide hybridization or wide cross, is defined as the distant crossing between two kinds of plants belonging to different species, genera, subfamilies, or families and so on. Among these crossings, interspecific and intergeneric hybridization between crop plants and between crops and their wild relatives has been widely carried out by breeders. Distant hybridization is a useful strategy for introducing useful genes, e.g., resistance genes from wild relatives or other cultivated species into target crops to widen the genetic base of crops or to construct stocks for genetic analysis (Li and Liu 2001).

The Brassicaceae (Cruciferae) family comprises approximately 330 genera, including more than 3,500 species, the majority of which grow in the Northern Hemisphere. Several genera, such as *Brassica*, *Sinapis*, and *Raphanus*, are the most important sources of edible vegetable oils, vegetables, forage, and spices. The gene pool of Brassicaceae family is quite large with extremely rich genetic type and extensive genetic variation, which provides solid material base and genetic background for wide cross between relative species or genera, breeding new varieties of crops, and creating new types of plants (Momotaz et al. 1998; Qian et al. 2003). In the classical triangle of U (U 1935), the three amphidiploids species (*Brassica napus*, 2n = AACC = 38; *B. juncea*, 2n = AABB = 36; *B. carinata*, 2n = BBCC = 34) were derived from natural spontaneous interspecific hybridization between *B. rapa* (2n = AA = 20), *B. oleracea* (2n = CC = 18), and *B. nigra* (2n = BB = 16) (Fig. 3.1).

A number of wide crosses are, however, often incompatible as a result of barriers to crossing such as the abortion of hybrid embryos (Liu 1984; Zhang et al. 2001). However, with the development of in vitro techniques such as ovary culture and embryo culture, it is now possible to utilize these genetic resources by distant hybridization (Agnihotri et al. 1991).

3.2 Major Barriers to Overcome in Distant Hybridization

3.2.1 Pre-fertilization Barriers

The barriers in distant hybridization may occur at both the stage from pollination to fertilization (pre-fertilization) and the stage from fertilization to development of the embryo into a seed and then a fertile plant (post-fertilization). It was suggested that the pre-fertilization barriers in wide cross among *Brassica* species might be mainly due to the incompatible reaction caused by pollen-stigma interaction (Meng 1990).

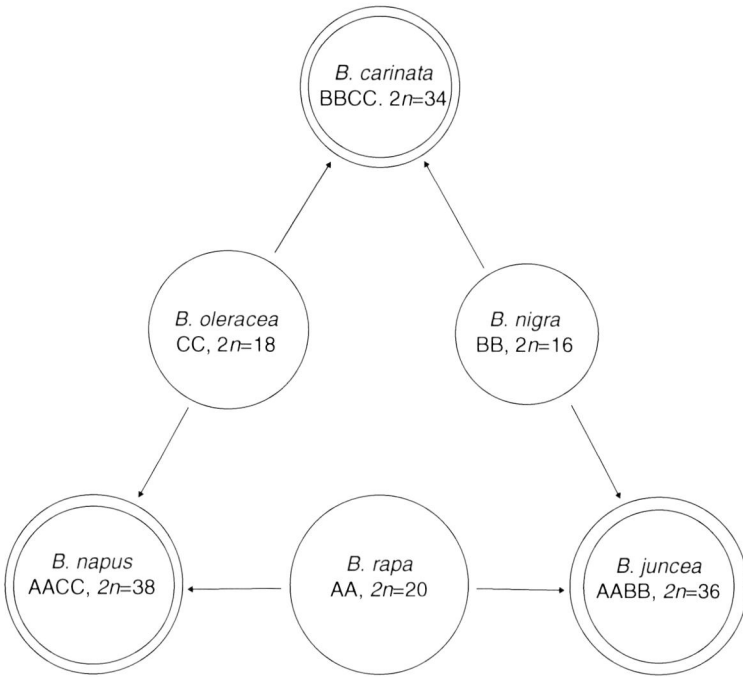

Fig. 3.1 Relationship of six natural *Brassica* species in U's triangle. (Adapted from U 1935)

Pre-fertilization barriers include failure of pollen adherence, hydration or germination, pollen tube growth, pollen tube growing into stigma, pollen tube growing into style, and pollen tube penetrating into ovule (Meng 1990; Bhat and Sarla 2004).

3.2.2   Post-fertilization Barriers

Successful formation of wide cross hybrids may also be limited by post-fertilization barriers, which include embryo and endosperm abortion, abnormal growth of result- ing hybrids and sterility of resulting hybrids. The possible mechanism of post- fertilization barriers is a dysfunctional embryo-endosperm relationship during seed development (Friedman and Ryerson 2009). Embryo and endosperm abortion may be overcome by sequential ovary culture or embryo rescue (Ayotte et al. 1987, 1989; Mohapatra and Bajaj 1988; Abel et al. 2005). Abnormal growth and sterility of hybrids may be overcome by selection or doubling the chromosome number of the hybrids (Pradhan et al. 2010).

3.3   Somatic Hybridization

Somatic cell fusion, which has been applied to transfer genes between various cruifers for increasing yield and disease resistance, is an important technique to bypass the sexual incompatibility barriers in distant hybridization. By somatic hybridization it is possible to transfer some useful genes such as disease resistance, nitrogen fixation, rapid growth rate, more product formation rate, protein quality, frost hardiness, drought resistance, herbicide resistance, heat and cold resistance from one species to another (Verma et al. 2004). Somatic hybridization generates novel hybrid and cybrid combinations of species that are sexually incompatible through new nuclear-cytoplasmic combinations (Davey et al. 2005).

3.3.1   Methods of Protoplast Fusion

In general, isolated protoplasts do not fuse with each other spontaneous because the surface of protoplasts carries negative charges around the outside of the plasma membrane. Thus, the fusion usually needs a fusion inducing chemicals which actually reduce the electronegativity of the isolated protoplasts and allow them to fuse with each other (Narayanswamy 1994). Currently, the most commonly used procedures of protoplast fusion are electrofusion, chemical treatment of protoplasts with polyethylene glycol (PEG) or a combination of these treatments.

Several chemicals have been used to induce protoplast fusion such as sodium nitrate, PEG, calcium ions (Ca^{2+}). Most somatic hybrid plants have been produced after the treatment of PEG. The PEG treatment induces agglutination of the protoplasts and fusion will occur after dilution of the PEG with a solution containing a high concentration of calcium ions at high pH. The resulting fusion frequency can vary roughly from 1 % to 20 % depending upon cell types and fusion conditions employed. Chemofusion is inexpensive, but it is cytotoxic and non selective, has low fusion frequency and the resulting products are not single ones (Waara and Glimelius 1995).

The electrofusion process is initiated by resuspending protoplasts in a medium of low conductivity in a chamber separated by two electrodes and applying a high alternating electric field. The protoplasts will move in the electric field by dielectrophoresis and will thereby become attached to each other like chains of pearls. A short pulse of direct current is then applied to induce fusion (Bates and Hasenkampf 1985). Electrofusion, having fusion frequency up to 100 %, is easy to control and has less cytotoxic. But the equipment is sophisticated and expensive (Senda et al. 1979).

3.3.2 Application of Somatic Hybridization in Brassicaceae Family

Since Schenck and Röbbelen (1982) obtained the hybrids of *Brassica oleracea* and *B. rapa* by using somatic cell fusion for the first time, considerable interspecific and intergeneric hybrids have been produced by somatic cell fusing in Brassicaceae. Qian et al. (2003) fused *B. rapa* and *B. napus* to transfer yield genes. *B. rapa* and *B. oleracea* have been fused to produce the somatic hybrids resistant to bacterial soft rot (Ren et al. 2000). Somatic hybridization has been used to produce very wide crosses. Skarzhinskaya et al. (1996) reported a wide somatic hybrid that was formed by the fusion of mesophyll protoplasts of *B. juncea* and *Lesquerella fendlert*. Two plants from symmetric fusions could be fertilized and set seeds after cross-pollination with *B. napus*. From the asymmetric fusions nine plants could be selfed as well as fertilized when backcrossed with *B. napus*. Somatic hybridization combinations between *Brassica* cultivated species and wild relatives of *Brassica, Sinapis, Moricandia, Raphanus, Eruca, Armoracia, Diplotaxis, Crambe,* and *Orichophragmus* are shown in Table 3.1. Intertribal somatic hybridization combinations between *Brassica* cultivated species and *Arabidopsis, Barbarea, Camelina, Lesqurella,* and *Thlaspi* are shown in Table 3.2.

Cytoplasmic male sterility (CMS) plays an important role in oilseed rape hybrid seed production. However, the natural occurrence of CMS has rarely been found in *Brassica* species. Recently, a number of new male sterile lines in *Brassica* have been developed by somatic hybridization within species or between species and genera. Barsby et al. (1987) reported that 'Polima' CMS, widely used in seed production of oilseed rape, was transferred in one step within *Brassica napus* L. by protoplast fusion. The 'Polima' CMS was transferred from a spring line Polima Regent to a winter line AWR of oilseed rape. The transfer was confirmed by restriction fragment length polymorphism (RFLP) analysis of mitochondrial DNA. All agronomic traits studied were stable through subsequent sexual generations. This introduction of CMS to winter lines was achieved in approximately only 9 months by protoplast fusion, which would take 3 years by a conventional backcross normally. Male-sterile cybrids have also been produced by the fusion of protoplasts of *B. rapa* and *B. oleracea* (Cardi and Earle 1997). The 'Anand' cytoplasm, which was derived from the wild species *B. tournefortii*, was transferred from *B. rapa* to *B. oleracea*. Prior to fusion with PEG, donor protoplasts were inactivated with 30 krad γ-rays and recipient ones with 3 mM iodoacetate, respectively. Sixty four percent of the cybrids were male-sterile. After crossed with fertile pollinations by both hand and insect pollinations, they found that some of sterile lines had good female fertility. Hu et al. (2002a, b) obtained the male-sterile cybrids of *B. napus* and *Sinapis arvensis* (Xinjiang wild mustard), *B. napus* and *Orychophragmus violaceus*, respectively, by somatic hybridization. Somatic hybridization between *Arabidopsis thaliana* and *B. napus* also produced the male-sterile hybrids (Leino et al. 2003). Hu and Li (2006) summarized the new male-sterile lines of *Brassica* which were produced by the fusion of protoplasts between genera or tribes (Table 3.3).

Table 3.1 Somatic hybridization combinations between species from different genera of the tribe Brassiceae

Somatic hybridization combination	Method of fusion	F1 hybrid[a]	Reference
Brassica napus (n=19) (+) *B. tournefortii* (n=10) irradiated with X-ray	PEG induced	+	Stiewe and Röbbelen (1994)
B. napus (n=19) (+) *Crambe abyssinica* (n=45) irradiated with UV	PEG induced	+	Wang et al. (2003)
B. napus (n=19) (+) *Diplotaxis harra* (n=13)	PEG induced	+	Klimaszewska and Keller (1988)
B. napus (n=19) (+) *D. muralis* (n=21)	PEG induced, Electrofusion	+	McLellan et al. (1987)
B. napus (n=19) (+) *Eruca sativa* (n=11)	PEG induced	+	Fahleson et al. (1988)
B. napus (n=19) (+) *Moricandia arvensis* (n=14)	PEG induced	+	O'Neill et al. (1996)
B. napus (n=19) (+) *M. nitens* (n=14)	PEG induced	+	Meng et al. (1999)
B. napus (n=19) (+) *Orychophragmus violaceus* (n=12) pre-treated with iodoacetate or irradiated with X-ray	PEG induced	+	Hu et al. (2002a)
B. napus (n=19) pre-treated with iodoacetate (+) *Raphanus sativus* (n=9)	PEG induced, PEG induced, PEG induced	+	Sundberg and Glimelius (1991), Lelivelt and Krens (1992), Sakai and Imamura (1990)
B. napus (n=19) (+) *Sinapis alba* (n=12)	PEG induced, Electrofusion	+	Lelivelt et al. (1993), Wang et al. (2005)
B. napus (n=19) pre-treated with iodoacetate (+) *S. arvensis* (n=9)	PEG induced	+	Hu et al. (2002b)
Brassica juncea (n=18) (+) *Diplotaxis catholia* (n=9)	PEG induced	+	Kirti et al. (1995b)
B. juncea (n=18) (+) *D. harra* (n=13)	PEG induced	+	Begum et al. (1995)
B. juncea (n=18) (+) *D. muralis* (n=21) irradiated or unirradiated with γ-ray	PEG induced	+	Chatterjee et al. (1988)
B. juncea (n=18) (+) *Eruca sativa* (n=11) irradiated or unirradiated with γ-ray	PEG induced	+	Sikdar et al. (1990)
B. juncea (n=18) (+) *Moricandia arvensis* (n=14)	PEG induced	+	Kirti et al. (1992b)

B. juncea (n=18)	(+)	*Sinapis alba* (n=12)		Electrofusion	+	Gaikwad et al. (1996)
B. juncea (n=18)	(+)	*Trachystoma ballii* (n=8)		PEG induced	+	Kirti et al. (1992a, 1995a)
B. oleracea (n=9)	(+)	*Armoracia rusticanna* (n=16)		PEG induced	+	Navratilova et al. (1997)
B. oleracea (n=9)	(+)	*M. nitens* (n=14)		PEG induced	+	Yan et al. (1999)
B. oleracea (n=9)	(+)	*M. arvensis* (n=14)		Dextran method, PEG induced	+	Toriyama et al. (1987a); Ishikawa et al. (2003)
B. oleracea (n=9)	(+)	*Raphanus sativus* (n=9)		Electrofusion, PEG induced	+	Hagimori et al. (1992); Yamanaka et al. (1992)
B. oleracea (n=9)	(+)	*S. alba* (n=12)		PEG induced	+	Hansen and Earle (1997)
B. oleracea (n=9)	(+)	*S. turgida* (n=9)		Dextran method	+	Toriyama et al. (1987b)
B. rapa (n=10)	(+)	*M. nitens* (n=14)		PEG induced	+	Meng et al. (1999)
B. rapa (n=10)	(+)	*Raphanus sativus* (n=9)		PEG induced	+	Pelletier et al. (1983)
B. nigra (n=8)	(+)	*S. turgida* (n=9)		Dextran method	+	Toriyama et al. (1987b)

[a]+: Hybrids more than one plant

Table 3.2 Intertribal somatic hybridization combinations between *Brassica* cultivated species and other species within crucifer family

Somatic hybridization combination		Method of fusion	F1 hybrid[a]	Reference
Brassica napus (n = 19)	(+) *Arabidopsis thaliana* (n = 5) irradiated or unirradiated with X-ray or UV	PEG induced	+	Forsberg et al. (1994; 1998); Bohman et al. (1999)
B. napus (n = 19)	(+) *Barbarea vulgaris* (n = 8)	PEG induced		Fahleson et al. (1994a)
B. napus (n = 19)	(+) *Camelina sativa* (n = 13)	Electrofusion	+	Jiang et al. (2009)
B. napus (n = 19)	(+) *Lesquerella fendleri* (n = 6) irradiated or unirradiated with X-ray	PEG induced	+	Skarzhinskaya et al. (1996)
B. napus (n = 19)	(+) *Thlaspi caerulescens* (n = 7)	Electrofusion	+	Brewer et al. (1999)
B. napus (n = 19)	(+) *T. perfoliatum* (n = 21) irradiated or unirradiated with X-ray	PEG induced	+	Fahleson et al. (1994b)
B. oleracea (n = 9)	(+) *A. thaliana* (n = 5) irradiated with γ-ray	PEG induced	+	Nitovskaya and Shakhovskii (1998)
B. oleracea (n = 9)	(+) *Camelina sativa* (n = 13)	PEG induced	+	Hansen (1998)
B. rapa (n = 10)	(+) *A. thaliana* (n = 5)	PEG induced	+	Gleba and Hoffmann (1978)
B. nigra (n = 8) irradiated with X-ray	(+) *A. thaliana* (n = 5)	PEG induced	+	Siemens and Sacristan (1995)
B. carinata (n = 17)	(+) *A. thaliana* (n = 5)	PEG induced	+	Gleba and Hoffman (1980)
B. carinata (n = 17)	(+) *C. sativa* (n = 13)	PEG induced	+	Narasimhulu et al. (1994)

[a]+: Hybrids more than one plant

3.4   Embryo Rescue to Overcome in Abortion of Hybrid Embryos

Embryo rescue is one of the earliest and most successful in vitro culture techniques that are used to assist in the development of plant embryos that might not survive to become viable plants (Sage et al. 1999). The basic premise for this technique is that integrity of the hybrid genome retained in a developmentally arrested or an abortive embryo and that its potential to resume normal growth may be realized if supplied with the proper growth substances. The technique depends on isolating the embryo without injury, formulating a suitable nutrient medium, and inducing continued embryogenic growth and seedling formation (Bridgen 1994).

Table 3.3 Alloplasmic male sterility systems established by somatic hybridization in oilseed *Brassica* crops (Adapted from Hu and Li 2006)

Brassica species	Kindred species	Mitochondrial type	Chloroplast type	Restorer	Reference
Brassica juncea (n = 18)	*Trachystoma ballii* (n = 8)	Recombinant	Recombinant	Found	Kirti et al. 1995a
B. juncea (n = 18)	*Moricandia arvensis* (n = 14)	*Moricandia arvensis*	Recombinant	Found	Kirti et al. 1998; Prakash et al. 1998
B. napus (n = 19)	*Raphanus sativus* (n = 9)	Recombinant	Recombinant	Found	Sakai and Imamura 1992
B. napus (n = 19)	*Brassica tournefortii* (n = 10)	Recombinant	*Brassica napus*	Found	Stiewe and Robbelen 1994
B. napus (n = 19)	*Arabidopsis thaliana* (n = 5)	Mix or Recombinant	*Brassica napus*	Not Known	Leino et al. 2003
B. napus (n = 19)	*Sinapis arvensis* (n = 9)	Recombinant	Not Determined	Found	Hu et al. 2002a
B. napus (n = 19)	*Orychophragmus violaceus* (n = 12)	Recombinant	Not Determined	Found	Hu et al. 2002b

The primary use of embryo rescue is considered to develop the interspecific and intergeneric hybrids. Since the hybrid embryos of distant hybridization are prone to abortion soon after they begin to develop because of the cross incompatible, rescue of the very-early stage embryos is often necessary to overcome the abnormal development of hybrid embryos (Cisneros and Tel-Zur 2010). Embryo rescue techniques also have been utilized to obtain progeny from hybrid seeds that are too weak to germinate normally.

Embryo rescue techniques considerably help in obtaining wide hybrids in which failure of endosperm to properly develop causes embryo abortion. Since its first use in *Brassica* by Nishi et al. (1959), extensive investigations have been carried out to improve the techniques for obtaining higher seed set in crucifers (Inomata 1976, 1978a, b, 1979, 2003; Zhang et al. 2003, 2004). It has been well summarized by Kaneko et al. (2009) that different embryo rescue techniques to overcome hybrid inviability and breakdown in early stages between *Brassica* crops and wild relatives of *Brassica*, *Sinapis*, *Diplotaxis*, *Moricandia*, *Eruca*, and *Orichophragmus*.

3.4.1 Procedures of Embryo Rescue

Embryo rescue may be classified as embryo, ovule, and ovary cultures depending on the organ cultured. The most commonly used embryo rescue procedure is embryo culture, in which embryos are excised and placed directly in culture medium. In some case, however, it is not technically possible to take embryos out of the ovules without injury. Thus, to rescue these embryos, the whole ovule or even ovary is cultured.

3.4.1.1 Embryo Culture

Since embryos are located in the sterile environment of the ovule, surface sterilization of embryos is not necessary. Instead, surface sterilization of whole ovules or ovaries is applied and then embryos are removed aseptically from the surrounding tissues. A harsh disinfestation procedure is usually enough since the embryo is well-protected by the surrounding tissue. Thus, axenic cultures of embryos are often easily established (Reed 2004). For example, Li et al. (1995) obtained the hybrid plants of *Brassica napus* and *Orychophragmus violaceus* by embryo culture with only seed coats sterilized (using 70 % ethanol for 5 min and then 0.15 % $HgCl_2$ for 15 min) before embryos were excised from the seeds.

Since the embryo is very small, micro-dissecting tools and a dissecting microscope are usually used to excise the embryo without injury. The intactness of the embryo is crucial to the embryo culture. The embryos should not be damaged either when they are dissected from the ovule or during culture (Rangan 1984). The point of incision into the ovule is also important. In some cases, the embryos are excised at the micropylar end of the ovule, while for heart-stage and younger embryos, it is important to keep the suspensors intact (Hu and Wang 1986). After its excision, the embryo should be placed directly into culture medium so that it does not become dry (Reed 2004).

The culture of immature embryos is difficult due to the micro-dissection operation and the complex nutrient medium requirements (Monnier 1978; Raghavan 1980). Thus, in practice embryo culture is sometimes preceded by ovule or ovary culture (ovary-embryo or ovary-ovule culture) to delay the embryo excision until the embryo becomes large enough to be removed without damage. Brown et al. (1997) obtained hybrid plants of *Sinapis alba* and *Brassica napus* by using ovary-embryo culture technique. Firstly, developing ovaries were excised from the plant 8 days after pollination and surface sterilized for 15 min in a 15 % bleach solution. The ovaries were then rinsed with sterile water and transferred to a sterile culture medium for a further 10–14 days. At this time the ovaries were removed from the media, under sterile conditions, and examined for developing ovules and subsequently developing embryos. These embryos were rescued and cultured in the fresh media. When hybridizing *B. rapa* with *Diplotaxis erucoides* and *B. maurourm*, the sequential ovary-ovule culture technique was used to rescue the hybrid embryos (Garg et al. 2007). For the ovary culture, pistils were excised 2–3 days after pollination, surface sterilized with mercuric chloride (0.01 %) for 8 min and cultured. Pollinated ovaries 8–9 days after the initial culture were dissected and enlarged ovules were re-cultured in a fresh culture medium.

3.4.1.2 Ovule Culture

The embryos are difficult to excise from the ovule when they are very young or from small-seeded species. To avoid embryos excision process, the whole ovules are sometimes cultured in a culture medium directly. Ovaries are collected and surface-sterilized, and then the ovules are removed from the ovaries and placed into culture media. For large-seeded species, this step is easy to accomplish, while for small-seeded polyovulate species, it is time-consuming and difficult to achieve (Reed 2004).

Some successful ovule culture systems have been established to save the hybrid embryos in the wide hybridization between crop brassicas and their wild relatives. Rawsthorne et al. (1998) reported that an intergeneric hybrid plant was produced between *Moricandia nitens* and *Brassica napus* by sexual hybridization and in vitro ovule culture. The developing ovules were dissected aseptically from the ovaries and placed into tissue culture media for 2–3 weeks after pod-set occurred. Seven interspecific hybrids were also produced between *B. maurorum*, a wild species resistant to *Alternaria* blight and white rust, and all the monogenomic (*B. campestris*, *B. nigra* and *B. oleracea*) and digenomic (*B. juncea*, *B. napus* and *B. carinata*) crop brassicas through embryo rescue (Chrungu et al. 1999). For ovule culture, pistils 10 days after pollination were surface-sterilized and the ovules were dissected and cultured.

3.4.1.3 Ovary Culture

Ovary culture, without dissecting ovule or embryo, is convenient to operate. In ovary culture, the entire ovary is placed into culture medium. When ovaries are sampled, the residual parts of flower should be cleared. The ovary should be surface-sterilized thoroughly without injury. Then, place the ovary into culture and make sure the cut end of the pedicel is in the medium (Reed 2004).

According to the review of Kaneko et al. (2009), ovary culture is more widely used in the distant hybridization between *Brassica* crops and their wild relatives. Rapeseed (*Brassica napus*) was synthesized from interspecific hybridization between *B. rapa* and *B. oleracea* through ovary culture techniques (Zhang et al. 2004; Wen et al. 2008). Intergeneric hybrids between *Moricandia arvensis* and *B. campestris* and *B. nigra* were also produced through ovary culture (Takahata and Takeda 1990). The ovaries were excised at 3–7 days after pollination and sterilized followed by being placed into agar-solidified culture medium.

3.4.2 Factors Affecting Embryo Rescue

3.4.2.1 Cross Combination

When attempting to rescue the hybrid embryos of incompatible crosses, the cross combination greatly influences success. Embryos of some cross combinations are easier to grow in culture medium than the others. Hybrids were obtained through either ovary culture from crosses *Brassica napus* × *Hirschfeldia incana*, *B. rapa* × *Moricandia arvensis*, and their reciprocal ones, or embryo culture from crosses *B. napus* × *Sinapis alba* and their reciprocal ones (Lefol et al. 1996; Takahata and Takeda 1990; Heyn 1977; Brown et al. 1997). However, Batra et al. (1990) and Inomata (1994) using ovary culture technique, produced hybrids from the crosses of *Diplotaxis siifolia* × *B. napus*, *Diplotaxis siifolia* × *B. rapa*, and *B. napus* × *Sinapis pubescens*, but not in the reciprocal ones. Similarly, the combination *Enarthrocarpus lyratus* × *B. oleracea* and *Eruca gallicum* × *B. juncea* also produced hybrids through ovule culture and ovary culture, respectively, but the reciprocal ones did not do (Gundimeda et al. 1992; Lefol et al. 1997).

3.4.2.2 Time of Sampling

As with the cross combination, the time of sampling of explants is also important. It is critical that the cultures be initiated prior to embryo abortion. The time of embryo abortion can be evaluated by histological examinations, but these evaluations are very laborious. Thus, the explants sampling is often started at various time points following pollination to maximize chances of obtaining viable plants (Reed 2004). The sampling time is ranged widely with different cross combinations and culture techniques. Wen et al. (2008) investigated the sampling time of both ovary culture and embryo culture using the reciprocal crosses between *Brassica oleracea* and *B. rapa*. They found that when ovaries were cultured at 4–7 days after pollination (DAP) the efficient of hybrid production was higher in *B. rapa* × *B. oleracea*. In embryo culture, the hybrid rate was significantly enhanced at 16–18 DAP, up to 48.1 % in *B. oleracea* × *B. rapa*.

3.4.2.3 Culture Media

The most widely used basal media in embryo rescue are Murashige and Skoog (MS) (Murashige and Skoog 1962) and Gamborg's B5 (Gamborg et al. 1968) culture media. The media supplements required depend on the development stage of embryos. In the heterotrophic phase, the young embryo depends on the endosperm and requires complex media supplemented with various vitamins, amino acids, growth hormones, and natural extracts such as coconut milk and casein hydrolysate. In the autotrophic phase, embryos usually can germinate on a simple inorganic medium supplemented with sucrose (Raghavan 1966; Sharma et al. 1996). The most commonly used energy source in embryo rescue is sucrose, which also serves as an osmoticum in the media (Bridgen 1994). Agar with the concentrations of 0.5–1.5 % is generally used to solidify culture media in embryo rescue (Hu and Wang 1986).

Although they are extensively used in embryo rescue studies, plant growth regulators seem to be not necessary for embryo rescue. Bridgen (1994) thought auxins and cytokinins were not generally used for embryo culture unless callus induction was needed. To rescue the hybrid embryos of the cross between *Sinapis alba* and *Brassica napus*, Brown et al. (1997) used the ovary-embryo technique. For ovary culture, the most productive media was MS basal medium added 30 mg/L sucrose and 8 mg/L agar. Subsequently, the embryos were excised from the ovaries and placed in fresh culture media in which the constituents were the same as for the ovary culture until small plantlets were produced. No plant growth regulators were referred to through all the culture process. Takahata (1990) also obtained many hybrid embryos in the *Moricandia arvensis* × *Brassica oleracea* cross through ovary culture using hormone-free MS medium. However, the embryos produced in the ovaries failed to develop into plantlets on hormone-free MS or B5 medium. When the embryos failed to develop into plantlets, hypocotyls were cultured on MS medium containing 1.0 mg/L 6-benzyladenine (BA) or no hormones in order to induce shoot regeneration.

3.5 Microspore Embryogenesis

Microspore culture technique is widely used in *Brassica* breeding and also in genetic research, biochemical and physiological studies for plant breeding and experimental purposes due to its simplicity, efficiency in haploid and doubled haploid (DH) production, mutation and germplasm generation, and gene transformation (Palmer et al. 1996; Xu et al. 2007; Pratap et al. 2009). Microspore culture is relatively simpler and more effective to produce a DH population, which helps breeders develop homozygous lines from heterozygous parents in a single generation and allows fixing the recombinant gametes directly as fertile homozygous lines (Pratap et al. 2009). The BC_2S_3 lines from a cross between *Brassica napus* and *B. carinata* were fixed by Navabi et al. (2010) using microspore culture. Besides shortening breeding cycle, microspore culture also can be helpful in experimental purposes. It was used by Nelson et al. (2009) and Mason et al. (2011) to analyze the male gametes of interspecific hybrids of *B. napus* × *B. carinata*. Mason (2010) also used microspore culture technique to obtain male gametes from a near-hexaploid individual developed by hybridizing three allotetraploid species. Chen et al. (2011) suggested that this technique may help stabilize the trigenomic *Brassica* hybrids (AABBCC).

3.5.1 Microspore Culture

The microspore culture technique is relatively simpler and easier than that of anther culture in *Brassica* (Pratap et al. 2009). The flower buds of *Brassica napus* were selected as the donor, and the growth conditions were set as follows: a 16 h photoperiod (approximately 350 μE m^{-2} s^{-1}), a day/night temperature of 20/15 °C and a high nutritive status (Xu et al. 2007). For winter *B. napus* genotypes, the

TULANE UNIVERSITY LIBRARY

vernalization was conducted under the conditions of 4 °C and 8 h photoperiod for 8 weeks. One week before microspore isolation the day/night temperature was adjusted to 12/10 °C (Zhou et al. 2002b).

For microspore isolation, flower buds with appropriate size, which contained microspores at the late uninucleate stage and early binucleate stage, were selected. The proper stage of microspore development was determined from a combination of flower bud size (usually 2–3 mm in length for *B. rapa*, 3–4 mm for *B. napus*, and 5–6 mm for *B. oleracea*) and the ratio of petal length to anther length (P/A, usually 0.5–0.75 for *B. rapa* and *B. napus*, and 1.0–1.2 for *B. oleracea*), followed by the identification of microspore development under a dissecting microscope (Gu et al. 2004a; Xu et al. 2007). These buds are surface sterilized in 0.5 % sodium hypochlorite or 0.1 % $HgCl_2$ prior to washing 3 times with sterilized water, and then macerated in cold B5-13 medium (Gamborg et al. 1968), supplemented with 13 % sucrose at pH 6.0, and subsequently filtered through a 40-µm nylon filter into a centrifuge tube. The filtered microspores were centrifuged at 750 rpm for 5 min, and then were resuspended in NLN-13 medium (Lichter 1982) with 13 % sucrose at pH 6.0. The microspore density was about $3–10 \times 10^4$ per ml (Pratap et al. 2009).

Activated charcoal is added prior to dispending the suspension into 60×15 mm Petri dishes, 4 ml in each. The dishes are sealed with Parafilm and incubated at 30 °C in the dark for 7 days in the case of *Brassica napus*, or at 32 °C for 2 days in the case of *B. rapa* and *B. oleracea*, and then moved to 24 °C still in the dark (Xu et al. 2007). After 1 day of incubation following microspore isolation, the medium should be refreshed, for which the suspension should be centrifuged at 750 rpm for 5 min and the medium should be then changed with the same amount of fresh NLN-13 medium, and cultured as described above (Gu et al. 2003, 2004b).

3.5.2   Embryogenesis and Plant Development

Once embryos are visible to the naked eye (approximately 9–15 days after isolation), cultures are transferred to a slow rotary shaker (45 rpm) in the dark at 24 °C (Zhang et al. 2006a, b). After cultured 2–3 weeks, various embryos at different developmental stages such as globular to torpedo to cotyledonary stage are observed (Pratap et al. 2009). For plantlet induction, large embryos at the late torpedo stage are transferred to solid MS regeneration medium (2 % sucrose, half-strength macro nutrients and 2 mg/L 6-benzylaminopurine, solidified with 0.9 % agar, pH 5.8). The cultures are incubated under a 16 h photoperiod at 24 °C with low light density (approximately 100 µmol m^{-2} s^{-1}) (Zhang et al. 2006a, b). When shoots developed they are cut off from any callus or hypocotyl tissue and transferred to plastic growth vessels with fresh solid MS medium (Murashige and Skoog 1962; 2 % sucrose, 0.9 % agar, pH 5.8) for roots induction. Plantlets are transferred to a soil-perlite mixture and kept for 2 weeks in a nursing room under 16 h daylength at 24 °C with low light intensity and high relative humidity. Gradual adaptation to greenhouse conditions followed (Zhou et al. 2002a; Zhang et al. 2006a, b; Xu et al. 2007).

The microspore embryogenesis and haploid production are influenced by various factors including donor plant genotype, anther and microspore developmental stage, donor plant physiology, culture media and pretreatments, culture conditions, culture environment (Gu et al. 2003; Zhang et al. 2006b; Pratap et al. 2009). These various factors affecting microspore embryogenesis have been reviewed by Xu et al. (2007) and Pratap et al. (2009).

3.6 Doubling Chromosomes

The chromosomes are usually unpaired in hybrids of distant hybridization, thereby causing the hybrids sterile. Intergeneric hybridization followed by chromosome doubling of sterile hybrids produce allopolyploids, which might be fertile (Momotaz et al. 1998). Chen et al. (2011) reported that the trigenomic allohexaploids could be obtained via crossing tetraploid × diploid species to produce ABC triploids, followed by chromosome doubling. In the other hand, 70–90 % of regenerated plants from microspore-derived embryoids are haploid from rapeseed microspore culture (Charne et al. 1988; Chen and Beversdorf 1992). To develop the DH lines, the chromosome doubling technique is needed.

Colchicine, which is an anti-microtubule agent to inhibit spindle formation during mitosis in actively-dividing cells, is the most widely-used chemical to double the chromosome (Xu et al. 2007). Chromosome doubling of sterile haploid plantlets is usually achieved by culturing plantlets in colchicine-containing medium in the greenhouse (Mathias and Röbbelen 1991), or soaking roots or whole plants in a colchicine solution (Fletcher et al. 1998; Zhang et al. 2004, 2006a). These methods as well as injecting colchicine into the secondary buds or applying a colchicine-soaked cotton swab to axillary buds often cause the development of chimeric with small and uneven sectors of diploidization (Lichter et al. 1988; Chen et al. 2011). During microspore culture, Zhou et al. (2002a) reported that a high doubling efficiency (83–91 %) of haploid microspores of spring *Brassica napus* was achieved by treating them immediately with colchicines at 500 mg/L for 15 h. Immediate colchicine treatment of isolated microspores is superior to colchicine treatments of microspore-derived plants and microspore-derived embryos because it has higher efficiency of embryogenesis and lower chimeric percentages in *Brassica napus* (Chen et al. 1994; Zhou and Hagberg 2000). Thus, colchicine application of isolated microspores is safer, quicker and more efficient (Pratap et al. 2009; Chen et al. 2011).

3.7 Conclusion

The Brassicaceae family is pretty huge, in which a great number of wild species are the potential sources of novel and valuable characteristics of economic importance. This enables breeders to use distant hybridization, a useful strategy for introducing

genes from wild relatives or other cultivated species into target crops, to breed new
varieties of crops or create new types of plants. Although there are cross-
incompatibility barriers in distant hybridization, it is possible to utilize these new
genetic resources to broaden the genetic diversity for the breeding of cruciferous
crops due to the development in in vitro techniques such as somatic hybridization
and embryo rescue technique. A synergy of conventional sexually cross and biotechnological approaches will certainly enhance Brassica breeding by using the
gene pool resource and transferring the novel quality and resistance traits from wild
crucifers.

Acknowledgments This work was supported by National Natural Science Foundation of China
(31000678, 31071698, 31170405), the Science and Technology Department of Zhejiang Province
(2012C12902-1, 2011R50026-5), China Postdoctoral Science Foundation (2011M501012,
2012T50555), Jiangsu Province Agricultural Scientific and Technological Autonomous Innovation
Project (No. CX (10) 463), and National High Technology Research and Development Program of
China (2011AA10A206). Weijun Zhou (the corresponding author) is grateful to the 985-Institute
of Agrobiology and Environmental Sciences of Zhejiang University for providing convenience in
using the experimental equipments.

References

Abel S, Mollers C, Becker HC (2005) Development of synthetic *Brassica napus* lines for the
    analysis of "fixed heterosis" in allopolyploid plants. Euphytica 146:157–163
Agnihotri A, Lakshmikumaran MS, Prakash S, Jagannathan V (1991) Embryo rescue of *B
    napus × Raphanobrassica* hybrids. Cruciferae Newslett 14–15:92–93
Ayotte R, Harney PM, Machado VS (1987) The transfer of triazine resistance from *Brassica napus*
    L. to *Brassica oleracea* L. 1. Production of F1 hybrids through embryo rescue. Euphytica
    36:615–624
Ayotte R, Harney PM, Machado VS (1989) The transfer of triazine resistance from *Brassica napus*
    L. to *Brassica oleracea* L. 4. 2nd and 3rd backcross to *Brassica oleracea* and recovery of an
    18-chromosome, triazine resistance BC3. Euphytica 40:15–19
Barsby TL, Yarrow SA, Kemble RJ, Grant I (1987) The transfer of cytoplasmic male-sterility to
    winter-type oilseed rape (*Brassica napus* L.) by protoplast fusion. Plant Sci 53:243–248
Bates GW, Hasenkampf CA (1985) Culture of plant somatic hybrids following electrical fusion.
    Theor Appl Genet 70:227–233
Batra V, Prakash S, Shivanna KR (1990) Intergeneric hybridization between *Diplotaxis siifolia*, a
    wild-species and crop brassicas. Theor Appl Genet 80:537–541
Begum F, Paul S, Bag N, Sikdar SR, Sen SK (1995) Somatic hybrids between *Brassica juncea* (L).
    Czern. and *Diplotaxis harra* (Forsk.) Boiss and the generation of backcross progenies. Theor
    Appl Genet 91:1167–1172
Bhat S, Sarla N (2004) Identification and overcoming barriers between *Brassica rapa* L. em.
    Metzg. and *B. nigra* (L.) Koch crosses for the resynthesis of *B. juncea* (L.) Czern. Genet Resour
    Crop Evol 51:455–469
Bohman S, Forsberg J, Glimelius K, Dixelius C (1999) Inheritance of *Arabidopsis* DNA in offspring
    from *Brassica napus* and *A. thaliana* somatic hybrids. Theor Appl Genet 98:99–106
Brewer EP, Saunders JA, Angle JS, Chaney RL, McIntosh MS (1999) Somatic hybridization
    between the zinc accumulator *Thlaspi caerulescens* and *Brassica napus*. Theor Appl Genet
    99:761–771
Bridgen MP (1994) A review of plant embryo culture. Hortscience 29:1243–1246

Brown J, Brown AP, Davis JB, Erickson D (1997) Intergeneric hybridization between *Sinapis alba* and *Brassica napus*. Euphytica 93:163–168

Cardi T, Earle ED (1997) Production of new CMS *Brassica oleracea* by transfer of 'Anand' cytoplasm from *B. rapa* through protoplast fusion. Theor Appl Genet 94:204–212

Charne DG, Pukacki P, Kott LS, Beversdorf WD (1988) Embryogenesis following cryo-preservation in isolated microspores of rapeseed (*Brassica napus* L.). Plant Cell Rep 7:407–409

Chatterjee G, Sikdar SR, Das S, Sen SK (1988) Intergeneric somatic hybrid production through protoplast fusion between *Brassica juncea* and *Diplotaxis muralis*. Theor Appl Genet 76:915–922

Chen JL, Beversdorf WD (1992) Cryopreservation of isolated microspores of spring rapeseed (*Brassica napus* L.) for in vitro embryo production. Plant Cell Tiss Org 31:141–149

Chen ZZ, Snyder S, Fan ZG, Loh WH (1994) Efficient production of doubled haploid plants through chromosome doubling of isolated microspores in *Brassica napus*. Plant Breed 113:217–221

Chen S, Nelson MN, Chevre AM, Jenczewski E, Li Z, Mason AS, Meng J, Plummer JA, Pradhan A, Siddique KHM, Snowdon RJ, Yan G, Zhou W, Cowling WA (2011) Trigenomic bridges for *Brassica* improvement. Crit Rev Plant Sci 30:524–547

Chrungu B, Verma N, Mohanty A, Pradhan A, Shivanna KR (1999) Production and characteriza-tion of interspecific hybrids between *Brassica maurorum* and crop brassicas. Theor Appl Genet 98:608–613

Cisneros A, Tel-Zur N (2010) Embryo rescue and plant regeneration following interspecific crosses in the genus *Hylocereus* (Cactaceae). Euphytica 174:73–82

Davey MR, Anthony P, Power JB, Lowe KC (2005) Plant protoplast technology: current status. Acta Physiol Plant 27:117–129

Fahleson J, Rahlen L, Glimelius K (1988) Analysis of plants regenerated from protoplast fusions between *Brassica napus* and *Eruca sativa*. Theor Appl Genet 76:507–512

Fahleson J, Eriksson I, Glimelius K (1994a) Intertribal somatic hybrids between *Brassica napus* and *Barbarea vulgaris* production of in vitro plantlets. Plant Cell Rep 13:411–416

Fahleson J, Eriksson I, Landgren M, Stymne S, Glimelius K (1994b) Intertribal somatic hybrids between *Brassica napus* and *Thlaspi perfoliatum* with high content of the *T. perfoliatum-specific* nervonic acid. Theor Appl Genet 87:795–804

Fletcher R, Coventry J, Kott LS (1998) Doubled haploid technology for spring and winter *Brassica napus* (revised edition). OAC Publication, University of Guelph, Ontario, p 42

Forsberg J, Landgren M, Glimelius K (1994) Fertile somatic hybrids between *Brassica napus* and *Arabidopsis thaliana*. Plant Sci 95:213–223

Forsberg J, Dixelius C, Lagercrantz U, Glimelius K (1998) UV dose-dependent DNA elimination in asymmetric somatic hybrids between *Brassica napus* and *Arabidopsis thaliana*. Plant Sci 131:65–76

Friedman WE, Ryerson KC (2009) Reconstructing the ancestral female gametophyte of angiosperms: insights from *Amborella* and other ancient lineages of flowering plants. Am J Bot 96:129–143

Gaikwad K, Kirti PB, Sharma A, Prakash S, Chopra VL (1996) Cytogenetical and molecular investigations on somatic hybrids of *Sinapis alba* and *Brassica juncea* and their backcross progeny. Plant Breed 115:480–483

Gamborg OL, Miller RA, Ojima K (1968) Nutrient requirements of suspension cultures of soybean root cells. Exp Cell Res 50:151–158

Garg H, Banga S, Bansal P, Atri C, Banga SS (2007) Hybridizing *Brassica rapa* with wild cruci-fers *Diplotaxis erucoides* and *Brassica maurorum*. Euphytica 156:417–424

Gleba YY, Hoffmann F (1978) Hybrid cell lines *Arabidopsis thaliana* + *Brassica campestris*: no evidence for specific chromosome elimination. Mol Gen Genet 165:257–264

Gleba YY, Hoffmann F (1980) "Arabidobrassica": a novel plant obtained by protoplast fusion. Planta 149:112–117

Gu HH, Zhou WJ, Hagberg P (2003) High frequency spontaneous production of doubled haploid plants in microspore cultures of *Brassica rapa* ssp. *chinensis*. Euphytica 134:239–245

Gu HH, Hagberg P, Zhou WJ (2004a) Cold pretreatment enhances microspore embryogenesis in oilseed rape (*Brassica napus L.*). Plant Growth Regul 42:137–143

Gu HH, Tang GX, Zhang GQ, Zhou WJ (2004b) Embryogenesis and plant regeneration from isolated microspores of winter cauliflower (*Brassica oleracea* var. *botrytis*). J Zhejiang Univ (Agric Life Sci) 30:34–38

Gundimeda HR, Prakash S, Shivanna KR (1992) Intergeneric hybrids between *Enarthrocarpus lyratus*, a wild-species, and crop brassicas. Theor Appl Genet 83:655–662

Hagimori M, Nagaoka M, Kato N, Yoshikawa H (1992) Production and characterization of somatic hybrids between the Japanese radish and cauliflower. Theor Appl Genet 84:819–824

Hansen LN (1998) Intertribal somatic hybridization between rapid cycling *Brassica oleracea* L. and *Camelina sativa* (L.) Crantz. Euphytica 104:173–179

Hansen LN, Earle ED (1997) Somatic hybrids between *Brassica oleracea* L. and *Sinapis alba* L. with resistance to *Alternaria brassicae* (Berk.) Sacc. Theor Appl Genet 94:1078–1085

Heyn FW (1977) Analysis of unreduced gametes in the *Brassiceae* by crosses between species and ploidy level. Z Pflanzenzüchtg 78:13–70

Hu Q, Andersen SB, Dixelius C, Hansen LN (2002a) Production of fertile intergeneric somatic hybrids between *Brassica napus* and *Sinapis arvensis* for the enrichment of the rapeseed gene pool. Plant Cell Rep 21:147–152

Hu Q, Hansen LN, Laursen J, Dixelius C, Andersen SB (2002b) Intergeneric hybrids between *Brassica napus* and *Orychophragmus violaceus* containing traits of agronomic importance for oilseed rape breeding. Theor Appl Genet 105:834–840

Hu Q, Li YC (2006) Induction and improvement of cytoplasmic male sterility in oilseed *Brassica* by somatic hybridization. Acta Agron Sin 32:138–143

Hu C, Wang P (1986) Embryo culture: technique and application. In: Evans DA, Sharp WR, Ammirato PV (eds) Handbook of plant cell culture, vol 4. Macmillan, New York, pp 43–96

Inomata N (1976) Culture in vitro of excised ovaries in *Brassica campestris* L. I. Development of excised ovaries in culture media, temperature and light. Jpn J Breed 26:229–236

Inomata N (1978a) Production of interspecific hybrids between *Brassica campestris* and *Brassica oleracea* by culture in vitro of excised ovaries .2. Effects of coconut milk and casein hydrolysate on development of excised ovaries. Jpn J Genet 53:1–11

Inomata N (1978b) Production of interspecific hybrids in *Brassica campestris* × *Brassica oleracea* by culture in vitro of excised ovaries .1. Development of excised ovaries in crosses of various cultivars. Jpn J Genet 53:161–173

Inomata N (1979) Production of interspecific hybrids in *Brassica campestris* × *B. oleracea* by culture in vitro of excised ovaries II. Development of excised ovaries in culture media. Jpn J Breed 29:115–120

Inomata N (1994) Intergeneric hybridization between *Brassica napus* and *Sinapis pubescens*, and the cytology and crossability of their progenies. Theor Appl Genet 89:540–544

Inomata N (2003) Production of intergeneric hybrids between *Brassica juncea* and *Diplotaxis virgata* through ovary culture, and the cytology and crossability of their progenies. Euphytica 133:57–64

Ishikawa S, Bang SW, Kaneko Y, Matsuzawa Y (2003) Production and characterization of intergeneric somatic hybrids between *Moricandia arvensis* and *Brassica oleracea*. Plant Breed 122:233–238

Jiang JJ, Zhao XX, Tian W, Li TB, Wang YP (2009) Intertribal somatic hybrids between *Brassica napus* and *Camelina sativa* with high linolenic acid content. Plant Cell Tiss Org Cul 99:91–95

Kaneko Y, Bang SW, Matsuzawa Y (2009) Distant hybridization. In: Gupta SK (ed) Biology and breeding of crucifers. CRC, London, pp 207–247

Kirti PB, Narasimhulu SB, Prakash S, Chopra VL (1992a) Production and characterization of intergeneric somatic hybrids of *Trachystoma ballii* and *Brassica juncea*. Plant Cell Rep 11:90–92

Kirti PB, Narasimhulu SB, Prakash S, Chopra VL (1992b) Somatic hybridization between *Brassica juncea* and *Moricandia arvensis* by protoplast fusion. Plant Cell Rep 11:318–321

Kirti PB, Mohapatra T, Baldev A, Prakash S, Chopra VL (1995a) A stable cytoplasmic male-sterile line of *Brassica juncea* carrying restructured organelle genomes from the somatic hybrid *Trachystoma ballii* + *B. juncea*. Plant Breed 114:434–438

Kirti PB, Mohapatra T, Khanna H, Prakash S, Chopra VL (1995b) *Diplotaxis catholica* + *Brassica juncea* somatic hybrids: molecular and cytogenetic characterization. Plant Cell Rep 14:593–597

Kirti PB, Prakash S, Gaikwad K, Kumar VD, Bhat SR, Chopra VL (1998) Chloroplast substitution overcomes leaf chlorosis in a *Moricandia arvensis* based cytoplasmic male sterile *Brassica juncea*. Theor Appl Genet 97:1179–1182

Klimaszewska K, Keller WA (1988) Regeneration and characterization of somatic hybrids between *Brassica napus* and *Diplotaxis harra*. Plant Sci 58:211–222

Lefol E, Fleury A, Darmency H (1996) Gene dispersal from transgenic crops. 2. Hybridization between oilseed rape and the wild heavy mustard. Sex Plant Reprod 9:189–196

Lefol E, SeguinSwarts G, Downey RK (1997) Sexual hybridisation in crosses of cultivated *Brassica* species with the crucifers *Erucastrum gallicum* and *Raphanus raphanistrum*: potential for gene introgression. Euphytica 95:127–139

Leino M, Teixeira R, Landgren M, Glimelius K (2003) *Brassica napus* lines with rearranged *Arabidopsis* mitochondria display CMS and a range of developmental aberrations. Theor Appl Genet 106:1156–1163

Lelivelt CLC, Krens FA (1992) Transfer of resistance to the beet cyst nematode (*Heterodear schachtii* Schm.) into the *Brassica napus* L. gene pool through intergeneric somatic hybridization with *Raphanus sativus* L. Theor Appl Genet 83:887–894

Lelivelt CLC, Leunissen EHM, Frederiks HJ, Helsper J, Krens FA (1993) Transfer of resistance to the beet cyst nematode (*Heterodear schachtii* Schm.) from *Sinpis alba* L. (white mustard) to the *Brassica napus* L. gene pool by means of sexual and somatic hybridization. Theor Appl Genet 85:688–696

Li ZY, Liu Y (2001) Cytogenetics of intergeneric hybrids between *Brassica* species and *Orychophragmus violaceus*. Progr Nat Sci 11:721–727

Li Z, Liu HL, Luo P (1995) Production and cytogenetics of intergeneric hybrids between *Brassica napus* and *Orychophragmus violaceus*. Theor Appl Genet 91:131–136

Lichter R (1982) Induction of haploid plants from isolated pollen of *Brassica napus*. Z Pflanzenphysiol 103:229–237

Lichter R, Degroot E, Fiebig D, Schweiger R, Gland A (1988) Glucosinolates determined by HPLC in the seeds of microspore-derived homozygous lines of rapeseed (*Brassica napus* L.). Plant Breed 100:209–221

Liu HL (1984) Genetics and breeding of oilseed rape. Shanghai Science and Technology Press, Shanghai

Mason A (2010) Meiotic behavior and chromosome inheritance in interspecific hybrids of allotetraploid *Brassica* species. Ph.D. Thesis, The University of Western Australia, Perth

Mason AS, Nelson MN, Castello M-C, Yan G, Cowling WA (2011) Genotypic effects on the frequency of homoeologous and homologous recombination in *Brassica napus* × *B. carinata* hybrids. Theor Appl Genet 122:543–553

Mathias R, Robbelen G (1991) Effective diploidization of microspore-derived haploids of rape (*Brassica napus* L.) by in vitro colchicine treatment. Plant Breed 106:82–84

McLellan MS, Olesen P, Power JB (1987) Towards the introduction of cytoplasmic male sterility (CMS) into *Brassica napus* through protoplast fusion. In: Puite KJ, Done JJM, Iuizing HJ, Kool AJ, Koornneef M, Krens FA (eds) Progress in plant protoplast research. Proceedings of the 7th international protoplast symposium, Wageningen, pp 187–188

Meng JL (1990) Studies on pollen-pistil interaction between *Brassica napus* and its relative species and genus. Acta Agron Sin 16:19–26

Meng JL, Yan Z, Tian Z, Huang R, Huang B (1999) Somatic hybrids between *Moricandia nitens* and three *Brassica* species. In: Proceedings of the 10th international rapeseed congress, Canberra, pp 4–8

Mohapatra D, Bajaj YPS (1988) Hybridization in *Brassica juncea* × *Brassica campestris* through ovary culture. Euphytica 37:83–88

Momotaz A, Kato M, Kakihara F (1998) Production of intergeneric hybrids between *Brassica* and *Sinapis* species by means of embryo rescue techniques. Euphytica 103:123–130

Monnier M (1978) Culture of zygotic embryos. In: Thorpe TA (ed) Frontiers of plant tissue culture. University of Calgary Press, Canada, pp 277–286

Murashige T, Skoog F (1962) A revised medium for rapid growth and bio assays with tobacco tissue cultures. Physiol Plant 15:473–497

Narasimhulu SB, Kirti PB, Bhatt SR, Prakash S, Chopra VL (1994) Intergeneric protoplast fusion between *Brassica carinata* and *Camelina sativa*. Plant Cell Rep 13:657–660

Narayanswamy S (1994) Plant protoplast: isolation, culture and fusion. In: Plant cells and tissue cultures. TATA MCGraw Hill Publishing, New Delhi

Navabi ZK, Parkin IAP, Pires JC, Xiong Z, Thiagarajah MR, Good AG, Rahman MH (2010) Introgression of B-genome chromosomes in a doubled haploid population of *Brassica napus* × *B. carinata*. Genome 53:619–629

Navratilova B, Buzek J, Siroky J, Havranek P (1997) Construction of intergeneric somatic hybrids between *Brassica oleracea* and *Armoracia rusticana*. Biol Plant 39:531–541

Nelson M, Mason A, Castello MC, Thomson L, Yan G, Cowling W (2009) Microspore culture preferentially selects unreduced (2n) gametes from an interspecific hybrid of *Brassica napus* L. × *Brassica carinata* Braun. Theor Appl Genet 119:497–505

Nishi S, Kawata J, Toda M (1959) In the breeding of interspecific hybrids between two genomes "c" and "a" of *Brassica* through the application of embryo culture techniques. Jpn J Breed 5:215–222

Nitovskaya IA, Shakhovskii AM (1998) Obtaining of axisymmetrical somatic hybrids between *Brassica oleracea* L. and *Arabidopsis thaliana* L. Tsitologiya i Genetika 32:72–81

O'Neill CM, Murata T, Morgan CL, Mathias RJ (1996) Expression of the C_3-C_4 intermediate character in somatic hybrids between *Brassica napus* and the C_3-C_4 species *Moricandia arvensis*. Theor Appl Genet 93:1234–1241

Palmer CE, Keller WA, Arnison PG (1996) Experimental haploidy in *Brassica* species. In: Jain SM, Sopory SK, Veilleux RE (eds) In vitro haploid production in higher plants, vol 3. Kluwer Academic, Dordrecht, pp 143–172

Pelletier G, Primard C, Vedel F, Chetrit P, Remy R, Rousselle R, Renard M (1983) Intergeneric cytoplasmic hybridization in cruciferae by protoplast fusion. Mol Gen Genet 191:244–250

Pradhan A, Plummer JA, Nelson MN, Cowling WA, Yan G (2010) Successful induction of trigenomic hexaploid *Brassica* from a triploid hybrid of *B. napus* L. and *B. nigra* (L.) Koch. Euphytica 176:87–98

Prakash S, Kirti PB, Bhat SR, Gaikwad K, Kumar VD, Chopra VL (1998) A *Moricandia arvensis*-based cytoplasmic male sterility and fertility restoration system in *Brassica juncea*. Theor Appl Genet 97:488–492

Pratap A, Gupta SK, Takahata Y (2009) Microsporogenesis and haploidy breeding. In: Gupta SK (ed) Biology and breeding of crucifers. CRC, London, pp 293–307

Qian W, Liu R, Meng J (2003) Genetic effects on biomass yield in interspecific hybrids between *Brassica napus* and *B. rapa*. Euphytica 134:9–15

Raghavan V (1966) Nutrition growth and morphogenesis of plant embryos. Biol Rev 41:1–58

Raghavan V (1980) Embryo culture. In: Vasil IK (ed) Perspectives in plant cell and tissue culture. Academic, New York, pp 209–240

Rangan TS (1984) Culture of ovule. In: Vasil IK (ed) Cell culture and somatic cell genetics of plants, Vol. 1: laboratory procedures and their application. Academic, New York, pp 227–231

Rawsthorne S, Morgan CL, O'Neill CM, Hylton CM, Jones DA, Frean ML (1998) Cellular expression pattern of the glycine decarboxylase P protein in leaves of an intergeneric hybrid between the C_3-C_4 intermediate species *Moricandia nitens* and the C_3 species *Brassica napus*. Theor Appl Genet 96:922–927

Reed S (2004) Embryo rescue. In: Trigiano RN, Gray DJ (eds) Plant development and biotechnology. CRC, London, pp 235–239

Ren JP, Dickson MH, Earle ED (2000) Improved resistance to bacterial soft rot by protoplast fusion between *Brassica rapa* and *B. oleracea*. Theor Appl Genet 100:810–819

Sage TL, Strumas F, Cole WW, Barrett SCH (1999) Differential ovule development following self- and cross-pollination: the basis of self-sterility in *Narcissus triandrus* (Amaryllidaceae). Am J Bot 86:855–870

Sakai T, Imamura J (1990) Intergeneric transfer of cytoplasmic male-sterility between *Raphanus sativus* (cms line) and *Brassica napus* through cytoplast-protoplast fusion. Theor Appl Genet 80:421–427

Sakai T, Imamura J (1992) Alteration of mitochondrial genomes containing ATPA genes in the sexual progeny of cybrids between *Raphanus sativus* CMS line and *Brassica napus* cv. westar. Theor Appl Genet 84:923–929

Schenk HR, Röbblen G (1982) Somatic hybrids by fusion of protoplasts from *Brassica oleracea* and *B. campestris*. Z Pflanzenzüchtg 89:278–288

Senda M, Takeda J, Abe S, Nakamura T (1979) Induction of cell-fusion of plant-protoplasts by electrical-stimulation. Plant Cell Physiol 20:1441–1443

Sharma DR, Kaur R, Kumar K (1996) Embryo rescue in plants – a review. Euphytica 89:325–337

Siemens J, Sacristan MD (1995) Production and characterization of somatic hybrids between *Arabidopsis thaliana* and *Brassica nigra*. Plant Sci 111:95–106

Sikdar SR, Chatterjee G, Das S, Sen SK (1990) Erussica, the intergeneric fertile somatic hybrid developed through protoplast fusion between *Eruca sativa* Lam. And *Brassica juncea* (L.) Czen. Theor Appl Genet 79:561–567

Skarzhinskaya M, Landgren M, Glimelius K (1996) Production of intertribal somatic hybrids between *Brassica napus* L. and *Lesquerella fendleri* (Gray) wats. Theor Appl Genet 93:1242–1250

Stiewe G, Robbelen G (1994) Establishing cytoplasmic male-sterility in *Brassica napus* by mitochondrial recombination with *B. tournefortii*. Plant Breed 113:294–304

Sundberg E, Glimelius K (1991) Production of cybrid plants within Brassicaceae by fusing protoplasts and plasmolytically induced cytplasts. Plant Sci 79:205–216

Takahata Y (1990) Production of intergeneric hybrids between a C3-C4 intermediated species *Moricandia arvensis* and a C3 species *Brassica oleracea* through ovary culture. Euphytica 46:259–264

Takahata Y, Takeda T (1990) Intergeneric (intersubtribe) hybridization between *Moricandia arvensis* and *Brassica*-A and *Brassica*-B genome species by ovary culture. Theor Appl Genet 80:38–42

Toriyama K, Hinata K, Kameya T (1987a) Production of somatic hybrid plants, 'Brassicomoricandia', through protoplast fusion between *Moricandia arvensis* and *Brassica oleracea*. Plant Sci 48:123–128

Toriyama K, Kameya T, Hinata K (1987b) Selection of a universal hybridizer in *Sinapis turgid* Del. and regeneration of plantlets from somatic hybrids with *Brassica* species. Planta 170:308–313

U N (1935) Genomic analysis in *Brassica* with special reference to the experimental formation of B. napus and peculiar mode of fertilization. Jpn J Bot 7:389–452

Verma N, Bansal MC, Kumar V (2004) Protoplast fusion technology and its biotechnological applications. Department of paper technology, Indian institute of technology, Roorkee/ Saharanpur

Waara S, Glimelius K (1995) The potential of somatic hybridization in crop breeding. Euphytica 85:217–233

Wang YP, Sonntag K, Rudloff E (2003) Development of rapeseed with high erucic acid content by asymmetric somatic hybridization between *Brassica napus* and *Crambe abyssinica*. Theor Appl Genet 106:1147–1155

Wang YP, Sonntag K, Rudloff E, Chen JM (2005) Intergeneric somatic hybridization between *Brassica napus* L. and *Sinapis alba* L. J Integr Plant Biol 47:84–91

Wen J, Tu JX, Li ZY, Fu TD, Ma CZ, Shen JX (2008) Improving ovary and embryo culture techniques for efficient resynthesis of *Brassica napus* from reciprocal crosses between yellow-seeded diploids *B. rapa* and *B. oleracea*. Euphytica 162:81–89

Xu L, Najeeb U, Tang GX, Gu HH, Zhang GQ, He Y, Zhou WJ (2007) Haploid and doubled haploid technology. In: Gupta SK (ed) Advances in botanical research, rapeseed breeding, vol 45.
Academic, San Diego, pp 181–216

Yamanaka H, Kuginuki Y, Kanno T, Nishio T (1992) Efficient production of somatic hybrids
between *Raphanus sativus* and *Brassica oleracea*. Jpn J Breed 42:329–339

Yan Z, Tian Z, Huang B, Huang R, Meng J (1999) Production of somatic hybrids between *Brassica
oleracea* and the C_3-C_4 intermediate species *Moricandia nitens*. Theor Appl Genet
99:1281–1286

Zhang GQ, Zhou WJ, Yao XL, Zhang ZJ (2001) Studies on distant hybridization in *Brassica*
plants. J Shanxi Agric Sci 29:25–30

Zhang GQ, Zhou WJ, Gu HH, Song WJ, Momoh EJJ (2003) Plant regeneration from the hybridization of *Brassica juncea* and *B. napus* through embryo culture. J Agron Crop Sci 189:347–350

Zhang GQ, Tang GX, Song WJ, Zhou WJ (2004) Resynthesizing *Brassica napus* from interspecific hybridization between *Brassica rapa* and *B. oleracea* through ovary culture. Euphytica
140:181–187

Zhang GQ, He Y, Xu L, Tang GX, Zhou WJ (2006a) Genetic analyses of agronomic and seed
quality traits of doubled haploid population in *Brassica napus* through microspore culture.
Euphytica 149:169–177

Zhang GQ, Zhang DQ, Tang GX, He Y, Zhou WJ (2006b) Plant development from microspore-
derived embryos in oilseed rape as affected by chilling, desiccation and cotyledon excision.
Biol Plant 50:180–186

Zhou WJ, Hagberg P (2000) High frequency production of doubled haploid rapeseed plants by
direct colchicine treatment of isolated microspores. J Zhejiang University (Agric Life Sci)
26:125–126

Zhou WJ, Hagberg P, Tang GX (2002a) Increasing embryogenesis and doubling efficiency by
immediate colchicine treatment of isolated microspores in spring *Brassica napus*. Euphytica
128:27–34

Zhou WJ, Tang GX, Hagberg P (2002b) Efficient production of doubled haploid plants by immediate
colchicine treatment of isolated microspores in winter *Brassica napus*. Plant Growth Regul
37:185–192

Chapter 4
Microspore Culture and Doubled Haploid Technology

Yoshihito Takahata, Yu Takahashi, and Ryo Tsuwamoto

Abstract Haploids and doubled haploids (DHs) produced by in vitro culture of gametophytic cells, especially male gametophyto, are of great importance to plant breeding and basic science. The routine protocol of isolated microspore culture of *Brassica* has been established, and used worldwide. However, recently, new protocols dealing with multiple samples have been developed. Although various factors influencing microspore embryogenesis, plant regeneration and diplodization have been examined, attempts to enhance these efficiencies have continued. With the advancement of molecular genetics, many approaches from both forward- and reverse-genetics for understanding the mechanism of microspore embryogenesis have been attempted. This chapter reviews the recent progress of microspore embryogenesis of *Brassica* in relation to culture protocol and factors affecting microspore embryogenesis, and recent achievements in elucidating the mechanism of microspore embryogenesis.

Keywords Microspore embryogenesis • Doubled haploid • Protocol development • Mechanism

4.1 Introduction

Haploids and doubled haploids (DHs) are of great importance to practical breeding and basic science. In plant breeding, homozygous lines are utilized as final varieties in self-pollinating crops and as parent lines in F1 hybrid varieties in

Y. Takahata, Ph.D. (✉) • Y. Takahashi, Ph.D.
Iwate University, 3-18-8 Ueda, Morioka, Iwate 020-8550, Japan
e-mail: ytakahata@iwate-u.ac.jp

R. Tsuwamoto, Ph.D.
Misato Agricultural Extension Centre, Miyagi prefecture, 5 Sasadate, Misato-machi,
Toda-gun, Miyagi 987-0005, Japan

S.K. Gupta (ed.), *Biotechnology of Crucifers*, DOI 10.1007/978-1-4614-7795-2_4,   45
© Springer Science+Business Media, LLC 2013

cross-pollinating crops. Since successful production of haploids via an embryogenesis from isolated microspore culture of *B. napus* was reported by Lichter (1982), a large number of studies have been performed to improve this technique in genus *Brassica* and allied genera (reviewed by Takahata 1997; Palmer and Keller 1999; Ferrie and Keller 2004; Pratap et al. 2009). Numerous studies have also been performed to apply to genetic manipulation such as mutation, in vitro selection, transformation and artificial seed because of the haploid and/or DH production system from a single cell (reviewed by Takahata et al. 2005; Ferrie and Mollers 2011). Furthermore, this system can be utilized as a model system of zygotic embryogenesis and cell differentiation. In addition, DH populations are utilized as materials for construction of linkage map and for QTL analysis (Pink et al. 2008).

In this chapter, we describe the recent progress of microspore embryogenesis of *Brassica* in relation to culture protocol and factors affecting microspore embryogenesis, as well as recent achievements elucidating the mechanism of microspore embryogenesis.

4.2 Improvement of Culture Technique

The methodology of microspore culture of *Brassica*, whose basic components (sterilization, isolation of microspores and culture medium) almost have barely changed, was established in 1980s, and has been employed routinely around the world (Custers 2003; Dias 2003; Ferrie 2003). Although the routine protocol of microspore culture has been established, genotypic variations remain as a non-dissolved factor. For breeding and basic science, it is important to identify high responsive genotypes. However, surveying the response of a large number of genotypes is laborious. Recently, we developed the microspore culture method for dealing with multiple samples (Takahashi et al. 2011), and are now developing the method for super-multiple samples (Takahashi et al. unpublished). Such improved protocols can deal with 2–10 times more genotypes than the conventional one during the same period. Figure 4.1 shows the comparison of these three methods. In the conventional method (Fig. 4.1 A–F), generally 4–6 genotypes at maximum have been used in one experiment (Ferrie 2003). On the other hand, in the improved method for multiple samples (Fig. 4.1 G–L), 12 or more genotypes are used in one experiment. In addition, the improved method is characterized by (1) easy handling of samples in a plastic tube, (2) easy microspore isolation using a beads cell disruptor, and (3) a simple method of sterilization of tools used. When compared between the improved method and the conventional method, the number of isolated microspores per bud and the number of embryos per dish were not significantly different between the two methods (Takahashi et al. 2011). This improved method provides some advantages in breeding and basic science in various phases such as the survey of high responsive genotypes and QTL analysis on microspore embryogenesis.

Furthermore, we are developing the protocol for super-multiple samples (Fig. 4.1 M–S). This method can deal with 96 genotypes at maximum in one experiment.

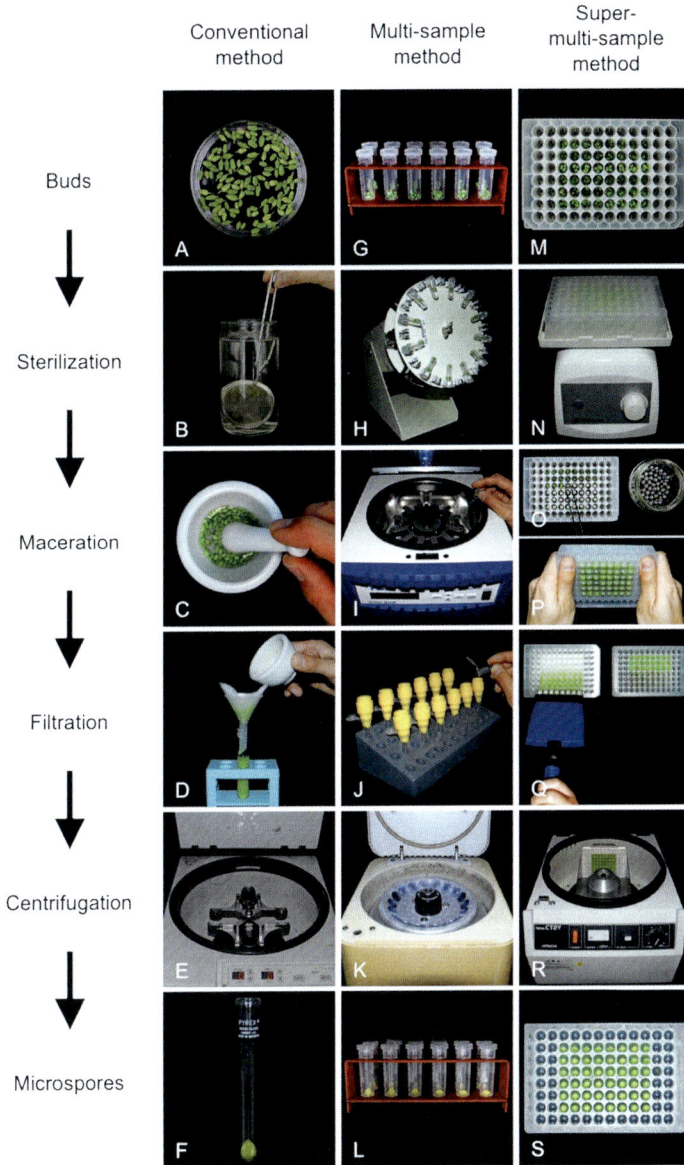

Fig. 4.1 The procedure of three microspore culture methods for *Brassica* and allied genera. The conventional method (**A–F**), multi-sample method (**G–L**) and supermulti-sample method (**M–S**). The buds in a Petri dish (**A**), 2 ml plastic tube (**G**) and 96-well plate (**M**). Sterilization of buds by sodium hypochlorite in tea egg (**B**), 2 ml plastic tube (**H**) and 96 well-plate on vortex mixer (**N**). Maceration of buds by a pestle and mortar (**C**), multi-beads shocker (**I**) and shake with stainless steel balls (**O**, **P**). Filtration of the microspore suspension through Miracloth (**D**), CellTrics filters (**J**) and 96-well filters plate (**Q**). Centrifuges (**E**, **K**, **R**). Isolated microspores (**F**, **L**, **S**)

This method is characterized by using a 96-well plate for bud sterilization and isolation of microspores. Microspore isolation is carried out in a 96-well plate with stainless steel balls. Preliminary experiments using *B. napus* showed that the number of isolated microspores per bud and the number of embryos per dish were not significantly different among the three methods.

4.3   Factors Influencing Microspore Embryogenesis

As mentioned by many reviews, various factors influence microspore embryogenesis, involving pre-isolation factors such as genotypes and physical conditions of donor plants, microspore developmental stage and pretreatment, and post-isolation factors such as culture media and culture conditions (Takahata 1997; Palmer and Keller 1999; Ferrie and Keller 2004; Pratap et al. 2009). In this part, we describe factors affecting microspore embryogenesis with a focus on recent progress.

4.3.1   Genotypes and Growth Conditions of Donor Plants

Although the routine protocol of microspore culture of *Brassica* has been established, studies on finding high responsive genotypes of recalcitrant species and subspecies and on overcoming genotypic variation have been attempted (Table 4.1).

As in various tissue culture and microspore embryogenesis of other crops, genotype is one of the most important factors. Large genotypic variations on embryo yield were reported not only within the same subspecies, but also among different species. It is unclear whether high responsive genotypes have some common characters or not. In *B. napus*, *B. rapa* ssp. *pekinensis* and *B. juncea*, genotypes of spring type, which bolts irrespective of chilling exposure, showed high response of embryogenesis (Ohkawa et al. 1987; Kuginuki et al. 1997; Hiramatsu et al. 1995; Ferrie 2003). On the other hand, Keller et al. (1987) reported that winter types had a high response in anther culture of *B. napus*. Despite recent progress in the study of molecular markers, molecular markers closely linked to microspore embryogenesis have not yet been found.

The ability of embryogenesis is known to be enhanced by selection. High responsive line Topas DH4079, which is frequently used as a model plant of *Brassica* microspore culture, is a selected line of cv. Topas of *B. napus* (Ferrie and Mollers 2011). Embryogenic Westar DH-2 was reported to be selected from cv. Westar, one of recalcitrant cultivars of *B. napus* (Malik et al. 2008). The ability of embryogenesis could be transferred from a high responsive genotype to recalcitrant genotypes by sexual crossing (Fig. 4.2). Zhang and Takahata (2001) reported genetic analysis of microspore embryogenic ability in rapeseed (*B. napus*) and Chinese cabbage (*B. rapa*) using diallel analysis. They indicated that (1) the dominant genes had positive effects on microspore embryogenesis, (2) both additive and dominant effects were

Table 4.1 High responses of microspore embryogenesis in *Brassica* and allied genera

Species	Subspecies	Embryo yields	Reference
Brassica carinata		325.6/bud	Ferrie (2003)
B. juncea		586.5/1.2 × 10⁵ microspores	Agarwal et al. (2006)
B. napus	*oleifera*	50 % of microspores	Pechan and Keller (1988)
	rapifera	81.4/bud	Hansen and Svinnset (1993)
B. nigra		22.3/bud	Ferrie and Keller (2007)
B. oleracea	*acephala*	43.4/6 × 10⁵ microspores	Zhang et al. (2008)
	alboglabra	42.0/bud	Ferrie and Keller (2007)
	botrytis	0.24/bud	Duijs et al. (1992)
	capitata	223.0/bud	Yuan et al. (2012)
	fimbriata	4.2/bud	Duijs et al. (1992)
	gemmifera	1.17/bud	Duijs et al. (1992)
	italica	0.99 % of microspores	Takahata et al. (1993)
	sabauda	12.1/bud	Duijs et al. (1992)
B. rapa (syn. *B. campestris*)	*broccoletto*	0.08/2 × 10⁵ microspores	Takahashi et al. (2012)
	chinensis	57.4/bud	Cao et al. (1994)
	niposinica	4.07/2 × 10⁵ microspores	Takahashi et al. (2012)
	oleifera	407.6/bud	Ferrie (2003)
	parachinensis	0.46/bud	Wong et al. (1996)
	pekinensis	227/1 × 10⁵ microspores	Kuginuki et al. (1997)
	perviridis	4.67/2 × 10⁵ microspores	Takahashi et al. (2012)
	rapa	108/2 × 10⁵ microspores	Takahashi et al. (2011)
Crambe abyssinica		0.16/bud	Ferrie and Keller (2007)
Eruca sativa		80.0/2 × 10⁵ microspores	Leskovsek et al. (2008)
Raphanus oleifera		10.0/bud	Ferrie and Keller (2007)
R. sativa		9.0/1 × 10⁵ microspores	Takahata et al. (1996)
Sinapis alba		0.42/bud	Ferrie and Keller (2007)

significant, (3) no significant maternal effects were observed, (4) the heritability was high, especially broad-sense heritability was higher than 0.9, and (5) narrow-sense heritability was higher in *B. napus* than in *B. rapa*, because of a larger contribution of dominant effects in *B. rapa*. Furthermore, from the results of microspore embryogenesis in the F2 population of 'Lisandra' (high responsive) × 'Kamikita' (recalcitrant), Zhang and Takahata (2001) considered that microspore embryogenesis in *B. napus* is mainly controlled by two multiple gene loci with additive effects.

Growth and physiological conditions of donor plants are an important factor in microspore embryogenesis. It is empirically known that the donor plants used for microspore culture are needed to grow healthfully and vigorously (Ferrie 2003). Although donor plants are grown in a controlled environment chamber, greenhouse or field conditions, the best conditions are in a controlled growth chamber. The growth temperature after bolting is especially important. Donor plants grown at low temperature are more productive and the productivity is maintained for long periods of time (Keller et al. 1987; Takahata et al. 1991; Ferrie 2003). This is supported by the result of Prem et al. (2012) who reported that donor plants grown in a

Fig. 4.2 Comparison of embryo production of *B. napus* among high responsive cultivar (cv. Topas), recalcitrant cultivar (cv. Kamikita) and their reciprocal F1 hybrids

growth chamber at 15/10 °C showed seven times higher microspore embryogenic ability compared to donor plants in a greenhouse at 18 °C. Although the influence of the age of donor plants and bud position (main vs. lateral raceme) were formerly reported, they do not influence embryogenesis when the suitable developmental stage of microspores was used (Takahata et al. 1991).

4.3.2 Culture Medium

NLN medium, which is a minor modification of Lichter's (1982) medium, has been widely used for microspore culture of *Brassica* and allied genera. One-half NLN medium, in which the concentration of major salts is reduced to half, was often used (Takahata and Keller 1999; Wakui et al. 1994; Dias 2003; Takahashi et al. 2012), and was reported to induce better embryo production than full strength NLN medium in several genotypes (Dias 2001; Wang et al. 2009a). Of major salts, reduction of NO_3 was considered to mainly relate to promotion of embryogenesis (Dias 2001). Ohkawa et al. (2000)

Table 4.2 Factors to enhance microspore embryogenesis in *Brassica* and allied genera

Treatments	Substance	Reference[a]
Addition to medium		
Anti-auxin	p-Chlorophenoxyisobutyric acid	Agarwal et al. (2006)
		Ahmadi et al. (2012)
		Zhang et al. (2011)
Auxin	Naphthalene acetic acid	Zhang et al. (2012)
Cytokinin	6-benzylaminopurine	Takahashi et al. (2012)
		Zhang et al. (2012)
Brassinosteroid	24-epibrassinolide, Brassinolide	Ferrie et al. (2005)
		Belmonte et al. (2010)
Buffer agent	2-(N-Morpholino) ethanesulfonic acid	Yuan et al. (2012)
Carbide	Activated charcoal	Prem et al. (2008)
High medium pH	pH6.2, 6.4	Yuan et al. (2012)
Inhibitors of ethylene biosynthesis	Aminoethoxyvinylglycine Cobalt chloride	Leroux et al. (2009)
	Silver nitrate	Prem et al. (2005)
Osmoticum	Polyethylene glycol	Ferrie and Keller (2007)
Plant proteoglycans	Arabinogalactan proteins	Yuan et al. (2012)
Stress and pretreatment		
Antimicrotuble	Cholchicine	Agarwal et al. (2006)
Glycopeptide antibiotic	Bleomycin	Ahmadi et al. (2012)
		Zeng et al. (2010)
Mutagen	Ethyl methane sulphonate	Ferrie et al. (2008)
Low temperature		Takahashi et al. (2012)
		Yuan et al. (2011)

[a]The recent paper is listed

reported that nitrate was not used for early embryogenesis in *B. napus* because of the lack of nitrate reductase activity, and was an ideal source of nitrogen at the later stage of embryogenesis.

High concentration of sucrose is essential for induction of embryogenesis. Although 10–13 % sucrose is used as a standard concentration in *Brassica* (Takahata 1997), 8–17 % sucrose is sometimes favorable depending on genotypes (Baillie et al. 1992; Lionneton et al. 2001; Ferrie and Keller 2007). Such high concentration of sucrose is considered to work as a nutrient and osmoticum. Ilic-Grubor et al. (1998) reported that 25 % polyethylene glycol (PEG) with a very low level of sucrose induced similar embryo yields as 13 % sucrose in *B. napus*. Similar results were obtained in many genotypes of *B. napus*, and several species of Brassicaceae (Ferrie and Keller 2007). In particular, they reported that substitution of PEG for sucrose enhanced embryo yields in *B. nigra*, *Crambe abyssinica*, *Raphanus oleifera* and several genotypes of *B. napus*, and produced embryos, which are morphologically more similar to zygotic embryos.

Although the exogenous plant growth regulators and other additives are not essential in the embryogenesis of *Brassica* (Keller et al. 1987), several kinds of additives were recently reported to enhance microspore embryogenesis (Table 4.2).

The anti-auxin, *p*-chlorophenoxyisobutyric acid (PCIB) improved microspore embryogenesis by 2–6 fold in *B. juncea*, *B. rapa* and *B. napus* (Agarwal et al. 2006; Zhang et al. 2011; Ahmadi et al. 2012). The methods of treatment were different among these studies; one was pretreatment (1 day) of PCIB (Agarwal et al. 2006; Ahmadi et al. 2012), and another was treatment of PCIB during culture period (Zhang et al. 2011). Brassinosteroids (BRs) such as 24-epibrassinolide (EBR) and brassinolide (BL) increased the production of microspore embryogenesis in *B. napus* and *B. juncea*, however, they had no effects in *B. carinata*, *B. nigra* and *B. rapa* (Ferrie et al. 2005). The positive effect of BRs was recently reported by Belmonte et al. (2010), who found that BL increased embryo yield in *B. napus*, whereas treatment with brssinazole, a BL biosynthetic inhibitor, decreased the embryo yield. Ferrie et al. (2005) speculated that BRs have a role in protecting microspores from initial heat shock treatment. Low concentration of 6-benzylaminopurine (BA) has sometimes been used to increase embryo yield in *B. napus* (reviewed by Takahata 1997). In *B. rapa*, BA was also recommended in microspore culture without data (Ferrie 2003), and recently Takahashi et al. (2012) demonstrated that BA enhanced embryo yields in all genotypes of *B. rapa*. Aminoethoxyvinylglycine (AVG) and $CoCl_2$, inhibitors of ethylene biosynthesis, and silver nitrate ($AgNO_3$), an inhibitor of ethylene action, led to increased embryo yields in *B. napus* and *B. juncea*, respectively (Prem et al. 2005; Leroux et al. 2009). In addtion, Leroux et al. (2009) reported that S-adenosylmethionine (SAM), an ethylene biosynthesis precursor, and ethephon, the ethylene-releasing agent, decreased embryo yields. Yuan et al. (2012) reported that though adding 2-(N-Morpholino) ethanesulfonic acid (MES) or the arabinogalactan proteins (AGPs) alone had small effects for embryogenesis of *B. oleracea*, the combination of AGPs and MES in medium at pH6.4 enhanced the embryo yield. The use of activated charcoal (AC) has been reported to improve microspore embryogenesis in *Brassica* spp. (reviewed by Palmer and Keller 1999). In contrast, AC gave a negative effect in some reports (Prem et al. 2005; Takahashi et al. 2012). In almost all reports showing positive effect, addition of AC with agarose was used, but in reports showing negative effect, AC was used without agarose.

4.3.3 Stress and Pretreatment

In order to induce microspore embryogenesis, it is known that exposure of microspores to stress is needed (Shariatpanahi et al. 2006; Islam and Tuteja 2012). Heat shock is generally used, but other stressing agents have been developed. A high temperature treatment (32–33 °C) for 1–4 days is required for induction of embryogenesis in *Brassica* microspore culture. The exposure periods depend on the species. Treatments for 3–4 days bring about effective embryogenesis in *B. napus*, *B. juncea* and *B. carinata*, while 1 day exposure is optimum in *B. rapa*, *B. oleracea* and *B. nigra* (Takahata 1997). Ferrie and Keller (2007) reported that microspore embryogenesis was induced in *B. napus* without 32 °C heat shock treatment (3-day treatment at 4 °C, 15 °C, 18 °C, 24 °C), when microspores were cultured in the medium containing PEG instead of sucrose. Recently, a milder (short) heat-stress treatment (32 °C for less than 24 h) induced a higher frequency of embryos with suspensor-like structure (Joosen et al. 2007; Supena et al. 2008; Dubas et al. 2011). Prem et al. (2012) reported that embryogenesis with a producing suspensor was also induced at 18 °C. In particular, they demonstrated that such a system at lower temperatures induced embryogenesis through two different development pathways: one involving the formation of a suspensor-like structure (52.4 %) and another producing embryos without a suspensor (13.1 %). In addition, they observed that 34.4 % of non-responsive microspores followed a gametophytic-like development leading to trinucleate mature pollen.

In order to stimulate embryogenesis, other treatments, which substitute heat shock or are used in conjunction with heat shock, have been reported (Table 4.2). It is known that treatment of colchicine instead of heat shock or with heat shock stimulates embryogenesis (Zhao et al. 1996; Zhou et al. 2002; Agarwal et al. 2006). Furthermore, bleomycin (0.1–0.2 µg/ml), a glycopeptide antibiotic, for 20–30 min pretreatment increased embryo yields in *B. napus*, (Zeng et al. 2010; Ahmadi et al. 2012). Some mutagens such as γ-rays irradiation, N-Nitroso-N-Methyl Urethane (NMU) and ethyl methane sulphonate (EMS) were reported to promote microspore embryogenesis in *B. napus*, *B. oleracea* and *B. juncea* (Jeung and Lee 1997; Pechan and Keller 1989; Ferrie et al. 2008).

Low temperature (4 °C) pretreatment of buds is known to increase microspore embryogenesis (reviewed by Palmer and Keller 1999; Pratap et al. 2009). Recently, such pretreatments for 1–2 days and for 10 days were reported to enhance embryo yields in *B. oleracea* and *B. rapa*, respectively (Yuan et al. 2011; Takahashi et al. 2012). In addition to positive effects, the storage of buds at a low temperature could avoid the concentration of works in limited time (Sato et al. 2002).

4.4 Plant Regeneration and DH Production

In order to effectively obtain haploids and DHs, not only effective induction of microspore embryogenesis, but also a high frequency of plant regeneration from embryos is needed. The frequency of direct plant regeneration from embryos is often low, depending on genotypes and culture conditions. Several treatments of embryos such as low temperature, ABA and desiccation enhanced direct plant regeneration (reviewed by Takahata 1997; Palmer and Keller 1999). These treatments are considered to involve embryo maturation, because ABA relates to seed maturation and accumulation of storage substances, and other stress treatments are known to increase ABA concentration. Instead of these treatments, culturing embryos on filter paper placed on top of medium and/or on the medium containing high concentration of gelling agents (1.6% agar) increased the frequency of plant regeneration in *B. oleracea*, *B. napus* and *B. rapa* (Takahata and Keller 1991; Peng et al. 1994; Takahashi et al. 2012). Peng et al. (1994) indicated that optimum matric potential for regeneration was from −3 to −4 kpa, which was caused by 1.6 % agar.

The embryos induced by PEG showed more morphologically similarity to zygotic embryos (Ilic-Grubor et al. 1998; Ferrie and Keller 2007; Supena et al. 2008) and higher plant regeneration ability than those induced by sucrose in *B. napus* (Ferrie and Keller 2007). Belmonte et al. (2010) reported that the embryos induced in the medium supplemented with brassinolide (BL) had higher regeneration ability, because BL-treated embryos developed zygotic-like shoot apical meristem. Further, they indicated that the application of BL switched the glutathione and ascorbate pools to the oxidized forms, and these metabolic alterations were related to embryo structure and performance. In contrast, Ferrie et al. (2005) observed that brassinosteroids did not affect the plant regeneration from embryos. Although the exogenous plant growth regulators are not essential in the plant regeneration from microspore-derived embryos, 0.05–0.1 mg/l GA_3, 2.0 mg/l BA or 0.1 mg/l BA+0.2 mg/l IAA were reported to enhance the regeneration in *B. napus* (Ahmadi et al. 2012).

Chromosome doubling is needed for both plant breeding and basic science. DHs can be obtained by treating microspores in vitro or plants with antimicrotubule agents such as colchicines and trifluralin (Palmer and Keller 1999; Ferrie 2003; Pratap et al. 2009). Spontaneous chromosome doubling is advantageous, because it omits the need of doubling treatments. Although the frequency of spontaneous diploids in *B. rapa* tends to be higher than those of *B. napus* and *B. oleracea* (Takahata 1997), they depend on genotypes. Recently, Ferrie and Keller (2007) demonstrated that in *B. napus* higher spontaneous diploids (64–92 %) were obtained from embryos developed in the medium containing PEG instead of sucrose, in comparison with 2–18 % spontaneous diploids from those in the sucrose medium. The embryos obtained in the medium supplemented with brassinosteroids increased the frequency of spontaneous diploids in *B. napus* (Ferrie et al. 2005).

4.5 Mechanism

Microspore embryogenesis is an efficient single-cell system and this developmental pathway is morphologically similar to zygotic embryogenesis. Various studies have been carried out to understand the mechanism of induction from gametophytic to sporophytic development. In particular, morphological, molecular biological and genetic studies have contributed to the understanding of microspore embryogenesis. Late uninucleate to early binucleate microspores are well known to be suitable for induction of embryogenesis in *Brassica*. In pollen development, late uninucleate microspores, which are characterized by vacuole, divide asymmetrically to form a small lens-shaped generative cell and a large vegetative cell. The former cell divides again to produce two sperm cells, resulting in a mature pollen grain (Fig. 4.3 A–C). In microspore embryogenesis, symmetrical division of uninucleate microspores is the first symptom (Zaki and Dickinson 1991, Fig. 4.3 D). Subsequently, random cell divisions repeat to produce proembryos (Fig. 4.3 E). In addition to uninucleate microspores, the vegetative cell of binucleate microspores was observed to contribute to embryogenesis (Fig. 4.3 F).

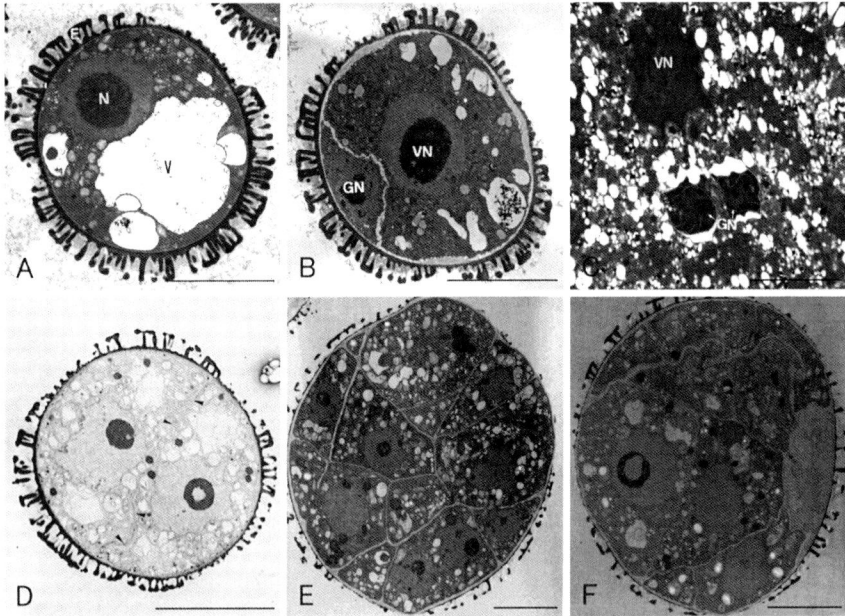

Fig. 4.3 Transmission electron micrographs of gametophytic (**A**–**C**) and sporophytic (**D**–**F**) developments of microspores of *B. napus*. (**A**) Uninucleate stage, *N* nucleus, *V* Vacuole, *E* Exine, (**B**) Binucleate stage, *GN* Generative nucleus, *VN* Vegetative nucleus, (**C**) Trinucleate stage, *GN* Sperms, *VN* Vegetative nucleus, (**D**) Symmetric devision, (**E**) Multicellular proembryo, (**F**) Multicellure derived from vegetative cell. Bar = 10 μm (**A**, **B**, **D**–**E**), 5 μm (Reprinted from Takahata and Nitta (1996) Ultrastructure of pollen embryogenesis. Tiss Cult 22: 178–181 (in Japanese) with permission)

Although morphological development of microspore embryogenesis is similar to zygotic embryos, early cell division is different between them. In zygotic embryos, cell linage during the early phase is obvious. At first, a fertilized cell undergoes asymmetric division to produce a large basal and a small proper cell. The former cell produces root apical meristem and suspensor, and the latter cell differentiates to produce the embryo proper (Tykarska 1976; Harada 1999). In microspore embryogenesis induced by the conventional method, such distinct establishment of apical-basal axis cannot be observed, but a series of random cell divisions produce embryos without suspensors. Recently, novel microspore culture systems producing embryos with a suspensor (zygotic-like microspore derived embryo) have been developed in *B. napus* (Joosen et al. 2007; Supena et al. 2008; Prem et al. 2012; Tang et al. 2013). Since these embryos follow the highly regular cell division pattern typical for zygotic embryos (Joosen et al. 2007), these culture systems provide a tool to clear zygotic embryogenesis, especially cell polarity in early embryogenesis. Supena et al. (2008) observed that in this culture system, the microspore first divided transversely to form a filamentous structure, from which the distal cell formed the globular embryo, while the low part resembled the suspensor. Dubas et al. (2011) also reported that this system induced microspores to elongate, to rearrange their micro-tubular cytoskeleton and to re-enter the cell cycle, and embryogenesis was charac-terized as being preceded by pre-prophase band formation and DNA synthesis. On the other hand, Tang et al. (2013) reported that the exine-dehiscing microspore showed cell polarity and subsequent asymmetric cell division resulting in the two daughter cells having different cell fate, which finally developed an embryo with a suspensor.

To understand the mechanism in induction of microspore embryogenesis and subsequent development, differential display, subtractive hybridization and micro-array analysis have been carried out to reveal changes in gene expression during microspore embryogenesis (Custers et al. 2001; Boutilier et al. 2002; Malik et al. 2007; Joosen et al. 2007; Tsuwamoto et al. 2007; Tsuwamoto and Takahata 2008; Malik and Krochko 2009). Of a large number of genes specific to microspore embryogenesis isolated in *B. napus*, some genes have been examined using ectopic expression of the genes in order to understand their roles in embryogenesis. Boutilier et al. (2002) reported that ectopic expression of *BABY BOOM 1* (*BBM1*) induced somatic embryogenesis from vegetative tissues. Tsuwamoto et al. (2010) found that ectopic expression of *EMBRYOMAKER* (*EMK*) also induced somatic embryos from cotyledons, and enhanced the efficiency of somatic embryogenesis in vitro culture. These phenomena are similar to ectopic embryogenesis caused from over-expression of *LEAFY COTYLEDON 1* (*LEC1*), *LEC2*, *WUSCHEL* (*WUS*), *PGA37/MYB118* and *MYB115*, and from disruption of *PICKEL* (*PKL*) and *VP1/ABSCISIC ACID INSENSITIVE 3-LIKE* (*VAL*) genes (Lotan et al. 1998; Stone et al. 2001; Zuo et al. 2002; Wang et al. 2009b; Ogas et al. 1999; Suzuki et al. 2007). Whether these factors act as critical triggers in the initiation of microspore embryogenesis has remained obscure. Although the expression of *LEC1*, *LEC2*, *BBM* and *EMK* in early embryogen-esis of *B. napus* has been found (Boutilier et al. 2002; Tsuwamoto et al. 2007; Malik et al. 2007), there is no direct evidence to combine them to the induction of microspore embryogenesis. Mutants, in which their gene activities are up and down regulated, are needed for more analysis. Tsuwamoto et al. (2008) demonstrated that *GASSHO* gene, which was isolated at early stage of microspore embryogenesis, are essential for the formation of a normal epidermal surface during embryogenesis. Apart from iso-lated genes, Elhiti et al. (2010) reported that *Brassica SHOOTMERISTEMLESS* (*STM*) gene gave effects on the microspore embryogenesis in *B. napus*, that is, the yield of embryos increased in lines overexpressing *BnSTM* and decreased in lines down-regulating *BnSTM*.

On the other hand, one of the problems in gene expression studies is the difficulty in selecting embryogenic microspores from non-embryogenic ones. When compara-tive analysis of gene expression between induced embryogenic microspores and non-induced microspores is carried out, a large number of non-embryogenic microspores is contained in the former, because of low frequency of embryogenesis (embryogenic microspores are less than 10 % of cultured microspores). Sakai et al. (2008) devel-oped the method for selection of embryogenic microspores using embryo specific promoter *22a1* (Fukuoka et al. 2003). When *B. napus* transformed with *p22a-1:sGFP*

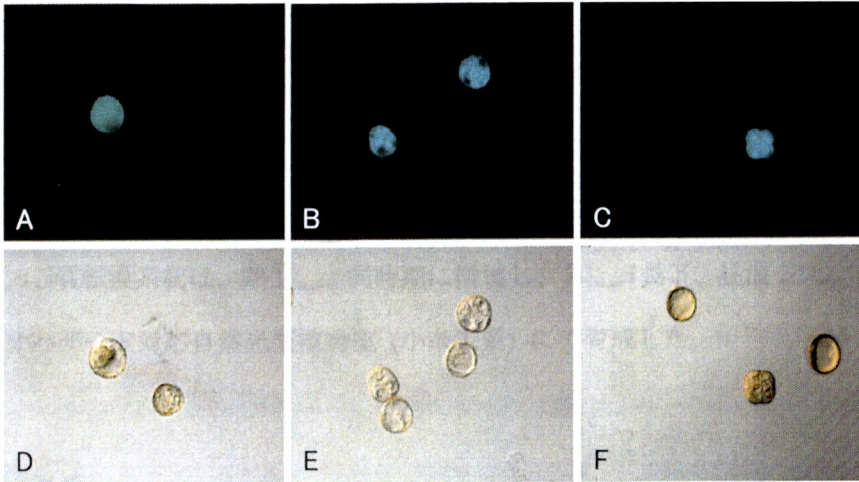

Fig. 4.4 Fluorescent (**A–C**) and light (**D–F**) microscopy in cultured microspores of *B.napus* transformed with *p22a1:sGFP*. (**A, D**) Single cell, (**B, E**) Two cells, (**C, F**) Four cells. (Sakai et al. unpublished)

chimeric gene was used for microspore culture, only embryogenic microspores showed GFP expression (Fig. 4.4). Using only GFP-expressed microspores, Sakai et al. (2008) isolated microspore embryogenesis-specific genes.

As described before, genetic analysis of microspore embryogenesis indicated that both additive and dominant effects contributed to embryogenesis with high heritability and that microspore embryogenesis was mainly controlled by two multiple gene loci (Zhang and Takahata 2001). Mapping of the genomic regions related to and identification of DNA marker linked to microspore embryogenesis have been carried out in *B. napus* and *B. rapa* (Cloutier et al. 1995; Ajisaka et al. 1999; Zhang et al. 2003). These analyses are based on the comparison of DNA marker segregation between a microspore-derived (MD) population and a F2 population derived from F1 of high responsive parent x low responsive parent. In MD populations, several DNA markers are known to deviate from an expected 1:1 Mendelian ratio. Of these distorted markers, those which are not distorted in the F2 population are considered to originate from selection pressure during microspore embryogenesis. Such markers with distorted segregation are speculated to link to gene(s) controlling microspore embryogenesis. Cloutier et al. (1995) reported that chromosome regions of linkage groups 1 and 18 were related to embryogenesis in *B. napus*. Ajisaka et al. (1999) found that a certain region of linkage group 6 played a role in embryogenesis in *B. rapa*. Zhang et al. (2003) identified 7 and 3 DNA markers associated with embryogenic ability in *B. rapa* and *B. napus*, respectively. Furthermore, they found that these markers had additive effects on embryo yield, that is, the plants having more alleles of high responsive parent produced higher embryo yields. Such analyses are based on a hypothesis that distorted marker segregation arises from the linkage to loci associated with microspore embryogenesis, but distorted segregation on the DH population is also able to be caused by linkage of loci related to ability of plant

regeneration from embryos and/or to diplodization ability. To precisely determine
the loci related to microspore embryogenesis, it is necessary that distortion analysis
is carried out in the phase of embryos (Ferrie and Mollers 2011).

4.6 Conclusion

Isolated microspore culture of Brassicas has been routinely used not only for practical
breeding, but also for genetic manipulation and basic science because of the effective haploid and/or DH production system from a single cell. However, there still
are some problems. In particular, large genotypic variations are present within the
identical species as well as among species. For example, we have been attempting
microspore culture of a large number of genotypes of *Rahpanus sativus*, but have
not yet obtained an effective system. Although routine protocol has already been
established and used worldwide, improvement of protocol will provide us with a
powerful tool for solving these problems. Recently, noble microspore culture systems were developed; one is a culture system for dealing with multiple samples
(Takahashi et al. 2011), and another is a culture system producing suspensor-bearing
embryos (Joosen et al. 2007). The former is utilized to effectively locate high
responsive genotypes, and a developing protocol dealing with super-multiple sample
will provide a useful tool for QTL analysis on microspore embryogenesis. The latter
has been utilized as a model system for studying early embryogenesis of zygotic
embryos.

Elucidation of the mechanism of microspore embryogenesis is fundamentally
needed for overcoming problems existing in this system. In order to molecular-
genetically reveal the mechanism of microspore embryogenesis, both analyses of
forward- and reverse-genetics have been carried out. Candidate genes isolated by
transcriptome analysis are not yet positioned in the chromosome region identified by
DNA marker segregation analysis. Recently, Kitashiba et al. (personal communication) found that several genes isolated using subtraction hybridization by Tsuwamoto
et al. (2007) are linked to chromosome regions showing distorted segregation in
the F1DH population.

References

Agarwal PK, Agarwal P, Custers JBM, Liu C, Bhojwani SS (2006) PCIB an antiauxin enhances
microspore embryogenesis in microspore culture of *Brassica juncea*. Plant Cell Tiss Org Cult
86:201–210
Ahmadi B, Alizadeh K, Teixeira da Silva JA (2012) Enhanced regeneration of haploid plantlets
from microspores of *Brassica napus* L. using bleomycin, PCIB, and phytohormones. Plant Cell
Tiss Org Cult 109:525–533
Ajisaka H, Kuginuki Y, Shiratori M, Ishiguro K, Enomoto S, Hirai M (1999) Mapping loci affecting the cultural efficiency of microspore culture of *Brassica rapa* L. syn. *campestris* L. using
DNA polymorphism. Breed Sci 49:187–192

Baillie AMR, Epp DJ, Hutcheson D, Keller WA (1992) In vitro culture of isolated microspores and regeneration of plants in *Brassica campestris*. Plant Cell Rep 11:234–237

Belmonte M, Elhiti M, Waldner B, Stasolla C (2010) Depletion of cellular brassinolide decreases embryo production and disrupts the architecture of the apical meristemes in *Brassica napus* microspore-derived embryos. J Exp Bot 61:2779–2794

Boutilier K, Offringa R, Sharma VK, Kieft H, Ouellet T, Zhang L, Hattori J, Liu CM, van Lammeren AA, Miki BL, Custers JB, van Lookeren Campagne MM (2002) Ectopic expression of *BABY BOOM* triggers a conversion from vegetative to embryonic growth. Plant Cell 14:1737–1749

Cao MQ, Li Y, Liu F, Dore C (1994) Embryogenesis and plant regeneration of pakchoi (*Brassica rapa* L. ssp. *chinensis*) via in vitro isolated microspore culture. Plant Cell Rep 13:447–450

Cloutier S, Cappadocia M, Landry BS (1995) Study of microspore-culture responsiveness in oil-seed rape (*Brassica napus* L.) by comparative mapping of a F2 population and two microspore-derived population. Theor Appl Genet 91:841–847

Custers JBM (2003) Microspore culture in rapeseed (*Brassica napus* L.). In: Maluszynski M, Kasha KJ, Forster BP, Szarejko I (eds) Doubled haploid production in crop plants. Kluwer, Dordrecht, pp 185–193

Custers JBM, Cordewener JHG, Fiers MA, Maassen BTH, van Lookeren Campagne MM, Liu CM (2001) Androgenesis in *Brassica*. A model system to study the initiation of plant embryogenesis. In: Bhojwani SS, Soh WY (eds) Current trends in the embryology of angiosperms. Kluwer, The Netherlands, pp 451–470

Dias JS (2001) Effect of incubation temperature regimes and culture medium on broccoli microspore culture embryogenesis. Euphytica 119:389–394

Dias JS (2003) Protocol for broccoli microspore culture. In: Maluszynski M, Kasha KJ, Forster BP, Szarejko I (eds) Doubled haploid production in crop plants. Kluwer, Dordrecht, pp 195–204

Dubas E, Custers J, Kieft H, Wedzony M, von Lammeren AAM (2011) Microtubule configurations and nuclear DNA synthesis during initiation of suspensor-bearing embryos from *Brassica napus* cv. Topas microspores. Plant Cell Rep 30:2105–2116

Duijs JG, Voorrips RE, Visser DL, Custers JBM (1992) Microspore culture is successful in most crop types of *Brassica oleracea* L. Euphytica 60:45–55

Elhiti M, Tahir M, Gulden RH, Khamiss K, Stasolla C (2010) Modulation of embryo-forming capacity in culture through the expression of *Brassica* genes involved in the regulation of the shoot apical meristem. J Exp Bot 61:4069–4085

Ferrie AMR (2003) Microspore culture of *Brassica* species. In: Maluszynski M, Kasha KJ, Forster BP, Szarejko I (eds) Doubled haploid production in crop plants. Kluwer Academic, Dordrecht, pp 205–215

Ferrie AMR, Keller WA (2004) *Brassica* improvement through microspore culture. In: Pua EC, Douglas CJ (eds) Biotechnology in agriculture and forestry, vol 54, Brassica. Springer, Berlin Heidelberg, pp 149–168

Ferrie AMR, Keller WA (2007) Optimization of methods for using polyethylene glycol as a non-permeating osmoticum for the induction of microspore embryogenesis in the Brassicaceae. In Vitro Cell Dev Biol Plant 43:348–355

Ferrie AMR, Mollers C (2011) Haploids and doubled haploids in *Brassica* spp. for genetic and genomic research. Plant Cell Tiss Org Cult 104:375–386

Ferrie AMR, Dirpaul J, Krishna P, Keller WA (2005) Effects of brassinosteroids on microspore embryogenesis in *Brassica* species. In Vitro Cell Dev Biol Plant 41:742–745

Ferrie AMR, Taylor DC, MacKenzie SL, Rakow G, Raney JP, Keller WA (2008) Microspore mutagenesis of *Brassica* species for fatty acid modifications: a preliminary evaluation. Plant Breed 127:501–506

Fukuoka H, Tsuwamoto R, Nunome T, Ohyama A, Takahata Y (2003) Isolation and characterization of an embryo-specific promoter 22a1 in Brassicaceae. In: Proceedings of 7th international congress of plant molecular biology. Barceiona, p 41

Hansen M, Svinnset K (1993) Microspore culture of sweda (*Brassica napus* ssp. *rapiferae*) and the effects of fresh and conditioned media. Plant Cell Rep 12:496–500

Harada JJ (1999) Signaling in plant embryogenesis. Curr Opin Plant Biol 2:23–27

Hiramatsu M, Odahara K, Matsu Y (1995) A survey of microspore embryogenesis in leaf musterd (*Brassica juncea*). Acta Hort 392:139–145

Ilic-Grubor K, Attree SM, Fowke LC (1998) Induction of microspore- derived embryos of *Brassica napus* L. with polyethylene glycol (PEG) as osmoticum in a low sucrose medium. Plant Cell Rep 17:329–333

Islam SMS, Tuteja N (2012) Enhancement of androgenesis by abiotic stress and other pretreatments in major crop species. Plant Sci 182:134–144

Jeung HW, Lee SS (1997) Influence of NMU on embryo induction and plant development of microspore culture in broccoli. J Korean Soc Hortic Sci 38:379–383

Joosen R, Cordewener J, Supena EDJ, Vorst O, Lammers M, Maliepaard C, Zeilmaker T, Miki B, America T, Custers J, Boutilier K (2007) Combined transcriptome and proteome analysis identifies pathways and markers associated with the establishment of rapeseed microspore-derived embryo development. Plant Physiol 144:155–172

Keller WA, Arnison PG, Cardy BJ (1987) Haploids from gametophytic cells: recent developments and future prospects. In: Green CE, Somers DA, Hackett WP, Biesboer DD (eds) Plant tissue and cell culture. Alan R Liss, New York, pp 223–241

Kuginuki Y, Nakamura K, Hida KI, Yosikawa H (1997) Varietal differences in embryogenic and regenerative ability in microspore culture of Chinese cabbage (*Brassica rapa* L. ssp. *pekinensis*). Breed Sci 47:341–346

Leroux B, Carmoy N, Giraudet D, Potin P, Larher F, Bodin M (2009) Inhibition of ethylene biosynthesis enhances embryogenesis of cultured microspores of *Brassica napus*. Plant Biotechnol Rep 3:347–353

Leskovsek L, Jakse M, Bohanec B (2008) Doubled haploid production in rocket (*Eruca sativa* Mill) through isolated microspore culture. Plant Cell Tiss Org Cult 93:181–189

Lichter R (1982) Induction of haploid plants from isolated pollen of *Brassica napus* L. Z Pflanzenphysiol 105:427–434

Lionneton E, Beuret W, Delaitre C, Ochatt S, Rancillac M (2001) Improved microspore culture and double-haploid plant regeneration in the brown condiment mustard (*Brassica juncea*). Plant Cell Rep 20:126–130

Lotan T, Ohto M, Yee KM, Westm MA, Lo R, Kwong RW, Yamagishi K, Fischer RL, Goldberg RB, Harada JJ (1998) Arabidopsis *LEAFY COTYLEDON1* is sufficient to induce embryo development in vegetative cells. Cell 93:1195–1205

Malik MR, Krochko JE (2009) Gene expression profiling of microspore embryogenesis in *Brassica napus*. In: Touraev A, Forster BP, Jain SM (eds) Advances in haploid production in higher plants. Springer, Dordrecht/London, pp 115–123

Malik MR, Wang F, Dirpaul JM, Zhou N, Polowick PL, Ferrie AMR, Krochko JE (2007) Transcript profiling and identification of molecular markers for early microspore embryogenesis in *Brassica napus*. Plant Physiol 144:134–154

Malik MR, Wang F, Dirpaul JM, Zhou N, Hammerlindl J, Keller W, Abrams S, Ferrie AMR, Krochko JE (2008) Isolation of an embryogenic line from non-embryogenic *Brassica napus* cv. Westar through microspore embryogenesis. J Exp Bot 59:2857–2873

Ogas J, Kaufmann S, Henderson J, Somerville C (1999) PICKLE is a CHD3 chromatin-remodeling factor that regulates the transition from embryonic to vegetative development in Arabidopsis. Proc Natl Acad Sci USA 96:13839–13844

Ohkawa Y, Nakajima K, Keller WA (1987) Ability to induced embryoid in *Brassica napus* cultivars. Japan J Breed 37(Suppl 2):44–55 (in Japanese)

Ohkawa Y, Fukuoka H, Ogawa T, Minami H (2000) Utilization of nitrate for early androgenic embryogenesis of *Brassica napus* L. Plant Biotechnol 17:225–233

Palmer CE, Keller WA (1999) Haploidy. In: Gomez C (ed) Biology of Brassica and coenospecies. Elsevier, Amsterdam, pp 247–286

Pechan PM, Keller WA (1988) Identification of potentially embryogenesis microspores in *Brassica napus*. Physiol Plant 74:377–384

Pechan PM, Keller WA (1989) Induction of microspore embryogenesis in *Brassica napus* by gamma irradiation and ethanol stress. In Vitro Cell Dev Biol Plant 25:1073–1075

Peng S, Takahata Y, Hara M, Shono H, Ito M (1994) Effects of the matric potential of culture medium on plant regeneration of embryos derived from microspore of *Brassica napus*. J SHITA 5/6:8–14 (in Japanese)

Pink D, Bailey L, McClement S, Hand P, Mathas E, Buchanan-Wollaston V, Astley D, King G, Teakle G (2008) Doubled haploids, markers and QTL analysis in vegetable Brassicas. Euphytica 164:509–514

Pratap A, Gupta SK, Takahata Y (2009) Microsporogenesis and haploidy breeding. In: Gupta SK (ed) Biology and breeding of Crucifers. CRC, New York, pp 293–307

Prem D, Gupta K, Agnihotri A (2005) Effect of various exogenous and endogenous factors on microspore embryogenesis in Indian mustard (*Brassica juncea* (L.) Czern and Coss). In Vitro Cell Dev Biol Plant 41:266–273

Prem D, Gupta K, Sankar G, Agnihotri A (2008) Activated charcoal induced high frequency microspore embryogenesis and efficient doubled haploid production in *Brassica juncea*. Plant Cell Tiss Org Cult 93:269–282

Prem D, Solis MT, Bárány I, Rodríguez-Sanz H, Risueño MC, Testillano PS (2012) A new microspore embryogenesis system under low temperature which mimics zygotic embryogenesis initials, expresses auxin and efficiently regenerates doubled-haploid plants in *Brassica napus*. BMC Plant Biol 12:127

Sakai M, Yokoi S, Takahata Y (2008) Isolation of microspore embryogenesis specific genes in rapeseed using embryo specific promoter. In: 16th Crucifer Genetic Workshop: Brassica 2008, Lillehammer, pp 150

Sato S, Katoh N, Iwai S, Hagimori M (2002) Effect of low temperature pretreatment of buds or inflorescence on isolated microspore culture in *Brassica rapa* (syn. *B. campestris*). Breed Sci 52:23–26

Shariatpanahi ME, Bal U, Heberle-Bors E, Touraev A (2006) Stress applied for the re-programming of plant microspores towards in vitro embryogenesis. Physiol Plant 127:519–534

Stone SL, Kwong LW, Yee KM, Pelletier J, Lepiniec L, Fischer RL, Goldberg RB, Harada JJ (2001) *LEAFY COTYLEDON2* encodes a B3 domain transcription factor that induces embryo development. Proc Natl Acad Sci USA 98:11806–11811

Supena EDJ, Winarto B, Riksen T, Dubas E, van Lammeren A, Offringa R, Boutilier K, Custers J (2008) Regeneration of zygotic-like microspore-derived embryos suggests an important role for the suspensor in early embryo patterning. J Exp Bot 59:803–814

Suzuki M, Wang HH, McCarty DR (2007) Repression of the *LEAFY COTYLEDON 1*/B3 regulatory network in plant embryo development by *VP1/ABSCISIC ACID INSENSITIVE 3-LIKE B3* genes. Plant Physiol 143:902–911

Takahashi Y, Yokoi S, Takahata Y (2011) Improvement of microspore culture method for multiple samples in *Brassica*. Breed Sci 61:96–98

Takahashi Y, Yokoi S, Takahata Y (2012) Effects of genotypes and culture conditions on microspore embryogenesis and plant regeneration in several subspecies of *Brassica rapa* L. Plant Biotechnol Rep 6:297–304

Takahata Y (1997) Microspore culture. In: Kalia HR, Guputa SK (eds) Recent advances in oilseed Brassicas. Kalyani, Ludhiana, pp 162–181

Takahata Y, Keller WA (1991) High frequency embryogenesis and plant regeneration in isolated microspore culture of *Brassica oleracea* L. Plant Sci 74:235–242

Takahata Y, Nitta T (1996) Ultrastructure of pollen embryogenesis. Tiss Cult 22:178–181 (in Japanese)

Takahata Y, Brown DCW, Keller WA (1991) Effect of donor plant age and inflorescence age on microspore culture of *Brassica napus* L. Euphytica 58:51–55

Takahata Y, Takani Y, Kaizuma N (1993) Determination of microspore population to obtain high frequency embryogenesis in broccoli (*Brassica oleracea* L.). Plant Tiss Cult Lett 10:49–53

Takahata Y, Komatsu H, Kaizuma N (1996) Microspore culture of radish (*Raphanus sativus* L.): influence of genotype and culture conditions on embryogenesis. Plant Cell Rep 16:163–166

Takahata Y, Fukuoka H, Wakui K (2005) Utilization of microspore-derived embryos. In: Palmer CE, Keller WA, Kasha KJ (eds) Biotechnology in agriculture and forestry, vol 56, Haploid in crop improvement II. Springer, Berlin Heidelberk, pp 153–169

Tang X, Liu Y, He Y, Ma L, Sun MX (2013) Exine dehiscing induces rape microspore polarity, which results in different daughter cell fate and fixes the apical-basal axis of the embryo. J Exp Bot 64:215–228

Tsuwamoto R, Takahata Y (2008) Identification of genes specifically expressed in androgenesis-derived embryo in rapeseed (*Brassica napus* L.). Breed Sci 58:251–259

Tsuwamoto R, Fukuoka H, Takahata Y (2007) Identification and characterization of genes expressed in early embryogenesis from microspores of *Brassica napus*. Planta 224:641–652

Tsuwamoto R, Fukuoka H, Takahata Y (2008) *GASSHO1* and *GASSHO2* encoding transmembrane-type receptor kinase are essential for the normal development of the epidermal surface in Arabidopsis embryos. Plant J 54:30–42

Tsuwamoto R, Yokoi S, Takahata Y (2010) Arabidopsis *EMBRYOMAKER* encoding an AP2 domain transcription factor plays a key role in developmental change from embryonic to vegetative phase. Plant Mol Biol 73:481–492

Tykarska T (1976) Rape embryogenesis. I. The proembryo development. Acta Soc Bot Poloniae 45:3–16

Wakui K, Takahata Y, Kaizuma N (1994) Effect of abscisic acid and high osmoticum concentration on the induction of desiccation tolerance in microspore-derived embryos of Chinese cabbage (*Brassica campestris* L.). Breed Sci 44:29–34

Wang T, Li H, Zhang J, Ouyang B, Lu Y, Ye Z (2009a) Initiation and development of microspore embryogenesis in recalcitrant purple flowering stalk (*Brassica campestris* ssp. *chinensis* var. *purpurea* Hort.) genotypes. Sci Hort 121:419–424

Wang X, Niu QW, Teng C, Li C, Mu J, Chua NH, Zuo J (2009b) Overexpression of *PGA37/MYB118* and *MYB115* promotes vegetative-to-embryonic transition in *Arabidopsis*. Cell Res 19:224–235

Wong RSC, Zee SY, Swanson EB (1996) Isolated microspore culture of Chinese flowering cabbage (*Brassica campestris* ssp. *parachinensis*). Plant Cell Rep 15:396–400

Yuan SX, Liu YM, Fang ZY, Yang LM, Zhuang M, Zhang YY, Sun PT (2011) Effect of combined cold pretreatment and heat shock on microspore cultures in broccoli. Plant Breed 130:80–85

Yuan S, Su Y, Liu Y, Fang Z, Yang L, Zhuang M, Zhang Y (2012) Effects of pH, MES, arabinogalactan-proteins on microspore cultures in white cabbage. Plant Cell Tiss Org Cult 110:69–76

Zaki M, Dickinson HG (1991) Microspore-derived embryos in *Brassica*: the significance of division symmetry in pollen mitosis to embryonic development. Sex Plant Reprod 4:48–55

Zeng X, Wen J, Wan Z, Yi B, Shen J, Ma C, Tu J (2010) Effects of Bleomycin on microspore embryogenesis in *Brassica napus* and detection of somaclonal variation using AFLP molecular markers. Plant Cell Tiss Org Cult 101:23–29

Zhang FL, Takahata Y (2001) Inheritance of microspore embryogenic ability in *Brassica* crops. Theor Appl Genet 103:254–258

Zhang FL, Aoki S, Takahata Y (2003) RAPD markers linked to microspore embryogenesis ability in *Brassica* crops. Euphytica 131:207–213

Zhang W, Fu Q, Dai X, Bao M (2008) The culture of isolated microspores of ornamental kale (*Brassica oleracea* var. *acephala*) and the importance of genotype to embryo generation. Sci Hort 117:69–72

Zhang Y, Wang A, Liu Y, Wang Y, Feng H (2011) Effects of the antiauxin PCIB on microspore embryogenesis and plant regeneration in *Brassica rapa*. Sci Hort 130:32–37

Zhang Y, Wang A, Liu Y, Wang Y, Feng H (2012) Improved production of doubled haploids in *Brassica rapa* through microspore culture. Plant Breed 131:164–169

Zhao JP, Simmonds DH, Newcomb W (1996) Induction of embryogenesis with colchicine instead of heat in microspores of *Brassica napus* L. cv. Topas. Planta 198:433–439

Zhou WJ, Hagberg P, Tang GX (2002) Increasing embryogenesis and doubling efficiency by immediate colchicines treatment of isolated microspores in spring *Brassica napus*. Euphytica 128:27–34

Zuo J, Niu QW, Frugis G, Chua NH (2002) The *WUSCHEL* gene promotes vegetative-to- embryonic transition in *Arabidopsis*. Plant J 30:349–359

Chapter 5
Biotechnological Strategies for Enhancing Phytoremediation

Bhawana Pathak, Razia Khan, Jyoti Fulekar, and M.H. Fulekar

Abstract Phytoremediation to clean up soil or sediments contaminated with metals and other pollutant compound has gained increasing attention as environmental friendly and cost effective. Achievements of the last decade suggest that genetic engineering of plants can be instrumental in improving phytoremediation. Members of the Cruciferae plant family have a key role in phytoremediation technology. Many wild crucifer species are known to hyperaccumulate heavy metals and possess genes for resistance or tolerance to the toxic effects of a wide range of metals. Many of these species are well adapted to a range of environmental conditions. Some species are tolerant to high levels of heavy metals, and there is the potential to select superior genotypes for phytoremediation. They are well suited to genetic manipulation and in vitro culture techniques and are attractive candidates for the introduction of genes aimed at phytoremediation. The use of genetic engineering to modify plants for metal uptake, transport and sequestration may open up new avenues for enhancing efficiency of phytoremediation. Metal chelator, metallothionein, phytochelatin and metal transporter genes have been transferred to plants for improved metal uptake and sequestration in crucifers. The purpose of this article is to review different biotechnological approaches to enhance phytoremediation in crucifers.

Keywords Crucifers • Phytoremediation • Genetic engineering • Metabolic pathways • PGPB

B. Pathak, Ph.D. • R. Khan, M.Phil. • J. Fulekar, M.Phil. • M.H. Fulekar, Ph.D. (✉)
School of Environment and Sustainable Development, Central University of Gujarat,
Sector 30, Gandhinagar, Gujarat 382030, India
e-mail: mhfulekar@yahoo.com

S.K. Gupta (ed.), *Biotechnology of Crucifers*, DOI 10.1007/978-1-4614-7795-2_5,
© Springer Science+Business Media, LLC 2013

5.1 Introduction

Over the last century, continued population growth and increased industrialization have resulted in the degradation of various ecosystems on which quality of life depends (Khan et al. 2012). In addition to this, mining, agriculture, metallurgy, combustion of fossil fuels, faulty waste disposal and military operations have released enormous amounts of toxic compounds, heavy metals and metalloids into the environment with a consequent impact on health (Wijnhoven et al. 2007; Kotrba et al. 2009). Of these, the metals may include lead, cadmium, zinc, selenium, chromium, nickel, cobalt, copper and mercury; the radioactive compounds may be uranium, strontium or cesium; and the other inorganic compounds might include arsenic, sodium, nitrate, ammonia or phosphate. In addition to the above-mentioned inorganic compounds, soils and water systems may also be contaminated with organic compounds including chlorinated solvents like trichloroethylene; explosives such as trinitrotoluene (TNT) and 1,3,5-trinitro-1,3,5-hexahydrotriazine (RDX); petroleum hydrocarbons including benzene, toluene and xylene (BTX), polyaromatic hydrocarbons (PAHs); and pesticides such as atrazine and bentazon (Glick 2003). Co-existence and persistence of these compounds and heavy metals in soils as multiple contaminants and human exposure to them through ingestion of heavy metal contaminated food or uptake of contaminated drinking water can lead to their accumulation in humans, plants and animals (Khan 2005). Cleaning up of the environment by removing these persistent and hazardous contaminants needs effective approaches that allow for a precise restoration of polluted sites. Established methods to remediate contaminated soils and groundwater are frequently expensive, environmentally invasive, labour intensive, and do not make cost-effective use of existing resources. Especially in case of large scale contaminated areas, phytoremediation is considered to be a cost-effective and sustainable remediation alternative, as it works in situ, is solar powered and demands minimal site disturbance and maintenance (Weyens et al. 2010).

As an environmentally-friendly alternative, the use of plants to remedy soils contaminated with inorganic and organic xenobiotics has gained increasing attention in recent years, giving rise to the phytoremediation concept (Eapen et al. 2007; Jadia and Fulekar 2009). Phytoremediation possesses some particularly important advantages over bioremediation using microorganisms: the capability of autotrophic plants to produce high biomass with low nutrient requirements; the capacity to reduce the spread of pollutants through water and wind erosion; and a larger public acceptance. Conventional soil and crop management methods such as increasing the soil pH, draining wet soils and applying phosphate can help prevent the uptake of heavy metals by plants, leaving them in the soil and the soil becomes the sink of these toxic metals in due course of time (Selvam and Wong 2009). Plants also produce various beneficial root exudates which support the proliferation of soil microflora, participating in remediation, especially at the rhizosphere, as well as specific chelating agents mobilizing elements in bioavailable forms (Kotrba et al. 2009). Phytoremediation covers several different strategies, of which bioremediation pollutant compounds employs phytoextraction, rhizofiltration, phytostabilization and phytovolatization. In phytoextraction, metal-accumulating plants are used that are able to concentrate pollutants in aboveground harvestable parts. Rhizofiltration uses plant roots to absorb, concentrate and/or precipitate pollutants from contaminated effluents. Phytostabilization aims at using plants to prevent the migration of pollutants, rendering them harmless. Phytovolatization is a process by which plants allow the accumulated pollutants to evaporate through their leaf surface when converted in plants to volatile forms (Kotrba et al. 2009; Doty 2008; Macek et al. 2008; Eapen and D'Souza 2005).

Plants with extensive root system help for absorption and transport of large amounts of contaminants which can be directed to the shoots and removed by biomass harvesting. The metal-enriched plant material can then be removed from the site, the contaminants concentrated, disposed or if possible, the metal element recovered and valuable metal recycled. In such cases, some species are very efficient at heavy metal accumulation in shoots and are being exploited for phytoremediation and phytomining (Baker et al. 1994; Anderson et al. 1998; Brooks et al. 1998).

Phytoextraction using hyperaccumulator plants has been proposed as a promising, environmental friendly, low-cost technology for decreasing the heavy-metal contents of contaminated soils and has emerged as an alternative to the engineering-based methods (Selvam and Wong 2009; McGrath et al. 2002). The uptake of heavy metals and metalloids by plants is influenced by many soil factors, including the presence of competitive ions in the rhizosphere (Schiavon 2012; Hopper and Parker 1999). Metalliferous soils, with abnormally high concentrations of some of the elements that are normally present only at minor or trace levels, vary widely in their effects on different plant species and some plant species have the inherent ability to sequester high concentrations of metals in the shoot tissues (Table 5.1). Hyperaccumulators are defined as plants with leaves able to accumulate at least 100 mgkg^{-1} of Cd; 1,000 mgkg^{-1} of As, Cu, Pb, Ni, Co, Se, or Cr; or 10,000 mgkg^{-1} of Mn or Zn (dry weight) when grown in a metal-rich environment (Reeves and Baker 2000; Brooks 1998; Hoang Ha et al. 2011). To date, there are approximately 400 known metal hyperaccumulators in the world (Reeves and Baker 2000) and the number is increasing. However, the remediation potential of many of these plants is limited because of their slow growth and low biomass. The ideal plant species for phytoremediation should have high biomass with high metal accumulation in the shoot tissues (Lasat 2002; McGrath et al. 2002). The advantages of phytoremediation compared to other approaches are: (1) it preserves the natural structure and texture of the soil; (2) energy is primarily derived from sunlight; (3) high levels of biomass in the soil can be achieved; (4) it is low in cost; and (5) it has the potential to be rapid. Although using plants for remediation of persistent contaminants may have advantages over other methods, many limitations exist for the large-scale application of this technology. For example, many plant species are sensitive to contaminants including PAHs so that they grow slowly and it is time consuming to establish sufficient biomass for meaningful soil remediation. In addition, in most contaminated soils, the number of microorganisms is depressed so that there are not enough bacteria either to facilitate contaminant degradation or to support plant growth.

Table 5.1 Concentration of heavy metals in soils and plants (Mukhopadhyay and Maiti 2010; Alloway 1990)

Element	Normal range in soil (ppm)	Critical soil total concentration[a] (ppm)	Normal range in plants (ppm)	Critical concentration in plants[b] (ppm)
As	0.1–40	20–50	0.02–7	5–20
Cd	0.01–2	3–8	0.1–2.4	5–30
Co	0.5–65	25–50	0.02–1	15–50
Cr	5–1500	75–100	0.03–14	5–30
Cu	2–250	60–125	5–20	2–100
Hg	0.01–0.5	0.3–5	0.005–0.17	1–3
Mn	20–10000	1500–3000	20–1000	300–500
Mo	0.1–40	2–10	0.03–5	10–50
Ni	2–750	100	0.02–5	10–100
Pb	2–300	100–400	0.2–20	30–300
Se	0.1–5	5–10	0.001–2	5–30
Zn	1–900	70–400	1–400	100–400

[a]The critical concentration in the range of values above which toxicity is considered to be possible
[b]The critical concentration in plants is the level above which toxicity effects are likely to occur

To remedy this situation, both degradative and plant growth-promoting bacteria may be added to the plant rhizosphere. The purpose of this review is to provide a summary of recent biotechnological advances in development of transgenic plants for remediation of pollutant compounds and heavy metals. The consequences of some biotechnological approaches are discussed here.

5.2 Application of Crucifers in Phytoremediation

Members of the Brassicaceae or Cruciferae plant family have a key role in phytoremediation technology. Many wild crucifer species are well known to hyperaccumulate heavy metals and possess genes for resistance or tolerance to the toxic effects of a wide range of metals (Table 5.2) (Palmer et al. 2001). Among Cruciferae, species of *Arabidopsis, Brassica, Hirschfeldia, Capsella, Thlaspi and Lepidium* have repeatedly been reported to well perform for phytoremediation (Davies et al. 2004; Fisherova et al. 2006; Gisbert et al. 2006; Jiménez-Ambriz et al. 2007; Madejon et al. 2005; Madejon et al. 2007). Members of the family Brassicaceae (Brussels sprout, cabbage, cauliflower, radish, rape seed, turnip) are the most sulfur-demanding plants and therefore excellent candidates for phyto-extraction. The enhanced sulfur demand of the members of the Brassicaceae is due to the occurrence of glucosinolates which have an array of function in plants (Schnug 1993) and play an important role in the economic use of these plants (Schnug 1997).

Among the plants of the Brassica species, the *Brassica juncea* deserve special attention because its relevance to the phytoextraction of heavy metals from soil was confirmed in many experiments.

Table 5.2 Promising crucifers for phytoremediation of various metals and radionuclides

Metal or Radionuclide	Plant species	References
Cd	*Brassica juncea*	Anamika et al. (2009), Nouairi et al. (2006)
Cr (VI)	*Brassica juncea*	Kumar et al. (1995)
Cs	*Brassica juncea*	Lasat et al. (1997)
	Brassica oleracea	Lasat et al. (1997)
Cu	*Brassica juncea*	Gisbert et al. (2006), Jordan et al. (2002)
	Brassica napus	Tappero et al. (2007), Marchiol et al. (2004)
Ni	*Brassica juncea*	Kumar et al. (1995)
Pb	*Brassica campetris*	Kumar et al. (1995)
	Brassica carinata	Gisbert et al. (2006)
	Brassica nigra	Kumar et al. (1995)
	Brassica juncea	Anamika et al. (2009), Gisbert et al. (2006), Zaier et al. (2010), Jordan et al. (2002)
	Brassica napus	Tappero et al. (2007), Marchiol et al. (2004)
Hg	*Brassica napus*	Tappero et al. (2007)
	A. thaliana	
Se	*Brassica napus*	Banuelos et al. (1997)
	Brassica juncea	Schiavon et al. (2012)
U	*Brassica chinensis*	Huang et al. (1998)
	Brassica juncea	Huang et al. (1998)
	Brassica narinosa	Huang et al. (1998)
Zn	*Brassica juncea*	Anamika et al. (2009), Gisbert et al. (2006), Jordan et al. (2002)
	Thlaspi caerulescens	
	Brassica napus	Marchiol et al. (2004)
	Brassica rapa	Ebbs and Kochian (1997)

B. juncea is a dry-land species which accumulates several metals (Pb, Cu, and Zn) from contaminated soils together with reasonable biomass yields (Zaier et al. 2010). It has been found that *B. juncea* exhibits a high capacity to accumulate Cd mainly in the shoots, where Cd level was recorded at level of 1,450 µg Cd/g dry wt. This is three times more than reported in *Brassica napus* (555 µg/g dry wt) (Nouairi et al. 2006). In addition, this plant exhibit a high removal efficiency of other metals such as Pb (28 % reduction) and Se (reduced between 13 % and 48 %) (Salt et al. 1998). *B. juncea* is more effective in removing Zn from soil than *Thlaspi caerulescens*, a known hyperaccumulator of zinc. This is due to the fact, that *B. juncea* produces ten-times more biomass than *T. cearullescens* (Gisbert et al. 2006). However *Brassica juncea* needs to be harvested shortly after the plant becomes mature, resulting in problems of disposal of obtained biomass. When these plants are dried, they easily crumble and flake off, greatly reducing the yield obtained, and the rest of the plant residues are a source of secondary emissions of toxic substances. Hyperaccumulator plants, such as *Brassica napus* are capable of concentrating trace metals (Cr, Cu, Hg, and Pb) in their harvestable biomass, thereby offering a sustainable treatment option for metal contaminated sites (Tappero et al. 2007). In the case of Chinese cabbage, the high cumulative capacity of lead was observed within the

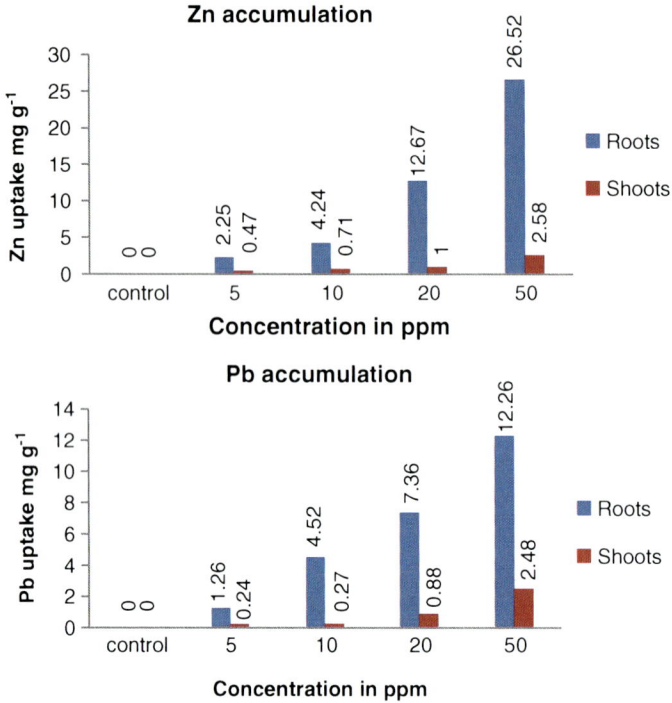

Fig. 5.1 Accumulation of Cd, Zn, and Pb in roots and shoots of *Brassica juncea*

limits of 4,620–5,010 mg/kg dry wt. During testing capacity of phyoextraction of Zn, Cu and Pb for three Brassica crop species: *B. oleracea* L., *B. carinata* A. Br. and *B. juneca* (L.) Czern., the highest concentration of Zn (381 mg/kg dry wt.) and Cu (834 mg/kg dry wt.) were recorded in the shoots of *B. oleracea* L. The Pb concentrations of all Brassica species were more or less constant over the tested range of soil Pb concentrations, with lower values than the other metals. The low bioaccumulation of lead is due to its extreme insolubility and not generally being available for plant uptake in the normal range of soil pH (Szczygłowska et al. 2011; Gruca-Królikowska et al. 2006; Gisbert et al. 2006). Metal uptake, sensitivity, and sequestration have been studied extensively in *Arabidopsis thaliana*, and a number of heavy metal sensitive and ion-accumulating mutants have been identified (Palmer et al. 2001). Robinson et al. (1998) examined the potential of *T. caerulescens* for phytoremediation under field conditions and in pot trials. The authors concluded that this species could reduce soil Cd levels by half in a single year, while a similar reduction in Zn levels required a much longer period.

The uptake of Cd, Pb and Zn by *Brassica juncea* was studied at various concentrations i.e. 0, 5, 10, 20 and 50 μg ml^{-1} in Steinberg's solution over a period of 21 days. The uptake of each metal was studied in the root and shoot separately (Fig. 5.1). The result showed that the heavy metal accumulated more in roots than in the shoots. When plants were exposed to the higher concentration (50 μg ml^{-1}) of Cd and Pb, the metals were present at an average of 18.42 and 12.27 mg g^{-1} tissue in the root, respectively and at 3.35 and 2.48 mg g^{-1} tissue in the shoots respectively. The average concentration of zinc was 26.52 mg g^{-1} in the root and shoot respectively, when exposed to 50 µg ml^{-1} of zinc (Anamika et al. 2009).

5.3   Genetic Engineering for Phytoremediation in Crucifers

Metal-hyperaccumulating plants and microbes with unique abilities to tolerate, accumulate and detoxify metals and metalloids, represent an important reservoir of unique genes (Danika and Norman 2005). These genes could be transferred to fast-growing plant species for enhanced phytoremediation (Fulekar et al. 2009; De Souza et al. 1998). The use of genetic engineering to modify plants for metal uptake, transport and sequestration may open new avenues for enhancing efficiency of phytoremediation. Generally, the processes to be considered influencing metal(loid) accumulations in plants (Fig. 5.2), and thus being targets for genetic modification, lay in such pathways as: (1) mobilization and uptake from the soil; (2) sequestration by metal complex formation and deposition in vacuoles for detoxification within roots; (3) competence of metal(loid) translocation to shoots via symplast or xylem (apoplast), including efficiency of xylem loading; (4) distribution to aboveground

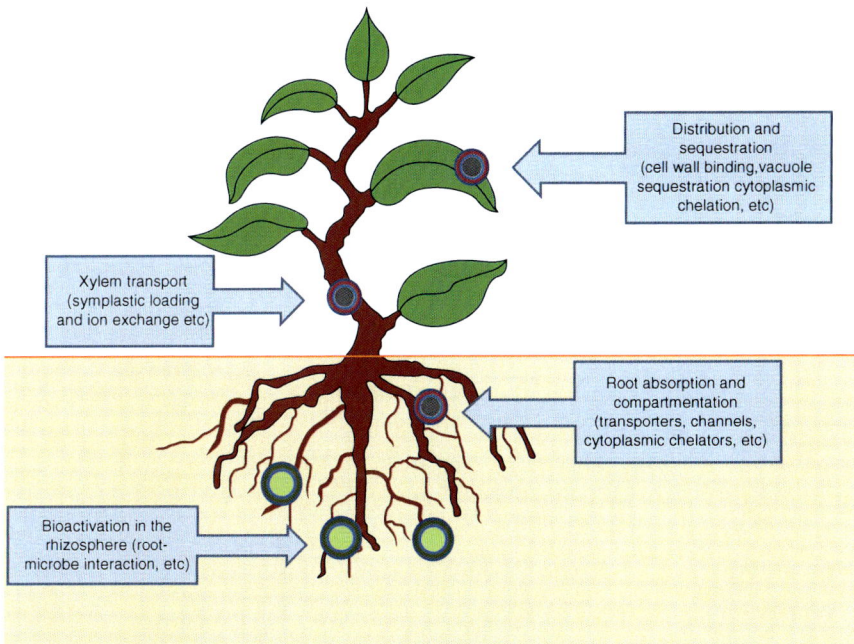

Fig. 5.2 Major processes proposed to be involved in heavy metal hyperaccumulation by plants

organs and tissues; (5) sequestration within tissue cells; and eventually (6) expulsion of accumulated metal(loid) to less metabolically-active cells, for example, to trichomes (Clemens et al. 2002; Kotrba et al. 2009). Many genes are involved in metal uptake, translocation and sequestration and transfer of any of these genes into candidate plants is a possible strategy for genetic engineering of plants for improved phytoremediation traits. Depending on the strategy, transgenic plants can be developed which will be engineered to accumulate high concentrations of metals in harvestable parts. Genetic engineering of plants for synthesis of metal chelators will improve the capability of plant for metal uptake (Pilon-Smits and Pilon 2002; Clemens et al. 2002). Classic genetic studies have shown that only a few genes are responsible for metal tolerance (Macnair et al. 2000). Recent research, including overexpression of genes whose protein products are involved in metal uptake, transport, and sequestration, or act as enzymes involved in the degradation of hazardous organics, have opened up new possibilities in field of phytoremediation.

For development of efficient genetically engineered plants for phytoremediation, genes can be transferred from hyperaccumulators or from other sources. Some of the possible areas of genetic manipulation are outlined below.

5.3.1 Metallothioneins

Metallothioneins (MTs) are cysteine-rich peptides capable of high affinity coordination of heavy metal ions via cysteine residues shared along the peptide sequence in Cys–X–Cys or Cys–Cys motifs. The role of plant MTs is generally attributed to the homeostasis of essential heavy metals and the transcription of their genes is controlled by signals instrumental during germination, organ development and senescence (Kotrba et al. 1999; Cobbet and Goldsbrough 2002; Clemens 2006; Kotrba et al. 2009). In the yeast *Saccharomyces cerevisiae* form 12 cysteine residues of CUP1, a 53 amino-acid MT variant, 8 binding centres for monovalent and 4 binding centres for divalent heavy metal ions. In these organisms, the intracellular sequestration of toxic heavy metal ions via MTs, represents one of the principal mechanisms conferring tolerance to particular heavy metal ions (Kotrba et al. 2009; Vašák 2005). Overproduction of recombinant MTs to enhance metallo resistance and support metal accumulation in plants may thus be considered as an attractive approach. Arabidopsis mutants have been useful in defining the role of metallothioneins in heavy metal tolerance in plants (Clemens et al. 1999; Ha et al. 1999; Vatamaniuk et al. 1999). Increased Cu^{2+} accumulation was reported also for roots of *A. thaliana*, overexpressing the plant MT gene PsMTA of pea *Pissum sativum* (Evans et al. 1992).

5.3.2 Phytochelatins

Phytochelatins (PCs) are small peptides having general structure (γ-Glu- Cys)nX (PCn; n = 2–11; X represents Gly, Ser, β-Ala, Glu, Gln or no residue) found in virtually all tested plants and in certain yeasts. These peptides are capable of an efficient sequestration of multiple metal and metalloid ions in metal(loid)-thiolate complexes and play a pivotal role in heavy metal detoxification in plants (Clemens 2006; Cobbett and Goldsbrough 2002; Kotrba et al. 1999). Unlike metallothioneins, phytochelatins are not primary gene products but are synthesized enzymatically in a transpeptidation reaction from glutathione (γ-glutamylcysteinylglycine, GSH) or its homologues (iso- PCs) by the constitutive PC synthase (PCS) in a metal or metalloid (e.g., arsenate) dependent manner (Leopold et al. 1999). When complexed with phytochelatins, the metals are less toxic and can be sequestered in the vacuoles (Kneer and Zenk 1992). Inside the vacuoles, the acidic conditions may cause the complex to disassemble, where the metals will then be complexed with organic acids and the phytochelatins degraded and the sulfur-containing cysteine recycled (Zenk 1996).

Gong et al. (2003) showed that the overproduction of heterologous PCS synthase of wheat *Triticum aestivum* (TaPCS1) in roots of *A. thaliana* supported the translocation of Cd^{2+} into the shoots, followed by a reduced metal accumulation, compared to wild-type control, in the roots. In contrast, the directed overproduction of intrinsic AtPCS1 in leaves (Peterson and Oliver 2006) and chloroplasts (Picault et al. 2006) of *A. thaliana* only enhanced Cd^{2+} tolerance, but had no effect on metal accumulation in shoots. The notion that PCs can be involved in long-distance metal transport predominantly via symplasmic passage is further supported by the high PC content and, compared to the xylem, four times higher Cd^{2+} levels in the phloem sap of Cd^{2+} exposed rapeseed *Brassica napus* (Mendoza-Cózatl et al. 2008). The inhibitors of phytochelatin biosynthesis, buthionine sulfoximine increased Zn tolerance in *Festuca rubra* roots (Davies et al. 1991). Also, inhibition of phytochelatin synthesis by sulfur starvation increased heavy metal sensitivity in nontolerant plants but not in tolerant ones (Schultz and Hutchinson 1998). Studies using Arabidopsis have revealed that metal-sensitive mutants are either deficient in the synthesis of phytochelatins or glutathione biosynthesis (Howden et al. 1995). Even though heavy metal challenge induces the synthesis of phytochelatins in many systems (Steffen 1990; Gekeler et al. 1998), this may not be a universal mechanism for metal tolerance (Grill 1989; Salt et al. 1998).

In *B. juncea*, heavy metals, such as Cu^{2+} and Cd^{2+}, increased the mRNA for γ glutamylcysteine synthetase (γECS) in roots and shoots and resulted in increases in phytochelatins and GSH levels even when growth was inhibited (Schäfer et al. 1997). In addition, Cd^{2+} increased the level of expression of sulfate transporters and SO_4^{-2} metabolizing enzymes, ATP sulfurylase and APS reductase, in roots and leaves (Heiss et al. 1999). Such changes are related to the increased demand for GSH for the synthesis of phytochelatins involved in metal sequestration. Also, expression of bacterial glutathione reductase (GR) gene in plastids of *B. juncea* enhanced Cd tolerance at the chloroplast level but not in the whole plant. This was related to increased GR levels in the transgenics (Pilon-Smits et al. 2000). The ability of plants to complex and transport heavy metals to the shoot is the key to accumulation and tolerance. In addition to organic acids, metal binding proteins and peptides, increases in free amino acids, such as histidine that complexes heavy metals (Krämer et al. 1996) should be targeted in plants used for phytoremediation.

5.3.3   Metal Transporters

Genetic manipulation of metal transporters is known to alter metal tolerance and accumulation in plants. Arabidopsis transporter proteins, such as the AtMRP2 gene product, are similar in function to the yeast YCF1 gene that confers Cd tolerance by transporting Cd/glutathione complexes to the vacuole (Li et al. 1996). Other Arabidopsis transporter proteins are similar to the ATP-binding cassette (ABC) transporters. These are vacuolar GS-x pumps that transport a variety of conjugates of glutathione (GSH) into the vacuole as a means of detoxification or maintenance of homeostasis (Rea et al. 1998; Thomasini et al. 1998). There are a number of other metal transporter genes identified in Arabidopsis and yeast. Some are expressed only under cation limiting conditions, while others act to prevent toxicity by regulating intracellular compartmentation (Guerinot and Eide 1999). Metal transporters of the Nramp family were isolated from *Arabidopsis thaliana*. These can transport Cd, Zn, Fe, and Mn and are expressed in both roots and shoots and inducible by Fe starvation (Thomine et al. 2000). Overexpression of an Arabidopsis zinc transporter (CDF) cation diffusion facilitator gene led to enhanced resistance and Zn accumulation (van der Zaal et al. 1999). Transgenic plants showed increased Zn uptake and tolerance and antisense of this gene led to wild type Zn tolerance in the transgenic plants. Some Zn transporters, such as the ZIP genes that are homologous to the ZRT Zn uptake genes from yeast, are expressed in roots of Arabidopsis under conditions of Zn deficiency (Grotz et al. 1998). These could be useful for genetic manipulation and studies of metal transport and tolerance. For example, mutations in the IRTI metal transporter gene of Arabidopsis eliminated Zn transport, while another mutation removed both Mn and Fe transport (Rogers et al. 2000). Therefore, these two transporters can be manipulated to increase selectivity and accumulation of metal ions.

Another mechanism of ion transport identified in Arabidopsis involves nonselective ion channel proteins, some of which are regulated by cyclic nucleotide and others by calmodulin (Köhler et al. 1999). These authors isolated a gene family of six members, which code for these proteins, two of which were involved in K^+ transport. A calmodulin-binding transporter protein similar to the ion channel proteins of Arabidopsis was shown to modulate Ni^{2+} tolerance and to enhance Pb^{2+} sensitivity in transgenic tobacco, *Nicotiana tabacum* L. (Arazi et al. 1999). In this case Ni^{2+} uptake was reduced, while Pb^{2+} uptake was enhanced. Lead was assumed to be transported via the nonselective ion channel as previously indicated (Rubio et al. 1995). As suggested by the authors these genes are involved in ion uptake and transport across the plasma membrane and are targets for gene manipulation aimed at increasing phytoremediation potential. The genes for channel proteins of Nicotiana (NtcBP4) and the Arabidopsis homologue CNGC1, when disrupted, conferred improved lead accumulation and tolerance (Sunkar et al. 2000). This offers the potential to improve plant tolerance and metal accumulation through the manipulation of channel proteins.

5.3.4 Alteration of Metabolic Pathways

New metabolic pathways can be introduced into plants for hyperaccumulation or phytovolatilization as in case of MerA and MerB genes which were introduced into plants which resulted in plants being several fold tolerant to Hg and volatilized elemental mercury (Bizily et al. 2000; Eapen and D'Souza 2005). In cysteine biosynthesis, inorganic sulphate after uptake is activated by ATP sulphydrylase to form adenosine phosphosulphate (APS), which is subsequently reduced to free sulphide by APS reductase. Sulphide is subsequently used by O-acetylserine (thiol) lyase (OAS-TL, cysteine synthase) to substitute the acetate of O-acetyl-L-serine (OAS), which is produced from L-serine and acetyl-CoA by serin-O-acetyltransferase (SAT). The sulphur assimilatory mechanism and subsequent production of the antioxidant and PC precursor GSH in plants is known to be highly induced by heavy metal exposure (Xiang and Oliver 1998; Howarth et al. 2003). In this pathway, OAS production has been shown to limit the overall rate of GSH biosynthesis and the maintenance of an elevated GSH pool (Barroso et al. 1995; Meyer and Fricker 2002). Overproduction of mitochondrial SAT encoded by TgSATm of *Thlaspi goesingense* promoted an accumulation of GSH in leaves of *A. thaliana*, providing increased tolerance to Ni^{2+}, Co^{2+}, Zn^{2+} and Cd^{2+}, attributed mainly to the acquired advantage of an improved antioxidative defense potential (Freeman and Salt 2007). Constitutive overexpression of Atcys-3A encoding intrinsic OAS-TL in *A. thaliana* also increased intracellular cysteine and GSH levels, allowing transgenes to survive at 400 μM Cd^{2+} stress (Domínguez-Solís et al. 2004). Over a 14-day period, OAS-TL Arabidopsis accumulated from media containing 250 μM Cd^{2+} 72 % more metal than WT control plants, the highest Cd^{2+} content being detected in the trichomes.

Moreover, due to highly improved biomass yields on media with 100 μM Cd^{2+}, shoots of a 3-week old transgenic plant accumulated 2.8 times higher amount of metal than did shoots of a WT plant. In an extensive study measuring the effect of ATP sulphydrylase overproduction on the accumulation of 12 metal and metalloid cations and oxyanions, Wangeline et al. (2004) demonstrated that the expression of the APS1 gene of *A. thaliana* in *B. juncea* seedlings markedly contributed to both tolerance and accumulation of certain metal and metalloid species. Although the authors did not address the mechanisms behind the observed phenotypes, it seems likely that the oxyanions MoO_4^{2-}, CrO_4^{2-}, WO_4^{2-} could be, as are sulphate analogues (Leustek 1996), accumulated via sulphate permease upregulated on virtual sulphate starvation caused by the removal of free sulphate by the overexpressed enzyme. The higher tolerance and accumulation of cations and arsenic oxyanions could be attributed to the ATP sulphydrylase-promoted increase in GSH levels reported by Pilon- Smits et al. (1999). In this study, transgenic APS1 *B. juncea* exhibited a doubling of both ATP sulphydrylase activity and GSH content in both roots and shoots. Moreover, the transgenes showed an improved tolerance to selenate, and enhanced both the reduction of selenate and production of selenomethionine (SeMet), allowing for a significant increase in Se shoot accumulation from hydroponic solutions and polluted soil (Pilon-Smits et al. 1999; Bañuelos et al. 2005). Natural Se hyperaccumulating plants use selenocysteine methyltransferase (SMT) to diminish the misincorporation of SeMet and selenocysteine (SeCys) by decreasing their intracellular concentration via a conversion to the nonprotein amino acid methylselenocysteine (MetSeCys) (Neuhierl et al. 1999). The overexpression of the SMT gene originating from the Se hyperaccumulating milkvetch *Astragalus bisulcatus* in *A. thaliana* and *B. juncea* (LeDuc et al. 2004) substantially improved the tolerance of transformants to selenate and selenite. Overall Se accumulation in shoots and Se volatization was better pronounced with SMT *B. juncea*, which exhibited a three-fold higher content of foliar MetSeCys than the WT control. The additional implementation of ATP sulphydrylase had no impact on the ability of double-transformed *B. juncea* to tolerate selenate, but further promoted Se accumulation in shoots (LeDuc et al. 2006).

5.3.5   Alteration in Biomass

The development of commercial phytoextraction technologies require plants that produce high biomass and that accumulate high metal concentration in organs that can be easily harvested, i.e. in shoots. It has been suggested that phytoremediation would rapidly become commercially available if metal-removal properties of hyperaccumulator plants, such as *Thlaspi caerulescens*, could be transferred to high-biomass producing species, such as Indian mustard (*Brassica juncea*) or maize (*Zea mays*) (Brown et al. 1995). In an effort to correct for small size of hyperaccumulator plants, Brewer et al. (1999) generated somatic hybrids between *T. caerulescens* (a Zn hyperaccumulator) and *Brassica napus* (canola), followed by hybrid selection for Zn tolerance. High biomass hybrids with superior Zn tolerance were recovered.

5.3.6   Alteration of Oxidative Stress Mechanisms

Alteration of oxidative stress related enzymes may also result in altered metal tolerance as in the case of enhanced Al tolerance by overexpression of glutathione-S-transferase and peroxidase (Ezaki et al. 2000). Overexpression of 1-aminocyclopropane-1-carboxylic acid (ACC) deaminase led to an enhanced accumulation of a variety of metals (Grichko et al. 2000; Eapen and D'Souza 2005). In heavy metal-hyperaccumulating *Thlaspi goesigense*, the natural overproduction of GSH is considered a trait sustaining tolerance to oxidative stress caused by Cd^{2+} and Ni^{2+} (Freeman and Salt 2007; Boominathan and Doran 2003).

5.4   Somatic Cell Hybridization

Most species are well adapted to in vitro culture and plants can be quickly regenerated from a range of explants (Sjödin 1992; Palmer and Keller 1994) with high plant regeneration frequency. Selection of somaclonal variants for metal and organic compound tolerance is possible, and indeed in vitro selection of calli derived from Brassica species identified somaclonal variants for Zn and Mn tolerance (Rout et al. 1999). Protoplast culture techniques are well developed for members of the Brassicaceae and plant regeneration frequencies can be high (Glimelius 1984; Palmer and Keller 1994). The technique of protoplast fusion has allowed the production of a number of interspecific, intergeneric, and intertribal somatic hybrids (Glimelius 1999). With this potential to produce somatic hybrids from widely divergent species, crucifers such as *Streptanthus polygaloides* with heavy metal accumulating traits can by hybridized with high biomass producing Brassicas. There is evidence that such hybridizations may be possible as Brewer et al. (1999) showed that somatic hybridization between the Zn hyperaccumulator *Thlaspi caerulescens* and *Brassica napus* resulted in substantial accumulation and tolerance to Zn in the hybrids. Asymmetric hybrids have been produced in members of the Brassicaceae, where there is potential loss of all or some of the chromosomes from one parent (Glimelius 1999). Manipulation of the nuclear and cytoplasmic genome combinations that lead to enhanced biomass production is useful for phytoremediation.

5.5   Potential Genes for Enhancing Phytoremediation

The genetics of heavy metal tolerance in nonaccumulating tolerant plants suggest that tolerance is polygenic, with major genes and gene modifiers involved (Table 5.3) (Macnair 1993). Microorganisms and plants contain a number of genes useful in phytoremediation (Field and Thurman 1996; Hughes et al. 1997; Salt et al. 1998). These can be transferred to Brassica to augment the phytoremediation capacity. Bacterial genes involved in detoxification are attractive because they can be improved by DNA shuffling techniques prior to their expression in plants (Crameri et al. 1997). Similarly, the non proteinogenic amino acid nicotianamine occurs widely in plants and is an efficient complexing agent for metal ions (Stephan and Scholz 1993). Manipulation of the genes involved in the biosynthesis of these compounds and introduction into Brassicas could enhance both heavy metal tolerance and their use in phytoremediation.

   Evidence from heavy metal tolerant *Silene vulgaris* indicated that there are two distinct major genes responsible for Zn tolerance, with the level of tolerance affected by two additional modifier genes (Schat and Ten-Bookum 1992). Co-segregation analyses have also indicated that Cu, Zn, and Cd tolerances in *Silene vulgaris* are controlled by different genes, whereas tolerance to Ni and Co seem to be linked

Table 5.3 Major genes involved in phytoremediation by crucifers

Gene	Source	Effect	Reference
ZnT1	T. caerulescens	Zn hyperaccumulation	Lasat et al. (2000)
ATP-sulfurylase (APS)	Brassica juncea	Se hyperaccumulation	Pilon-Smits et al. (1999)
Arabidopsis IRT1	A. thaliana	Uptake of iron and other metals	Eide et al. (1996)
CAX2	A. thaliana	Accumulation of Cd, Ca and Mn	Hirschi et al. (2000)
		Accumulation of Cd, Mn and Zn	Korenkov et al. (2007)
Zn transporters ZAT(At MTPI)	Arabidopsis	Zn accumulation	Van der Zaal et al. (1999)
Selenocysteine methyl transferase	A. bisculatus	Resistance to selenite	Ellis et al. (2004)
CAX4	A. thaliana	Accumulation of Cd, Mn and Zn.	Korenkov et al. (2007)
CGS1	A. thaliana	Volatilization of Se	Van Huysen et al. (2003)
APS1	A. thaliana	Accumulation of Se	Yang et al. (2005)

pleiotropically to the tolerance allele of one of the loci for Zn tolerance (Schat and Vooijs 1997). Recent studies indicate that Zn hyperaccumulation in *T. caerulescens* is due to enhanced Zn loading into the xylem and influx into root and leaf cells (Lasat et al. 2000). These authors reported the cloning and analysis of a high-affinity Zn transporter gene, ZnT1, from *T. caerulescens* and demonstrated high constitutive expression in roots and shoots compared with low expression in *T. arvense*. Recently, other studies have revealed a number of Zn transporters homologous to the Arabidopsis ZIP gene family members (Pence et al. 2000; Assuncao et al. 2001). In *T. caerulescens* these genes were highly expressed in roots and shoots with either normal or Zn deficiency conditions. However, in *T. arvense*, a nonaccumulator, Zn transporter genes were only expressed under Zn-deficient conditions, leading to the conclusion that gene expression in the hyperaccumulator *T. caerulescens* is much less susceptible to Zn down-regulation (Assuncao et al. 2001). These results underscore the molecular basis for Zn hyperaccumulation.

Others have indicated that real tolerance that develops over time may be monogenic or oligogenic, while detoxification is more likely to be under polygenic control (Sanita-di Toppi and Gabbrielli 1999). Tolerance is constitutive and a high level of contamination over a long period is needed to elicit specific monogenic/oligogenic change. When Zn tolerance and accumulation were compared in *Arabidopsis halleri* populations from metallicolous and non metallicolous sites, similar tolerance was observed (Bert et al. 2000). This points to tolerance being a constitutive trait, and the greater tolerance observed in the metallicolous population was explained as the emergence of gene modifiers under metal stress conditions to augment the constitutive trait (Bert et al. 2000). A similar constitutive trait may operate in *Thlaspi caerulescens* where plant populations from normal soils accumulated more Zn than populations from contaminated soils (Escarré et al. 2000). However, the latter population accumulated more Cd and variations existed in both populations for Cd and Zn accumulation. Heavy metal accumulation and tolerance are probably independent traits as in crosses between Zn-tolerant, hyper accumulating *Arabidopsis halleri* and the non-tolerant non accumulating *A. petraea* there was segregation for both the Zn tolerant and -accumulating traits (Macnair et al. 1999). The authors concluded that a single major gene was involved in tolerance, although the number for hyper accumulation could not be assessed.

As a consequence, foreign genes of value in phytoremediation can be evaluated in Arabidopsis before incorporation into species suitable for phytoremediation. Because of the close relationship to Brassica, Arabidopsis can be viewed as a reservoir of genes for introgression into cultivated Brassica. With complete Arabidopsis genome sequence now available, genes useful in phytoremediation may be uncovered and Brassica species are prime subjects for the expression of such genes. The introduction of bacterial genes Mer A and Mer B, which code for enzymes that metabolize mercury, into Arabidopsis conferred mercury resistance (Rugh et al. 1996; Bizily et al. 1999; Bizily et al. 2000). In these transgenic plants organic mercury was converted to a less toxic form and, by modification of the plasmid construct, the plants were able to volatilize the mercury (Rugh et al. 1996; Pilon-Smits and Pilon 2000). Transgenic Arabidopsis plants with a bacterial dehalogenase gene (Dhla) were able to degrade, 1, 2 dichloroethane (Naested et al. 1999). This is a common degradative system in bacteria (Janssen et al. 1994), and it may be possible to engineer plants for the remediation of organic contaminants.

Rugh et al. (1998) modified yellow poplar trees with two bacterial genes, merA and merB, to detoxify methyl-Hg from contaminated soil. In transformed plants, merB catalyzes the release of Hg^{2+} from methyl-Hg, which is then converted to Hg^0 by merA. Elemental Hg is less toxic and more volatile than the mercuric ion, and is released into the atmosphere. Pilon-Smits et al. (1999) overexpressed the ATP-sulfurylase (APS) gene in Indian mustard. The transgenic plants had four-fold higher APS activity and accumulated three times more Se than wild-type plants. Recently, Dhankher et al. (2002) reported a genetics-based strategy to remediate As from contaminated soils. They overexpressed two bacterial genes in Arabidopsis. One was the E. coli AsrC gene encoding arsenate reductase that reduces arsenate to arsenite coupled to a light-induced soybean rubisco promoter. The second gene was the E. coli g-ECS coupled to a strong constitutive actin promoter. The AsrC protein, expressed strongly in stem and leaves, catalyzes the reduction of arsenate to arsenite, whereas g-ECS, which is the first enzyme in the PC-biosynthetic pathway, increases the pool of PCs in the plant. The transgenic plants expressing both AsrC and g-ECS proteins showed substantially higher As tolerance; when grown on As, these plants accumulated a 4–17-fold greater fresh shoot weight and accumulated 2–3-fold more as than wild-type plants.

5.6 Plant Growth-Promoting Bacteria in Phytoremediation

Phytoremediation (i.e., degradation of organics in the presence of plants) alone is not significantly faster than bioremediation (i.e., where biodegradation of the organics is by microorganisms independent of plants) for removal of PAHs that include three rings or less, although phytoremediation outperformed bacterial treatment with respect to removal of the larger, more strongly soil bound PAHs. Beneficial free-living soil bacteria are generally referred to as plant growth-promoting rhizobacteria and are found in association with the roots of many different plants (Glick et al. 1999). Cultivating plants together with plant growth-promoting bacteria allowed the plants to germinate to a much greater extent, and then to grow well and rapidly accumulate a large amount of biomass. The plant growth-promoting bacteria increased seed germination and plant survival in heavily contaminated soils, decreased the plant dry weight to fresh weight ratio, increased the plant water content, helped plants to maintain their chlorophyll contents and chlorophyll a/b ratio, and promoted plant root growth. As a consequence of the treatment of plants with plant growth-promoting bacteria, the plants provide a greater sink for the contaminants since they are better able to survive and proliferate. These bacteria can positively influence plant growth and development in two different ways: indirectly or directly (Glick et al. 1999). The indirect promotion of plant growth occurs when these bacteria decrease or prevent some of the deleterious effects of a phytopathogenic organism. Bacteria can directly promote plant growth by providing the plant with a compound that is synthesized by the bacterium or by facilitating the uptake of nutrients from the environment by the plant. Plant growth promoting bacteria may: fix atmospheric nitrogen and supply it to plants; synthesize siderophores which can solubilize and sequester iron from the soil and provide it to plant cells; synthesize several different phytohormones including auxins and cytokinins which can enhance various stages of plant growth; have mechanisms for the solubilization of minerals such as phosphorus which then become more readily available for plant growth; and contain enzymes that can modulate plant growth and development (Brown 1974; Davison 1988; Kloepper et al. 1989; Lambert and Joos 1989; Patten and Glick 1996; Glick et al. 1998). A particular bacterium may affect plant growth and development using any one, or more, of these mechanisms and a bacterium may utilize different mechanisms under different conditions. For example, bacterial siderophore synthesis is likely to be induced only in soils that do not contain sufficient levels of iron. Similarly, bacteria do not fix nitrogen when sufficient fixed nitrogen is available.

When the wild-type bacterium and the siderophore overproducing mutant were tested in the laboratory, as expected both of them were observed to promote the growth of tomato, canola and Indian mustard plants in the presence of inhibitory levels (generally 2 mM) of nickel, lead or zinc. In addition, the siderophore overproducing mutant decreased the inhibitory effect of the added metal on plant growth significantly more than the wildtype bacterium. Heavy metal contamination of soil is often associated with iron-deficiency in a range of different plant species (Mishra and Kar 1974).

There were reports in the scientific literature that indicated that *Brassica juncea* was a nickel-hyperaccumulating plant and could be used for this purpose. However, preliminary laboratory experiments indicated that the growth of Indian mustard, and the related plant *Brassica campestris* (canola), which could also accumulate high levels of nickel and other metals, was significantly inhibited by the presence of moderate amounts of nickel in the soil. In an effort to overcome the inhibition of plant growth by nickel, a bacterium was isolated from a nickel contaminated soil sample; the bacterium was (i) nickel-resistant, (ii) able to grow at the cold temperatures (i.e., 5–10 °C) that one expects to find in nickel contaminated soil environments in Canada and (iii) an active producer of ACC deaminase (Burd et al. 1998).

When grown in the presence of arsenate, the fresh and dry weights of roots and shoots of transgenic canola, especially when they were treated with the ACC deaminase-containing plant growth-promoting bacterium *Enterobacter cloacae* CAL2, were much higher than with non-transformed canola. Other properties of this bacterial strain, in addition to ACC deaminase activity, may contribute to this result the bacterium synthesizes IAA, siderophores and antibiotics, all of which may stimulate plant growth. In this regard, antibiotic-secreting plant growth-promoting bacterial strains can inhibit the proliferation and subsequent invasion of phytopathogens, hence protecting plants, already debilitated by arsenate in the soil, from further damage.

Another rather promising approach appears to be the development of engineered endophytic bacteria that improve the phytoremediation of water-soluble, volatile organic compounds (Barac et al. 2004). Trichloroethylene (TCE)-degrading bacteria have been proven to protect host plants against the phytotoxicity of TCE and to contribute to a significant decrease in TCE evapotranspiration. Plant associated bacteria can be exploited to overcome constrains such as phytotoxicity, a limited contaminant uptake, and evapotranspiration of volatile organic contaminants (Weyens et al. 2010). In case of phytoremediation of organic contaminants, endophytes equipped with the appropriate degradation pathway can diminish phytotoxicity and evapotranspiration (Barac et al. 2004; Taghavi et al. 2005). To increase plant availability of metals plant-associated bacteria that are capable of producing siderophores and/or organic acids can be used (Weyens et al. 2009). To reduce internal metal bioavailability and by consequence metal phytotoxicity, endophytes equipped with a metal resistance/sequestration system (e.g. ncc-nre) leading to bioprecipitation of metals on the bacterial cell wall can be inoculated (Weyens et al. 2009). Combining increased plant availability and reduced internal bioavailability of metals will allow plants to accumulate higher amounts of metals without increasing phytotoxicity.

5.7   Transgenic Approaches for Phytoremediation in Crucifers

Transgenic plants, which detoxify/accumulate cadmium, lead mercury, arsenic and selenium have been developed (Table 5.4). The most spectacular application of biotechnology for environmental remediation has been the bioengineering of plants capable of volatilizing mercury from soil contaminated with methyl mercury. Methyl-mercury, a strong neurotoxic agent, is biosynthesized in Hg-contaminated soils. To detoxify this toxin, transgenic pants (Arabidopsis and tobacco) were engineered to express bacterial genes merB and merA. In these modified plants, merB catalyses the protonolysis of the carbon mercury bond with the generation of Hg^{+2}, a less mobile mercury species. Subsequently, merA converts Hg(II) to Hg(0) a less toxic, volatile element which is released into the

Table 5.4 Properties of transgenic crucifers engineered for phytoremediation of metals

Gene	Source	Target plant	Phenotype	Reference
CAX2	A. thaliana	N. tabacum	2.8, 2.5 and 1.3 times higher biomass when grown, respectively, on media with 3 μM Cd^{2+}, 500 μMMn^{2+} and 150 μM Zn^{2+}, then in roots 1.5 and 1.3 times higher Cd and Zn levels. Amount of metal accumulated per plant growing on media with 3 μM Cd^{2+}, 500 μMMn^{2+} and 150 μM Zn^{2+} was higher 3.4, 2.3 and 1.9 times, respectively	Korenkov et al. (2007)
AtPSC1 and MTL4	A. thaliana and H. sapiens	M. huakuii/A. sinicus	Roots of A. sinicus colonized with rhizobia M. huakuii producing AtPCS1 and AtPCS1+MT4 accumulated, respectively, 2.5 and 3 times more Cd from soil containing 1 ppm Cd. Colonized nodules increased Cd concentration [only] by 30 %	Ike et al. (2007)
gshI	E. coli	B. juncea	2.1 times longer roots in media with 200 μM Cd^{2+}. By 90 % higher shoot Cd levels when grown in media with 50 μM Cd^{2+}	Zhu et al. (1999a)
			When grown on polluted soilb, shoots showed 1.5, 2.0, 2.0 and 3.1 times higher Cd, Zn, Cu and Pb levels, respectively.	Bennett et al. (2003)
AtPCS1	A. thaliana	B. juncea	1.9 and 1.4 times longer roots on media with 100 μM Cd^{2+} and 500 μM AsO_4^{3-}, respectively	Gasic and Korban (2007)
gshI and arsC	E. coli	A. thaliana	Six times higher biomass yield from medium with 200 μM AsO_4^{3-}. Three times higher As accumulation from medium with 125 μM AsO_4^{3-}	Dhankher et al. (2002)
gshII	E. coli	B. juncea	1.5 times longer roots on medium with 200 μM Cd^{2+}. By 20 % enhanced Cd^{2+} accumulation from media with 50 μM Cd^{2+}	Zhu et al. (1999b)
			When grown on polluted soilb, shoots showed 1.5 higher Cd and Zn levels.	Bennett et al. (2003)
merC	A. ferrooxidans	A. thaliana	Hg^{2+} hypersensitivity biomass reduced by 6.4 times when grown on medium with 3 μM Hg^{2+}. Leaves submersed into test solution with 100 μM Hg2+ accumulated over 3 h period 3.2 more Hg	Sasaki et al. (2006)
merP	Bacillus megaterium	A. thaliana	Capable of germination and growth on media with 12.5 μM Hg^{2+} accumulating 5.35 μg Hg^{2+}/g of fresh seedling weight	Hsieh et al. (2009)
GSH1	S. cerevisiae	A. thaliana	No effect on Cd^{2+}, AsO_4^{3-} a AsO^{2-} tolerance. Increased accumulation of Cd (four times from media with 30 ppm Cd^{2+}) and As (2.5 and 4.4 times from media with 28 ppm AsO_4^{3-} and AsO^{2-}, respectively)	Guo et al. (2008)

GSH1 and AsPCS1	S. cerevisiae and A. sativum	A. thaliana	Two times longer roots on media with 50 μM Cd^{2+}, 150 μM AsO_4^{3-} or 50 μM AsO_2^{-}. Increased accumulation of Cd (ten times from media with 30 ppm Cd^{2+}) and As (three and ten times from media with 28 ppm AsO_4^{3-} and AsO_2^{-}, respectively)	Guo et al. (2008)
APS1	A. thaliana	B. juncea	1.5 times longer roots and 1.4 times higher biomass with plantlets grown on medium with 400 μM SeO_4^{2-} Improved accumulation of Se and S: three times higher Se levels in shoots when plantlets grown on medium with 40 μM SeO_4^{2-}. Doubled levels of glutathione, both in shoots and roots	Pilon-Smits et al. (1999)
			On metal(loid) containing mediac showed plantlets increased tolerance (as root elongation) to Cd^{2+} (2.2 times), Cu^{2+} (by 30 %), Hg^{2+} (by 20 %), Zn^{2+} (by 15 %), AsO_4^{3-} (by 35 %), AsO_2^{-} (by 20 %). In shoots increased accumulation of Cd^{2+} (1.9 times), VO_4^{3-} (2.5 times), CrO_4^{2-} (1.5 times), WO_4^{2-} (1.7 times), MoO_4^{2-} (1.4 times) from mediad	Wangeline et al. (2004)
SMT	A. bisulcatus	B. juncea	No phytotoxicity of 25 μM SeO_3^{2-} in medium (97 % growth inhibition with WT). By 40 % reduced growth on medium with 25 μM SeO_4^{2-} (60 % inhibition with WT). Se accumulation from media with 200 μM SeO_4^{2-} and 100 μM SeO_3^{2-} increased four and two times, respectively	LeDuc et al. (2004)
SMT and APS1	A. bisulcatus and A. thaliana	B. juncea	Se accumulation from media with 200 μM SeO_4^{2-} increased nine times (six times compared to single-transformed APS1 plant)	LeDuc et al. (2006)

atmosphere (Bazirmakenga et al. 1995). Hg reductase has also been successfully transferred to Brassica, tobacco and yellow poplar trees (Meager et al. 2000).

Metabolic modification and degradation of a xenobiotic molecule may depend on a single enzyme. A transgenic approach for modifying or improving this enzyme with benefit for the relevant phytotechnology, is therefore conceivable. Examples reported in literature concern engineering herbicide tolerance, since these compounds are completely assimilable to environmental xenobiotics. For instance, Diderjean et al. (2002) reported of a successful transgenic approach with a gene for cytochrome P450, involved in Phase I of the metabolism. The gene chosen is inducible by chemical stress (metals and drugs) in Jerusalem artichoke, and it conferred resistance to phenyl urea upon transfer in the sensitive species tobacco and Arabidopsis. This gene may further be considered a useful tool for phytotransformation application in case of contamination by herbicides in soils and water. The following example is not concerned strictly phytoremediation, but rather phytomonitoring of organic compounds with a transgenic approach. The Danish company Aresa Biotechnology has developed a GM plant of *Arabidopsis thaliana* which can detect nitrogen dioxide emitted by explosives and signal this contact by changing to red color (Anonymous 2002). The proposed application would be that of growing plants in areas affected by anti-personnel mines in order to contribute to decontamination of the site. The performance of a transgenic system coupling bacterial arsenate reductase arsC and gshI genes to improve arsenate removal from soil was inspected in A. thaliana (Dhankher et al. 2002). The ArsC/γ- ECS plants showed substantially greater tolerance to arsenate and accumulation of arsenic oxyanions in shoots (predominantly as [glutathione]3AsIII) from arsenate-containing hydroponic solutions than did the control WT and/or gshI-only-transformed plants. In transgenic plants of Indian mustard (*Brassica juncea*), and increase in the expression of the GSH biosynthetic pathways led to an increase in PC biosynthesis and Cd tolerance (Yang et al. 2005; Zhu et al. 1999).

Uptake and evaporation of Hg is achieved by some bacteria. The bacterial genes responsible have already been transferred to Brassica species and these transgenic plants may become useful in cleaning Hg- contaminated soils (Meager et al. 2000). The biodegradation of explosives by transgenic plants expressing pentaerythritol tetranitrate reductase (French et al. 1999) is the classical example of the exploitation of a bacterial gene for phytoremediation. More recently, plants have been constructed that express bacterial enzymes capable of TNT transformation and RDX (hexahydro-1,3,5-trinitro-1,3,5 triazine, an explosive nitroamine widely used in military and industrial applications) degradation (Bruce 2007).

5.8   Future Research Perspectives

Heavy metal hyperaccumulators have received increased attention in recent years, due to the potential of using these plants for phytoremediation of soil contaminated with metals. However, there are some limitations for this technology to become efficient and cost-effective on a commercial scale, as most of the metal hyperaccumulating plants identified have small biomass, and are not very adaptable to harsh environment. These limitations can be overcome by using biotechnological approaches for enhancing phytoremediation in plants. The problem of low biomass phytoremediators can be overcome by increasing plant yield and metal uptake by engineering common plants with hyperaccumulating genes. Transgenic plants which can convert toxic mercury compounds to volatile and less toxic forms and crucifer plant species with tolerance to high concentration of metals have been produced. Overexpression of proteins involved in intracellular metal sequestration may significantly increase metal accumulation and subcellular storage. Transgenic plants may enhance remediation of contaminated soil with obvious benefits, yet some question arises about their techno economic perspective and environmental safety. The potential of engineered plants for phytoremediation should be thus demonstrated in field trials. The ecological impact and underlying economics of phytoremediation with transgenics should be carefully evaluated and weighted against known disadvantages of conventional remediation techniques. A multidisciplinary research effort that integrates the work of plant biologists, microbiologists, soil chemists and environmental engineers is essential for greater success of phytoremediation technique.

References

Alloway BJ (1990) Heavy metals in soil. Blackie and Son, London, pp 1–339

Anamika S, Eapen S, Fulekar MH (2009) Phytoremediation of cadmium, lead and zinc by Brassica juncea L. Czern and Coss. J Appl Biosci 13:726–736

Anderson CWN, Brooks RR, Stewart RB, Simcock R (1998) Harvesting a crop of gold in plants. Nature 395:553–554

Anonymous (2003) Phytoremediation session at the 19th annual international conference on soils, sediments, and water, phytoremediation session, Amherst, MA, USA, October 20–23, 2003. Int J Phytoremediation 5:399–404

Arazi T, Sunkar R, Kaplan B, Fromm H (1999) A tobacco plasma membrane calmodulin-binding transporter confers Ni2+ tolerance and Pb2+ hypersensitivity in transgenic plants. Plant J 20:171–182

Assuncao AGL, DaCosta Martin P, De Folter S, Voolis R, Schat H, Aarts MGM (2001) Elevated expression of metal transporter genes in three accessions of the metal hyperaccumulator *Thlaspi caerulescens*. Plant Cell Environ 24:217–226

Baker AJM, McGrath SP, Sidoli CMD, Reeves RD (1994) The possibility of insitu heavy metal decontamination of polluted soils using crops of metal-accumulating plants. Resour Conserv Recycl 11:41–49

Bañuelos G, Terry N, Leduc DL, Pilon-Smits EAH, Mackey B (2005) Field trial of transgenic Indian mustard plants shows enhanced phytoremediation of selenium-contaminated sediment. Environ Sci Technol 39:1771–1777

Bañuelos GS, Ajwa HA, Terry N, Zayed A (1997) Phytoremediation of selenium laden soils: a new technology. J Soil Water Conserv 52(6):426–430

Barac T, Taghavi S, Borremans B, Provoost A, Oeyen L, Colpaert JV, Vangronsveld J, vander Lelie D (2004) Engineered endophytic bacteria improve phytoremediation of water soluble, volatile, organic pollutants. Nat Biotechnol 22:583–588

Barroso C, Vega J, Gotor C (1995) A new member of the cytosolic O-acetylserine(thiol)lyase gene family in Arabidopsis thaliana. FEBS Lett 363:1–5

Bazirmakenga R, Siomard RR, Leroux GD (1995) Determination of organic acids in soil extracts by ion chromatography. Soil Biol Biochem 27:349–356

Bennett LE, Burkhead JL, Hale KL, Terry N, Pilon M, Pilon-Smits EAH (2003) Analysis of transgenic Indian mustard plants for phytoremediation of metal contaminated mine tailings. J Environ Qual 32:432–440

Bert V, Macnair MR, de Laguerie P, Saumitou-Laprade P, Petit D (2000) Zinc tolerance and accumulation in metallicolous and nonmetallicolous populations of *Arabidopsis halleri* (Brassicaceae). New Phytol 146:225–233

Bizily SP, Rugh CL, Summers AO, Meagher RB (1999) Phytoremediation of methyl mercury pollution: merB expression in Arabidopsis thaliana confers resistance to organomercurials. Proc Natl Acad Sci U S A 96:6808–6813

Bizily SP, Rugh CL, Meagher RB (2000) Phytodetoxification of hazardous organomercurials by genetically engineered plants. Nat Biotechnol 18:213–217

Boominathan R, Doran PM (2003) Cadmium tolerance and antioxidative defenses in hairy roots of the cadmium hyperaccumulator, Thlaspi caerulescens. Biotechnol Bioeng 83:158–167

Brewer EP, Saunders JA, Angle JS, Chaney RL, McIntosh MS (1999) Somatic hybridization between the zinc accumulator *Thlaspi caerulescens* and *Brassica napus*. Theor Appl Genet 99:761–771

Brooks RR (1998) General introduction. In: Brooks RR (ed) Plants that hyperaccumulate heavy metals. CABI, Wallingford, pp 1–14

Brooks RR (1998) Plants that hyperaccumulate heavy metals. CAB International, Wallingford

Brown ME (1974) Seed and root bacterization. Ann Rev Phytopathol 12:181–197

Brown SL, Chaney RL, Angle JS, Baker AJM (1995) Zinc and cadmium uptake by hyperaccumulator *Thlaspi caerulescens* grown in nutrient solution. Soil Sci Soc Am J 59:125–133

Burd GI, Dixon DG, Glick BR (1998) A plant growth promoting bacterium that decreases nickel toxicity in plant seedlings. Appl Environ Microbiol 64:3663–3668

Clemens S (2006) Toxic metal accumulation, responses to exposure and mechanisms of tolerance in plants. Biochimie 88:1707–1719

Clemens S, Kim EJ, Neumann D, Schroeder JI (1999) Tolerance to toxic metals by a gene family of phytochelatin synthases from plants and yeast. EMBO J 18:3325–3333

Clemens S, Palmgren M, Krämer U (2002) A long way ahead: understanding and engineering plant metal accumulation. Trends Plant Sci 7:309–315

Cobbett C, Goldsbrough P (2002) Phytochelatins and metallothioneins: roles in heavy metal detoxification and homeostasis. Annu Rev Plant Physiol Plant Mol Biol 53:159–182

Crameri A, Dawes G, Rodriguez E, Silver S, Stemmer WPC (1997) Molecular evolution of an arsenate detoxification pathway by DNA shuffling. Nat Biotechnol 15:436–438

Davies KL, Davies MS, Francis D (1991) The influence of an inhibitor of phytochelatin synthesis on root growth and root meristematic activity in *Festuca rubra* L. in response to zinc. New Phytol 118:565–570

Davison J (1988) Plant beneficial bacteria. Bio/technol 6:282–286

De Souza MP, Pilon-Smits EAH, Lytle CM, Hwang S, Tai J, Honma TSU, Yeh L, Terry N (1998) Rate-limiting steps in selenium assimilation and volatilization by Indian mustard. Plant Physiol 117:1487–1494

Dhankher OP, Li Y, Rosen BP, Shi J, Salt D, Senecoff JF et al (2002) Engineering tolerance and hyperaccumulation of arsenic in plants by combining arsenate reductase and γ- glutamylcysteine synthetase expression. Nat Biotechnol 20:1140–1145

Diderjean L, Gondet L, Perkins R, Lau SMC, Schaller H, O'Keefe DP, Werck-Reickhart D (2002) Engineering herbicide metabolism in tobacco and Arabidopsis with CYP76B1, a cytochrome P450 enzyme from Jerusalem artichoke. Plant Physiol 130:179–189

Domínguez-Solís JR, López-Martín MC, Ager FJ, Ynsa MD, Romero LC, Gotor C (2004) Increased cysteine availability is essential for cadmium tolerance and accumulation in Arabidopsis thaliana. Plant Biotechnol J 2:469–476

Doty SL (2008) Enhancing phytoremediation through the use of transgenics and endophytes. New Phytol 179:318–333

Eapen S, D'Souza SF (2005) Prospects of genetic engineering of plants for phytoremediation of toxic metals. Biotechnol Adv 23:97–114

Eapen S, Singh S, D'Souza S (2007) Advances in development of transgenic plants for remediation of xenobiotic pollutants. Biotechnol Adv 25:442–451

Ebbs SD, Kochian LV (1997) Toxicity of zinc and copper to Brassica species. Implication for phytoremediation. J Environ Qual 26:776–781

Eide D, Broderius M, Fett JM, Guerinot ML (1996) A novel iron- regulated metal transporter from plants identified by functional expression in yeast. Proc Natl Acad Sci 93(11):5624–5628

Ellis DR, Sors TG, Brunk DG, Albrecht C, Orser C, Lahner B et al (2004) Production of Se-methylselenocysteine in transgenic plants expressing selenocysteine methyltransferase. BMC Plant Biol 4:1–11

Escarré J, Lefebvre C, Gruber W, Lablanc M, Leport J, Riviere Y, Delay B (2000) Zinc and cadmium hyperaccumulation by *Thlaspi caerulescens* from metalliferous and nonmetalliferous sites in the Mediterranean area: implications for phytoremediation. New Phytol 145:429–437

Evans KM, Gatehouse JA, Lindsay WP, Shi J, Tommey AM, Robinson NJ (1992) Expression of the pea metallothionein like gene PsMTA in Escherichia coli and Arabidopsis thaliana and analysis of trace metal ion accumulation: implications for gene PsMTA function. Plant Mol Biol 20:1019–1028

Ezaki B, Gardner RC, Ezaki Y, Matsumoto H (2000) Expression of aluminium induced genes in transgenic Arabidopsis plants can ameliorate aluminium stress and/or oxidative stress. Plant Physiol 122:657–665

Field JA, Thurman EM (1996) Glutathione conjugation and contamination transformation. Environ Sci Technol 30:1413–1418

Fischerová Z, Tlustoš P, Száková J, Šichorová K (2006) A comparison of phytoremediation capability of selected plant species for given trace elements. Environ Pollut 144:93–100

Freeman JL, Salt DE (2007) The metal tolerance profile of *Thlaspi goesingense* is mimicked in *Arabidopsis thaliana* heterologously expressing serine acetyl-transferase. BMC Plant Biol 7:63

French CE, Rosser SJ, Davies GJ, Nicklin S, Bruce NC (1999) Biodegradation of explosives by transgenic plants expressing pentaerythritol tetranitrate reductase. Nat Biotechnol 17:491–494

Fulekar MH, Singh A, Bhaduri AM (2009) Genetic engineering strategies for enhancing phytoremediation of heavy metals. Afr J Biotechnol 8(4):529–535

Gasic K, Korban SS (2007) Transgenic Indian mustard (Brassica juncea) plants expressing an Arabidopsis phytochelatin synthase (AtPCS1) exhibit enhanced As and Cd tolerance. Plant Mol Biol 64:361–369

Gekeler W, Grill E, Winnacker EL, Zenk MH (1998) Algae sequester heavy metals via synthesis of phytochelatin complexes. Arch Microbiol 105:197–202

Gisbert C, Clemente R, Navarro-Aviño JP, Baixauli C, Gines A, Serrano R, Walker DJ, Bernal MP (2006) Tolerance and accumulation of heavy metals by Brassicaceae species grown in contaminated soils from Mediterranean regions of Spain. Environ Exp Bot 56:19–27

Glick BR (2003) Phytoremediation: synergistic use of plants and bacteria to clean up the environment. Biotechnol Adv 21:383–393

Glick BR, Penrose DM, Li J (1998) A model for the lowering of plant ethylene concentrations by plant growth promoting bacteria. J Theor Biol 190:63–68

Glick BR, Patten CL, Holguin G, Penrose DM (1999) Biochemical and genetic mechanisms used by plant growthpromoting bacteria. Imperial College, London

Glimelius K (1984) High growth rate and regeneration capacity of hypocotyl protoplasts in some Brassicaceae. Plant Physiol 61:38–44

Glimelius K (1999) Somatic hybridization. In: Gomez- Campo C (ed) Biology of *Brassica Coenospecies*. Elsevier Science, Amsterdam, pp 107–148

Gong J, Lee DA, Schroeder JI (2003) Long-distance root-to-shoot transport of phytochelatins and cadmium in Arabidopsis. Proc Natl Acad Sci U S A 100:10118–10123

Grichko VP, Filby B, Glick BR (2000) Increased ability of transgenic plants expressing the enzyme ACC deaminase to accumulate Cd, Co, Cu, Ni, Pb and Zn. J Biotechnol 81:45–53

Grill E (1989) Phytochelatins in plants. In: Hammer DH, Winge DR (eds) Metal ion homeostasis: molecular biology and chemistry. Alan R. Liss, New York, pp 283–300

Grotz M, Fox T, Connolly E, Park W, Guerinot ML, Eide D (1998) Identification of a family of Zn transporter genes from *Arabidopsis* that responds to zinc deficiency. Proc Natl Acad Sci (U S A) 95:7220–7224

Guerinot ML, Eide D (1999) Zeroing in on zinc uptake in yeast and plants. Curr Opin Plant Biol 2:244–249

Guo JB, Dai XJ, Xu WZ, Ma M (2008) Overexpressing GSH1 and AsPCS1 simultaneously increases the tolerance and accumulation of cadmium and arsenic in Arabidopsis thaliana. Chemosphere 72:1020–1026

Ha SB, Smith AP, Howden R, Dietrich WM, Bugg S, O'Connell MJ, Goldsbrough PB, Cobbett CS (1999) Phytochelatin synthase genes from *Arabidopsis* and the yeast *Schizosaccharomyces pombe*. Plant Cell 11:1153–1163

Heiss S, Schäfer HJ, Haag-Kerwer A, Rausch T (1999) Cloning sulfur assimilation genes of *Brassica juncea* L.: cadmium differentially affects the expression of a putative low-affinity sulfate transporter and isoforms of ATP sulfurylase and APS reductase. Plant Mol Biol 39:847–857

Hirschi KD, Korenkov VD, Wilganowski NL, Wagner GJ (2000) Expression of Arabidopsis CAX2 in tobacco. Altered metal accumulation and increased manganese tolerance. Plant Physiol 124:125–133

Hopper JL, Parker DR (1999) Plant availability of selenate and selenate as influenced by the competing ions phosphate and sulfate. Plant Soil 210:199–207

Howarth JR, Domínguez-Solís JR, Gutiérrez-Alcalá G, Wray JL, Romero LC, Gotor C (2003) The serine acetyltransferase gene family in Arabidopsis thaliana and the regulation of its expression by cadmium. Plant Mol Biol 51:589–598

Howden R, Andersen CR, Goldsbrough PB, Cobbett CS (1995) A cadmium-sensitive, glutathione-deficient mutant of *Arabidopsis thaliana*. Plant Physiol 107:1067–1073

Hsieh JL, Chen CY, Chiu MH, Chein MF, Chang JS, Endo G, Huang CC (2009) Expressing a bacterial mercuric ion binding protein in plant for phytoremediation of heavy metals. J Hazard Mater 161(2–3):920–925

Huang JW, Blaylock MJ, Kapulnik Y, Ensley BD (1998) Phytoremediation of uranium-contaminated soils: role of organic acids in triggering uranium hyperaccumulation in plants. Environ Sci Technol 32:2004–2008

Hughes JB, Shanks J, Vanderford M, Lauritzen J, Bhadra R (1997) Transformation of TNT by aquatic plants and tissue cultures. Environ Sci Technol 31:266–271

Ike A, Sriprang R, Ono H, Murooka H, Yamashita M (2007) Bioremediation of cadmium contaminated soil using symbiosis between leguminous plant and recombinant rhizobia with the MTL4 and the PCS genes. Chemosphere 66:1670–1676

Jadia CD, Fulekar MH (2009) Phytoremediation of heavy metals: recent techniques. Afr J Biotechnol 8:921–928

Janssen DB, Pries F, van der Ploeg JR (1994) Genetics and biochemistry of dehalogenating enzymes. Annu Rev Microbiol 48:163–191

Jiménez-Ambriz G, Petit C, Bourrié I, Dubois S, Olivieri I, Ronce O (2007) Life history variation in the heavy metal tolerant plant Thalaspi caerulescens growing in a network of contaminated and noncontaminated sites in southern France: role of gene flow, selection and phenotypic plasticity. New Phytol 173:199–215

Jordan FL, Robin-Abbott M, Maier RM, Glenn EP (2002) A comparison of chelator-facilitated metal uptake by a halophyte and a glycophyte. Environ Toxicol Chem 21:2698–2704

Khan AG (2005) Role of soil microbes in the rhizospheres of plants growing on trace metal contaminated soils in phytoremediation. J Trace Elem Med Biol 18:355–364

Khan R, Bhawana P, Fulekar MH (2012) Microbial decolorization and degradation of synthetic dyes: a review. Rev Environ Sci Biotechnol. doi:10.1007/s11157-012-9287-6

Kloepper JW, Lifshitz R, Zablotowicz RM (1989) Free-living bacterial inocula for enhancing crop productivity. Trends Biotechnol 7:39–43

Kneer R, Zenk MH (1992) Phytochelatins protect plant enzymes from heavy metal poisoning. Phytochemistry 31:2662–2667

Köhler C, Merkle T, Neuhaus G (1999) Characterization of a novel gene family of putative cyclic nucleotides and calmodulin-regulated ion channels in *Arabidopsis thaliana*. Plant J 18:97–104

Korenkov V, Hirschi K, Crutchfield JD, Wagner GJ (2007) Enhancing tonoplast Cd/H antiport activity increases Cd Zn, and Mn tolerance, and impacts root/shoot Cd partitioning in Nicotianatabacum L. Planta 226:1379–1387

Kotrba P, Macek T, Ruml T (1999) Heavy metal-binding peptides and proteins in plants. A review. Collect Czech Chem Commun 64:1057–1086

Kotrba P, Najmanova J, Macek T, Ruml T, Mackova M (2009) Genetically modified plants in phytoremediation of heavy metal and metalloid soil and sediment pollution. Biotechnol Adv 27:799–810

Krämer U, Cotter-Howells JD, Charnock JM, Baker AJM, Smith AC (1996) Free histidine as metal chelator in plants that accumulate nickel. Nature 379:635–638

Kumar NPBA, Dushenkov V, Motto H, Raskin I (1995) Phytoextraction: the use of plants to remove heavy metals from soils. Environ Sci Technol 92:1232–1238

Lambert B, Joos H (1989) Fundamental aspects of rhizobacterial plant growth promotion research. Trends Biotechnol 7:215–219

Lasat MM, Pence NS, Garvin DF, Ebbs SD, Kochian LV (2000) Molecular physiology of zinc transport in the Zn hyperaccumulator *Thlaspi caerulescens*. J Exp Bot 51:71–79

Lasat MM (2002) Phytoextraction of toxic metals: a review of biological mechanisms. J Environ Qual 31:109–120

Lasat MM, Ebbs SD, Kochian LV (1997) Potential for phytoextraction of 137Cs from contaminated soils. Plant Soil 195:99–106

LeDuc DL, Tarun AS, Montes-Bayon M, Meija J, Malit MF, Wu CP et al (2004) Overexpression of selenocysteine methyltransferase in Arabidopsis and Indian mustard increases selenium tolerance and accumulation. Plant Physiol 135:377–383

LeDuc DL, Norman T (2005) Phytoremediation of toxic trace elements in soil and water. J Ind Microbiol Biotechnol 32:514–520

LeDuc DL, AbdelSamie M, Móntes-Bayon M, Wu CP, Reisinger SJ, Terry N (2006) Overexpressing both ATP sulfurylase and selenocysteine methyltransferase enhances selenium phytoremediation traits in Indian mustard. Environ Pollut 144:70–76

Leopold I, Guenther D, Schmidt J, Neumann D (1999) Phytochelatins and heavy metal tolerance. Phytochemistry 50:1323–1328

Leustek T (1996) Molecular genetics of sulfate assimilation in plants. Physiol Plant 97:411–419

Li ZS, Szcypka M, Lu YP, Thiele DJ, Rea PA (1996) The yeast cd factor protein (Y FC1) is a vacuolar glutathione-*S*-conjugate pump. J Biol Chem 271:6509–6517

Macek T, Kotrba P, Svatos A, Novakova M, Demnerova K, Mackova M (2008) Novel roles for genetically modified plants in environmental protection. Trends Biotechnol 26:146–152

Macnair MR (1993) The genetics of metal tolerance in vascular plants. New Phytol 124:541–559

Macnair MR, Bert V, Huitson SB, Saumitou-Laprade P, Petit D (1999) Zn tolerance and hyperaccumulation are genetically independent characters. Proc R Soc Lond B 266:2175–2179

Madejon P, Murillo JM, Maranon T, Valdes B, Rossini Oliva S (2005) Thallium accumulation in floral structures of *Hirschfeldiaincana* (L.) Lagrèze-Fossat (Brassicaceae). Bull Environ Contam Toxicol 74:1058–1064

Madejon P, Murillo JM, Maranon T, Lepp NW (2007) Factors affecting accumulation of thallium and other trace elements in two wild Brassicaceae spontaneously growing on soils contaminated by tailings dam waste. Chemosphere 67:20–28

Marchiol L, Asssolari S, Sacco P, Zerbi G (2004) Phytoremediation of heavy metals by canola (Brassica napus) and radish (Raphanussativus) grown on multi contaminated soil. Environ Pollut 132:21–27

McGrath SP, Zhao FJ, Lombi E (2002) Phytoremediation of metals, metalloids and radionuclides. Adv Agron 75:1–56

McNair MR, Tilstone GH, Smith SS (2000) The genetics of metaltolerance and accumulation in higher plants. In: Terry N, Bañuelos GS (eds) Phytoremediation of contaminated soil and water. Lewis Publishers, Boca Raton, pp 235–250

Meagher RB, Rugh CL, Kandasamy MK et al (2000) Engineered phytoremediation of mercury pollution in soil and water using bacterial genes. In: Terry N, Bañuelos G (eds) Phytoremediation of contaminated soil and water. CRC Press/Lewis, Boca Raton

Mendoza-Cózatl DG, Butko E, Springer F, Torpey JW, Komives EA, Kehr J et al (2008) Identification of high levels of phytochelatins, glutathione and cadmium in the phloem sap of Brassica napus. A role for thiol-peptides in the long-distance transport of cadmium and the effect of cadmium on iron translocation. Plant J 54:249–259

Meyer A, Fricker M (2002) Control of demand-driven biosynthesis of glutathione in green Arabidopsis suspension culture cells. Plant Physiol 130:1927–1937

Mishra D, Kar M (1974) Nickel in plant growth and metabolism. Bot Rev 40:395–452

Mukhopadhyay S, Maiti SK (2010) Phytoremediation of metal enriched mine waste: a review. Global J Environ Res 4(3):135–150

Naested H, Fennema M, Hao L, Anderson M, Janssen DB, Mundy J (1999) A bacterial haloalkane dehydrogenase gene as a negative selectable marker in *Arabidopsis*. Plant J 18:571–576

Neuhierl B, Thanbichler M, Lottspeich F, Böck A (1999) A family of S-methylmethionine dependent thiol/selenol methyltransferases. Role in selenium tolerance and evolutionary relation. J Biol Chem 274:5407–5414

Nouairi I, Ammar WB, Youssef NB, Daoud DBM, Ghorbal MH, Zarrouk M (2006) Comparative study of cadmium effects on membrane lipid composition of Brassica juncea and Brassica napus leaves. Plant Sci 170:511–519

Palmer CE, Keller WA (1994) In vitro culture of oilseeds. In: Vasil K, Thorpe TA (eds) Plant cell and tissue culture. Kluwer, Dordrecht, pp 413–455

Palmer CE, Warwick S, Keller W (2001) Brassicaceae (Cruciferae) family, plant biotechnology, and phytoremediation. Int J Phytoremediation 3:245–287

Patten CL, Glick BR (1996) Bacterial biosynthesis of indole-3-acetic acid. Can J Microbiol 42:207–220

Pence HS, Larsen PB, Ebbs SD, Letham DLD, Lasat MM, Garvin DF, Eide D, Kochian LV (2000) The molecular physiology of metal transporter in the Zn/Cd hyperaccumulator, *Thlaspi caerulescens*. Proc Natl Acad Sci U S A 97:4956–4960

Peterson AG, Oliver DJ (2006) Leaf-targeted phytochelatin synthase in Arabidopsis thaliana. Plant Physiol Biochem 44:885–892

Picault N, Cazalé AC, Beyly A, Cuiné S, Carrier P, Luu DT et al (2006) Chloroplast targeting of phytochelatin synthase in Arabidopsis: effects on heavy metal tolerance and accumulation. Biochimie 88:1743–1750

Pilon-Smits EAH, Pilon M (2000) Breeding mercury-breathing plants for environmental cleanup. Trends Plant Sci 5:235–236

Pilon-Smits EAH, Pilon M (2002) Phytoremediation of metals using transgenic plants. Crit Rev Plant Sci 21:439–456

Pilon-Smits EA, Hwang S, Mel Lytle C, Zhu Y, Tai JC, Bravo RC et al (1999) Overexpression of ATP sulfurylase in Indian mustard leads to increased selenate uptake, reduction, and tolerance. Plant Physiol 119:123–132

Pilon-Smits EAH, Zhu YL, Sears T, Terry N (2000) Overexpression of glutathione reductase in Brassica juncea: effects on cadmium accumulation and tolerance. Physiol Plant 110:455–460

Rea PA, Li ZS, Lu YP, Drozdowicz YM, Martinoia E (1998) From vacuolar GSX pumps to multi-specific ABC transporters. Annu Rev Plant Physiol Plant Mol Biol 99:727–760

Reeves R, Baker A (2000) Metal accumulating plants. In: Raskin I, Ensley BD (eds) Phytoremediation of toxic metals: using plants to clean up the environment. Wiley, New York, pp 193–229

Robinson BH, LeBlanc M, Petit D, Brooks RR, Kirkman JH, Gregg PEH (1998) The potential of *Thlaspi caerulescens* for phytoremediation of contaminated soils. Plant Soil 203:47–56

Rogers EE, Eide DJ, Guerinot ML (2000) Altered selectivity in an *Arabidopsis* metal transporter. Proc Natl Acad Sci U S A 97:12356–12360

Rout GR, Samantaray S, Das P (1999) In vitro selection and biochemical characterization of zinc and manganese adapted callus lines in *Brassica* spp. Plant Sci 137:89–100

Rubio F, Gassmann W, Schroeder JI (1995) Sodium driven potassium uptake by the plant potassium transporter HKT1 and mutations conferring salt tolerance. Science 270:1660–1663

Rugh CL, Wilde HD, Stack NM, Thompson DM, Summers AO, Meagher RB (1996) Mercuric ion reduction and resistance in transgenic Arabidopsis thaliana plants expressing a modified bacterial merA gene. Proc Natl Acad Sci U S A 93:3182–3187

Rugh CL, Senecoff JF, Meagher RB, Merkle SA (1998) Development of transgenic yellow poplar for mercury phytoremediation. Nat Biotechnol 16:925–928

Salt DE, Smith RD, Raskin I (1998) Phytoremediation. Annu Rev Plant Physiol Plant Mol Biol 49:643–668

Sanita-di Toppi L, Gabbrielli R (1999) Response to cadmium in higher plants. Environ Expt Bot 41:105–130

Sasaki Y, Hayakawa T, Inoue C, Miyazaki A, Silver S, Kusano T (2006) Generation of mercury hyperaccumulating plants through transgenic expression of the bacterial mercury membrane transport protein MerC. Transgenic Res 15:615–625

Schäfer HJ, Greiner S, Rausch T, Haag-Kerwer A (1997) In seedlings of the heavy metal accumulator *Brassica juncea* Cu2+ differentially affects transcript amounts for γ-glutamylcysteine synthetase (γ-ECS) and metallothionein (MT2). FEBS Lett 404:216–220

Schat H, Ten-Bookum WM (1992) Genetic control of copper tolerance in *Silene vulgaris*. Heredity 68:219–229

Schat H, Vooijs R (1997) Multiple tolerance and co-tolerance to heavy metals in *Silene vulgaris*: a co-segregation analysis. New Phytol 136:489–496

Schiavon M, Pittarello M, Pilon-Smits EAH, Wirtz M, Hell R, Malagoli M (2012a) Selenate and molybdate alter sulfate transport and assimilation in Brassica juncea L. Czern.: implications for phytoremediation. Environ Exp Bot 75:41–51

Schiavon M, Galla G, Wirtz M, Pilon-Smits EA, Telatin V, Quaggiotti S, Hell R, Barcaccia G, Malagoli M (2012b) Transcriptome profiling of genes differentially modulated by sulfur and chromium identifies potential targets for phytoremediation and reveals a complex S-Cr interplay on sulfate transport regulation in B. juncea. J Hazard Mater 239–240:192–205

Schultz CL, Hutchinson TC (1998) Evidence for a key role for metallothionein-like protein in the copper tolerance of *Deschampsia caespitosa* (L.) Beauv. New Phytol 110:163–172

Schnug E (1993) Physiological functions, environmental relevance of sulfurcontaining secondary metabolites. In: De Kok LJ, Stulen I, Rennenberg H, Brunold C, Rauser W (eds) Sulfur nutrition and sulfur assimilation in higher plants: regulatory agricultural, environmental aspects. SPB Academic Publishing, The Hague, pp 179–190

Schnug E (1997) Significance of sulphur for the quality of domesticated plants. In: Cram WJ, De Kok LJ, Brunold C, Rennenberg H (eds) Sulphur metabolism in higher plants: molecular ecophysiological, nutritional aspects. Backhuys Publishers, Leiden, pp 109–130

Selvam A, Wong JWC (2009) Cadmium uptake potential of Brassica napus cocropped with Brassica parachinensis and Zea mays. J Hazard Mater 167:170–177

Sjödin C (1992) Brassicaceae, a family well suited for modern biotechnology. Acta Agric Scand 42:197–207

Steffen JC (1990) The heavy metal building peptides of plants. Annu Rev Plant Physiol Plant Mol Biol 41:553–575

Stephan UW, Scholz G (1993) Nicotianamine: mediator of transport of iron and heavy metals in the phloem. Physiol Plant 88:522–529

Sunkar R, Kaplan B, Bouché N, Arazi T, Dolev D, Talke IN et al (2000) Expression of a truncated tobacco NtCBP4 channel in transgenic plants and disruption of the homologous Arabidopsis CNGC1 gene confer Pb2+ tolerance. Plant J 24:533–542

Szczygłowska M, Piekarska A, Konieczka P, Namiesnik J (2011) Use of Brassica plants in the phytoremediation and biofumigation processes. Int J Mol Sci 12:7760–7771

Taghavi S, Barac T, Greenberg B, Borremans B, Vangronsveld J, vanderLelie D (2005) Horizontal gene transfer to endogenous endophytic bacteria from poplar improves phytoremediation of toluene. Appl Environ Microbiol 71:8500–8505

Tappero R, Peltier E, Grafe M, Heidel K, Ginder-Vogel M, Livi KJT et al (2007) Hyperaccumulator Alyssum murale relies on a different metal storage mechanism for cobalt than for nickel. New Phytol 175:641–654

Thomasini R, Vogt E, Fromenteau M, Hortensteiner S, Matile P, Amrhein N, Martinoia E (1998) An ABC transporter of *Arabidopsis thaliana* has both glutathione conjugate and chlorophyll catabolite transporter activity. Plant J 13:773–780

Thomine S, Wang RC, Ward JM, Crawford NM, Schroeder JI (2000) Cadmium and iron transport by members of a plant metal transporter family in *Arabidopsis* with homology to Hramp genes. Proc Natl Acad Sci (U S A) 97:4991–4996

Van der Zaal BJ, Neuteboom LW, Pinas JE, Chardonnens AN, Schat H, Verkleij JAC, Hooykaas PJJ (1999) Overexpression of a novel *Arabidopsis* gene related to putative zinc-transporter genes from animals can lead to enhanced zinc resistance and accumulation. Plant Physiol 119(1047):1056

Van Huysen T, Abdel-Ghany S, Hale KL, LeDuc D, Terry N, Pilon-Smits EA (2003) Overexpression of cystathionine-gamma-synthase enhances selenium volatilization in Brassica juncea. Planta 218:71–78

Vasak M (2005) Advances in metallothionein structure and functions. J Trace Elem Med Biol 19:13–17

Vatamaniuk OK, Mari S, Lu YP, Rea PA (1999) AtPCS1, a phytochelatin synthase from *Arabidopsis*: isolation and in vitro reconstitution. Proc Natl Acad Sci (U S A) 96:7110–7115

Wangeline AL, Burkhead JL, Hale KL, Lindblom SD, Terry N, Pilon M et al (2004) Overexpression of ATP sulfurylase in Indian mustard: effects on tolerance and accumulation of twelve metals. J Environ Qual 33:54–60

Weyens N, vanderLelie D, Taghavi S, Vangronsveld J (2009) Phytoremediation: plant endophyte partnership stake the challenge. Curr Opin Biotechnol 20:248–254

Weyens N, Croes S, Dupae J, Newmanb L, vanderLelie D, Carleer R, Vangronsveld J (2010) Endophytic bacteria improve phytoremediation of Ni and TCE co-contamination. Environ Pollut 158:2422–2427

Wijnhoven S, Leuven R, Van Der Velde G, Jungheim G, Koelemij E, De Vries F et al (2007) Heavy-metal concentrations in small mammals from a diffusely polluted floodplain: importance of species- and location-specific characteristics. Arch Environ Contam Toxicol 52:603–613

Xiang C, Oliver DJ (1998) Glutathione metabolic genes coordinately respond to heavy metals and jasmonic acid in Arabidopsis. Plant Cell 10:1539–1550

Yang X, Jin XF, Feng Y, Islam E (2005) Molecular mechanisms and genetic bases of heavy metal tolerance/hyperaccumulation in plants. J Integr Plant 47:1025–1035

Zaier H, Tahar G, Abelbasset L, Rawdha B, Rim G, Majda M, Souhir S, Stanley L, Chedly A (2010) Comparative study of Pb-phytoextraction potential in Sesuvium portulacastrum and Brassica juncea: tolerance and accumulation. J Hazard Mater 183(1–3):609–615

Zenk MH (1996) Heavy metal detoxification in higher plants: a review. Gene 179:21–30

Zhu YL, Pilon-Smits EAH, Tarun AS, Weber SU, Jouanin L, Terry N (1999) Cadmium tolerance and accumulation in Indian mustard is enhanced by overexpressing glutamylcysteine synthetase. Plant Physiol 121:1169–1177

Zhu YL, Pilon-Smits EA, Jouanin L, Terry N (1999a) Overexpression of glutathione synthetase in Indian mustard enhances cadmium accumulation and tolerance. Plant Physiol 119:73–80

Zhu YL, Pilon-Smits EA, Tarun AS, Weber SU, Jouanin L, Terry N (1999b) Cadmium tolerance and accumulation in Indian mustard is enhanced by overexpressing γ-glutamylcysteine synthetase. Plant Physiol 121:1169–1178

Chapter 6
Genome Analysis

Graham J. King

Abstract Cruciferous crops are now cultivated throughout the world beyond their natural centres of diversity. The close evolutionary relationship of Arabidopsis has underpinned genome analysis and our current understanding of genome organisation and evolution within the Brassicaceae. Translating this reference information into crop species such as *Brassica*, of more recent polyploid origin and containing multiple paralogous and homoeologous gene copies, remains a challenge. *Brassica* experimental resources have enabled analysis of whole genome sequence data in combination with cytological approaches, This has revealed a series of large and small-scale duplication events driving a pattern of recursive paleopolyploidization, with genome triplication elegantly demonstrated in the *B. rapa* genome sequence. Progressive resolution and integration of genetic, physical and sequence maps underpins assignment of functional genes and regulatory networks to agronomic traits. High throughput sequencing now enables transcriptome mapping, associative transcriptomics and advanced eQTL analysis. Prospects for the future of genome analysis in cruciferous crops are considered. It is now possible to explore the role of epigenetic variation in modulating gene function and plasticity associated with phenotypic response to the environment.

Keywords Genomics • Brassica • Crops • Arabidopsis • Comparative genomics • Transcriptomics

G.J. King, B.Sc., Ph.D. (✉)
Southern Cross Plant Science, Southern Cross University,
Military Road, Lismore, PO Box 157 2480, Australia
e-mail: graham.king@scu.edu.au

S.K. Gupta (ed.), *Biotechnology of Crucifers*, DOI 10.1007/978-1-4614-7795-2_6,
© Springer Science+Business Media, LLC 2013

6.1   Introduction

Cruciferous species are geographically widely distributed, with various crop types now cultivated throughout the world beyond their natural centres of diversity. As outlined elsewhere in this volume, they are typified by their wide range of morphological forms, especially within the genus *Brassica*. This diversity (Gomez-Campo 1999), with a repertoire incorporating proliferation of many different tissues, has been the subject of much speculation in terms of the inherent plasticity within the genus, and so it has been noted that *Brassica* may represent the domestic dog of plants (Brown 2001). Darwin (1875) described the 'great differences in the shape, size, colour, arrangement and manner of growth of the leaves and stem, and of the flower-stem in the broccoli and cauliflower'. He also highlighted the morphotypic plasticity within the genus in terms of genotype x environment interactions, reporting that 'cabbages will not form heads in hot countries', and that 'extremely poor soil also affects the characters of certain varieties'.

Due to their relative economic value, and consequent investment of research effort, examples cited in this chapter will be associated with the genus *Brassica*, which includes distinct vegetable, oilseed, condiment, fodder and forage crops. Much of this information is directly relevant for crops of closely related species such as *Raphanus* (radish) and *Sinapis* (white mustard). Additional minor cruciferous crops, where genomic insights are also likely to benefit considerably from research carried out in *Brassica,* include *Camelina, Eruca* (rocket salad), and *Isatis*.

A range of reviews and books elsewhere have reflected the growing knowledge base and insights gained from genomic analysis of *Brassica* crops (e.g. King 2007; Edwards et al. 2012; Ramichiary and Lim 2011; Snowdon 2007; Quiros and Farnham 2011), which builds on the previous long history of *Brassica* genetics. The aim of this chapter is not to reiterate this information, but to provide a brief summary, and place in context an updated overview of the changing role of genomics, particularly in translational research and development from the model species Arabidopsis. Throughout the chapter, the aim will be to cite examples of relevant recent research, and to point readers to the richness of resources and information that have been accumulated for this globally important group of crops.

At the time of writing, the complete and annotated genome sequence of the diploid species *B. rapa* has been published (The *Brassica rapa* Genome Consortium 2011). Other genomes, notably *B. oleracea* and *B. napus* are in preparation. However, compared with other major crops, progress has been relatively hindered due to national and institutional decisions to embark on five 'reference' genome projects for the major oilseed crop and amphidiploid *B. napus*.

6.2   Cruciferous Crop Genomes and the Model Plant Arabidopsis

Arabidopsis thaliana was adopted by the international plant research community as the model reference plant species, with the annotated genome sequence released in 2000 (The Arabidopsis Genome Initiative 2000). Over the past 20 years sequencing and functional analysis of the Arabidopsis genome has generated fundamental and deep insights into many aspects of plant biology, and in particular the role of genes and regulatory networks modulating development, and adaptation to environment. The close evolutionary relationship between Arabidopsis and cruciferous crops has underpinned much of our current understanding of genome organisation and evolution within the Brassicaceae. The annotated Arabidopsis genome and functional role of individual genes provides a particularly valuable resource in terms of candidate genes underlying many crop agronomic traits. The challenge remains in translating this reference information into species that are of more recent polyploid origin, and so inherently contain multiple paralogous and homoeologous gene copies.

Compared with Arabidopsis, there has been relatively slow progress in functional analysis of individual genes, proteins, metabolites and complex traits of cruciferous crops, and less in characterising the details of relevant genomic regulatory networks. The increased complexity of genome organisation, driven by genome duplications and divergence, is exacerbated by the relatively close sequence homology between paralogous and orthologous genes. Thus for many single genes characterised within inbred Arabidopsis, there may be as many as 12 similar transcripts in an outcrossing amphidiploid such as *B. napus* (viz 3 paralogues × 2 genomes × heterozygotic alleles). These circumstances have until recently often been an impediment to unequivocal gene identification in relation to genetic locus.

6.3 Genomic Experimental Resources

Limitations to functional gene analysis in *Brassica* are rapidly being overcome through the availability of genomic tools and data. Over the past decade a valuable set of experimental resources has been assembled in laboratories worldwide, to facilitate genomic analysis of *Brassica* and related species. The accumulation and interpretation of genomic information has unsurprisingly been assisted by the ready access to experimental resources from Arabidopsis. In earlier analyses, genomic tools developed for Arabidopsis were used directly to probe the *Brassica* genomes (Gao et al. 2005; Ryder et al. 2001; Town et al. 2006). As well as large insert libraries for physical and chromosome in situ mapping (Mun et al. 2008; Wang et al. 2011a) and a series of EST libraries, the availability of *Brassica* transcriptome and whole genome sequence data has enabled generation of analysis platforms such as microarrays (Love et al. 2010; Parkin et al. 2010; Trick et al. 2009a) and SNP chips (Hayward et al. 2012).

Development and use of experimental plant populations has underpinned development of *Brassica* reference linkage maps. These have been particularly valuable when assembling the corresponding complex crop genome sequences, where it is essential to align and orientate scaffolds to high density and integrated linkage maps (Bancroft et al. 2011; Choi et al. 2007; Wang et al. 2011b; Xu et al. 2010). It has already been noted that the pace at which functional identification and characterisation of genes underlying agronomic traits has lagged behind the advances made in Arabidopsis. As well as issues related to genome duplication, this is due in part to the quantitative and polygenic complexity of many crop traits. For these to be resolved, there remains a need for generation of comprehensive genetic maps that survey a high density of recombination break points throughout the genome. This has been addressed elegantly in crops such as tomato, where alien substitutions and micro-introgression lines provide the tools for fine resolution of QTL (Eshed and Zamir 1994; Chapman et al. 2012). Such resources have been developed on a small scale in *B. oleracea* (Rae et al. 1999) and *B. napus* (Burns et al. 2003), but to date this approach has not been widely adopted for gene discovery. The generation and maintenance of EMS mutagenised populations (Wang et al. 2008; Stephenson et al. 2010) has long-term value as genomic information becomes available, particularly for reverse-genetic screening of functional genes. Finally, a series of online databases and other information resources are available, for which www.brassica.info and http://brassica.nbi.ac.uk/ provide a useful starting point.

6.4   Cytological Context

The relationship between karyotypes amongst *Brassica* species has long been of interest to cytologists (Prakash et al. 2009), and has provided the grounding for our current appreciation of genome evolution within the Brassiceae. Within the genus *Brassica*, the work of Nagaharu (1935) established the now emblematic view of the genus. The 'Triangle of U' provides the framework for understanding the overall relationships between the constituent genomes, with the three distinct 'cytodemes' (diploid genomes) represented by *B. rapa* (A, n=10), *B. nigra* (B, n=8) and *B. oleracea* (C, n=9), that are then combined in the amphidiploids *B. juncea* (AB, n=18), *B. napus* (AC, n=19) and *B. carinata* (BC, n=17). These cytodemes are effectively conserved for the diploid species, and account for all other described *Brassica* species. Thus many C genome species such as *B. cretica*, *B. insularis* and *B. hilarionis* not only share the same chromosome number of n=9, but are interfertile with *B. oleracea*, although they do contain more diverse chloroplast haplotypes (Allender et al. 2007).

6.5   Nomenclature

The description and comparison of genomes requires an appreciation of the underlying cytological framework of chromosome number, morphology and behaviour. Due to the relatively diminutive size of the smallest chromosomes within the karyotype of *Brassica* species, a number of different numbering systems had been developed. Moreover, the independent generation of genetic linkage groups resulted in a plethora of additional nomenclature. Since 2007 there has been international agreement (http://www.brassica.info/resource/maps/lg-assignments.php) to adopt the linkage group nomenclature outlined by Parkin et al. (1995, 2005) and Sharpe et al.

(1995) for *B. napus* (AC genomes), by Bohuon et al. (1998) for *B. oleracea* (C) and by Lagercrantz and Lydiate (1995) for *B. nigra* (B). This nomenclature scheme enables linkage groups for the amphidiploid genomes to be aligned with their constituent diploid linkage groups. The A1-A10 linkage groups of *B. rapa* thus correspond to the A1-A10 (previously N1-N10) of the *B. napus* A genome. Similarly the C1-C10 (previously O1-O9) linkage groups of *B. oleracea* correspond to the C1-C9 groups of the *B. napus* C genome. Moreover, firstly for *B. oleracea* (Howell et al. 2002) and then *B. rapa* (Koo et al. 2004; Lim et al. 2005) the linkage groups have been aligned and oriented with respect to the chromosomes of the karyotype, making use of fluorescent in situ hybridisation (FiSH) with sequence-tagged probes. The use of genomic in situ hybridisation (GISH) has also allowed the homoeologous A and C genome chromosomes to be distinguished in the amphidiploid *B. napus* (Howell et al. 2008). More recently Xiong and Pires (2011) consolidated this knowledge by identifying all homoeologous chromosomes of the allopolyploid *B. napus* and its diploid progenitors. Monosomic alien addition lines of individual C chromosomes in an A genome background have also enabled an unequivocal assignment of mapped SSR markers to specific C chromosomes (Geleta et al. 2012).

A nomenclature system for description of functional genes within the genus *Brassica* has been established (Ostergaard and King 2008). This system enables the assignment of a specific gene to the A, B or C genome, and provides a means of assigning a distinct identity to paralogous genes within any one genome. At present, a consistent nomenclature system for other genera within the Brassicaceae has not yet emerged. As *Brassica* spp. and other cruciferous genomes become sequenced and annotated, a series of reference gene models will become available in the context of chromosome position, as has been made available for *B. rapa* (The *Brassica rapa* Genome Consortium 2011). However, in terms of functional loci, gene name assignment remains a valuable functional identifier, as it does for Arabidopsis.

6.6   Whole Genome and Segmental Polyploidy

Genome sizes for *Brassica* species vary with relative chromosome lengths and numbers (Arumuganathan and Earle 1991). Whole-genome sequencing (The *Brassica rapa* Genome Consortium 2011) has provided updated estimates of the euchromatic gene space for *B. rapa* of 290 Mbp, representing approximately 60 % of the A genome. The contemporary structure and organisation of Brassicaceae genomes is a product of processes and events affecting chromosomes over evolutionary history. The analysis of genome sequence data in combination with cytological approaches has revealed a series of large and small-scale duplication events (Mun et al. 2009; Howell et al. 2008; Wang et al. 2011a) that over the past 5–43 million years (The *Brassica rapa* Genome Consortium 2011; Beilstein et al. 2010) have driven a pattern of recursive paleopolyploidizations (Wang et al. 2011a). The complete genome sequence of *B. rapa* elegantly demonstrates the genome triplication within the diploid genomes (The *Brassica rapa* Genome Consortium 2011) when compared with Arabidopsis. However, this overview needs to be placed in the context of variability in the degree of gene loss and fractionation found within each of the triplicated segments. The general pattern appears to consist of one out of the three paralogous segments retaining a disproportionately high proportion of genes that are inferred to have been present in the common ancestor (The *Brassica rapa* Genome Consortium 2011). It has been suggested that the variation in the numbers of paralogous genes, and gene families within each genome, have contributed to the well described plasticity within the genus.

6.7   Patterns of Genome Divergence

Our understanding of the evolutionary paths for different Brassicaceae genomes is based on integration of knowledge at cytological, genomic sequence and gene level. Current models for chromosome evolution have primarily been informed by comparison with Arabidopsis. The physical organisation of the Arabidopsis genome provided the basis for an experimental strategy that used fluorescent in situ hybridisation (FiSH) of sets of BACs representing segments of Arabidopsis chromosomes to probe other genomes within the Brassicaceae (Lysak et al. 2005; Howell et al. 2005). This has enabled a clear understanding to develop of the overall pattern of genomic duplications and rearrangements that led to the current chromosome numbers, and is based on the concept of a common ancestral karyotype (Barker et al. 2009; Lysak et al. 2005; Schranz et al. 2005, 2007). The prevailing model (Schranz et al. 2005, 2007) describes a series a crucifer genomic blocks (K blocks) in the ancestral karyotype of $n = 8$, which incorporates the 'At-alpha' duplication which appears to be Brassicaceae specific (Barker et al. 2009).

Comparative genetic mapping using probes or other sequence-tagged markers orthologous to Arabidopsis and *Brassica* provided the initial coarse-grain view of the relationships between segments of the Arabidopsis genome and the triplicated blocks within *Brassica* diploid genomes (Parkin et al. 2005; Panjabi et al. 2008). The use of sequence-tagged genetic markers is now routine (Bancroft et al. 2006; Kim et al. 2006; Wang et al. 2011b), and has greatly facilitated the anchoring of cruciferous crop genomes to the reference Arabidopsis. These tools have made a significant contribution to identifying candidate genes for trait loci.

The ability to generate high density linkage maps is limited by cross-over frequency and distribution in segregating populations, as well as the degree of allelic variation in the available marker set. Early restriction fragment length polymorphism (RFLP) markers were valuable for comparative mapping through the retrospective sequencing of previously arbitrarily selected cloned probes. The detailed work of Parkin et al. (1995, 2005) resulted in the description of a series of conserved collinearity blocks within the A and C genomes of *B. napus*, which has guided interpretation in subsequent studies. The need to integrate independent genetic linkage maps is also important, in order to navigate between genome sequence and trait loci. Suwabe et al. (2008) carried out a map integration of the A genomes of *B. rapa* and *B. napus* using a common set of microsatellite markers, and Wang et al. (2011b) integrated the A and C genomes based on the linkage maps generated from several independent and widely used segregating populations. This approach also facilitated direct integration of chromosome sequence scaffolds of *B. rapa* and the A genome linkage maps of *B. napus* (Wang et al. 2011a). The use of digital transcriptome mapping now greatly enhances this approach (Bancroft et al. 2011; Higgins et al. 2012).

The density and resolution of break-point positions that border conserved collinearity blocks has increased through a series of comparative genomic studies. This has enabled local and global patterns of genome duplication and divergence to be documented (Mun et al. 2009; Park et al. 2005; Rana et al. 2004; Ryder et al. 2001; Town et al. 2006; Trick et al. 2009b). This background work has now confirmed and guided the assembly of triplicated genomic regions observed in complete genome sequence.

The use of Illumina-based transcriptome mapping has been pioneered in *Brassica* (Trick et al. 2009c), where sequence variation and transcript abundance has been combined to construct SNP maps of *B. napus* containing up to 23,000 markers (Bancroft et al. 2011). This analysis has allowed the detection of genome rearrangements and greatly enhances the ability to track inheritance of genomic segments, including those arising from inter-specific crosses. More recently the same technology has been used successfully to assign transcripts to each of the A and C genomes in amphidiploid *B. napus* (Higgins et al. 2012). This greatly enhances the ability to analyse and assign gene expression to specific chromosomal segments, and to follow these in hybrid breeding programs.

6.8 Genome Sequencing

Generation of contiguous DNA sequences for each of the nuclear chromosomes in the widely grown *Brassica* crops has been the long-term goal of the Multinational Brassica Genome Project. This consortium was established in 2002 (http://www.brassica.info/info/about-mbgp.php#meetings) and has progressively led the development of genomic resources and sequencing initiatives, with *B. rapa* being the first genome annotated and published in 2011 (The *Brassica rapa* Genome Consortium 2011). The major challenge for generating high quality sequence assemblies in these complex crop genomes has been to generate data that may be unequivocally assigned and oriented with respect to known chromosomes and linkage groups. The initial approaches had been based on Sanger sequencing of bacterial artificial chromosome clones (BACs), with end-sequencing distributed amongst international partners. Analysis of these seed and end-sequenced BACs provided an initial view of the distribution of genes, transposons and SSRs within the genome (Hong et al. 2006). This stage was followed by seed-BAC sequencing and chromosome by chromosome assignment of tasks to different national consortia. The timing of this latter stage coincided with the availability of high throughput Illumina data generation, and thereby became superseded by this approach. However, there are valuable comparisons to be made between the two approaches, and the largest A genome chromosome (A3) was first completed by BAC sequencing (Mun et al. 2010).

There are currently ongoing efforts by different consortia under the umbrella of the Multinational Brassica Genome Project to sequence and annotate the complete genomes of *B. oleracea* and *B. napus*.

6.9   Genome Analysis of Specific Gene Families

One of the primary drivers for genome analysis in crop genomes is to identify and characterise genes associated with particular agronomic traits. In *Brassica* crops there are particular problems that need to be overcome, in terms of distinguishing between the multiple paralogous (within diploid genomes) and homoeologous (between genomes in amphidiploids) genes. A number of targeted studies (eg. Cardenas et al. 2012; Irwin et al. 2012; Wang et al. 2009; Wang et al. 2012; Yang et al. 2010) have focused on characterising members of paralogous and homoeolo-gous gene families, making use of accumulated sequence data to identify single nucleotide polymorphisms (SNPs) to distinguish between loci. These findings have started to provide an account of the role of genome evolution in divergence and selection of functional loci. For many paralogous genes there may be relative conservation of coding sequence, but divergence of promoters and other regulatory sequences (Wang et al. 2012). The power of comparative genomics to identify conserved non-coding sequences was elegantly demonstrated from early shotgun DNA data (Ayele et al. 2005) by a comparison of (*B. oleracea*) and Arabidopsis sequences upstream of 13 genes (Colinas et al. 2002). A more extensive approach taken by Haberer et al. (2006) involved analysing the same set of shotgun genome sequence data for *B. oleracea* and identifying conserved *cis* motifs of functionally characterised genes in Arabidopsis. This approach enabled discovery of candidate transcription factor binding sites in 64 % of the Arabidopsis genes analyzed.

6.10   Genomic Approaches to Resolution of Trait Loci

The primary driver for genome analysis in cruciferous crops is to understand the relationship between agronomic traits, any associated genes and their regulatory networks. Given the level of genome complexity compared with Arabidopsis, this remains a significant challenge. Many studies have focused on identifying quantitative trait loci (QTL) that account for many polygenic traits and may have differential contributions under different environmental conditions (Shi et al. 2009; Basunanda et al. 2010).

Traditional approaches to resolution of QTL have involved use of segregating populations and linkage maps defined by molecular markers. The availability of genomic information provides the opportunity to derive more detailed and functional information associated with such trait loci. Expression quantitative trait loci (eQTL) are regions of a genome associated with variation in gene expression amongst individuals (Kliebenstein 2009). Such variation may arise due to sequence polymorphisms in target genes, their *cis*-regulatory (proximal) or trans-regulatory (distal) regions, and lead to phenotypic differences. The value of this approach is the ability to infer genes and regulatory networks that underly complex traits, which is of particular importance for crop plants. Hammond et al. (2011) have used the reference *B. rapa* genome sequence and an exon-array to define *cis*- and *trans*-eQTL, and so characterise their environmental response to low phosphorus availability in this species. This analysis was also able to identify specific *trans*-eQTL within the genome that are candidates for potential regulatory hotspot of phosphorous nutrition.

The ability to generate whole transcriptome data using Illumina sequencing has revolutionised the capacity to carry out genome scanning of polyploid genomes such as *B. napus*. Harper et al. (2012) have demonstrated the approach of associative transcriptomics, which modifies the approach of eQTL mapping. By using transcriptome sequencing it is possible to identify and score molecular markers that represent variation both in gene sequences and gene expression, and then to correlate this with trait variation. These authors were able to identify genomic deletions in *B. napus* associated with two QTL for seed glucosinolate content, that corresponded to a transcription known to control aliphatic glucosinolate biosynthesis in Arabidopsis.

Metabolite QTL (mQTL) provide a further characterisation of trait variation in the context of genome organisation. By screening variation within a segregating population across particular metabolite classes it is possible to generate an overview of genomic regulation of particular pathways. Early analyses in *Brassica* focused on small numbers of metabolites within the context of well described pathways such as those involved in fatty acid biosynthesis and modification (Barker et al. 2007; Smooker et al. 2010). These studies were able to infer location of genes affecting (in *cis* or *trans*) specific enzymes, by mapping the adjacent substrate and products in the relevant biochemical pathways (Barker et al. 2007). More intensive analyses have shown the power of large-scale metabolite analysis. The dissection of glucosinolate concentration and individual components in different tissues of *B. napus* has enabled identification of 105 mQTL in *B. napus* (Feng et al. 2012). Further correlation analysis allows prediction of gene function underlying each mQTL and construction of an advanced metabolic network, with identification of epistatic interactions responsible for glucosinolate composition.

6.11 Transposon Analysis

Transposable elements (TEs) form a significant proportion of the genome within many Brassicaceae, and are likely to play important roles in genome divergence and plasticity. Due to their prevalence in such genomes, along with other repetitive sequences, they are often more difficult to compile accurately into contiguous genome sequences. However, they may account for much of the difference observed in genome size beween different *Brassica* diploid genomes, since it is apparent that the increase in genome size between Arabidopsis and *Brassica* is not solely accounted for by the effective segmental triplication within the latter (Gao et al. 2005; The *Brassica rapa* Genome Consortium 2011). Initial studies of *B. oleracea* shotgun genome data (Ayele et al. 2005) enabled identification of some of the patterns of TE amplification, diversification and loss since divergence from a common ancestor with Arabidopsis (Zhang and Wessler 2004). It thus appeared that nearly all TE lineages are shared between the genera, with the number of elements in each lineage being greater for *B. oleracea*. Based on genome sequence, it is estimated that transposons represent 39.5 % of the *B. rapa* genome (The *Brassica rapa* Genome Consortium 2011). The preferential proliferation of particular TE families within each species is well documented (Alix et al. 2005, 2008; Koo et al. 2011; Lim et al. 2007; The *Brassica rapa* Genome Consortium 2011). For example, the proliferation of the CACTA Bot1 transposon family (Alix et al. 2008) within *B. oleracea*, accounting for 2.3 % of the genome, has played a major role in the recent divergence of the A and C genomes. Lim et al. 2007 have also characterised the centromere and pericentromeric retrotransposons in *B. rapa*, and shown that many of these are species-specific and not found in the B or C genomes.

6.12   Epigenetic Variation and Plasticity

Within many cruciferous crops their evolutionary history, based on whole genome and intra-genomic polyploidy, has given rise to overlapping sets of duplicated genes, which then have the ability to diverge in sequence and function. When the divergent *Brassica* genomes that share a common history combine in hybrids, they often undergo a period of instability and re-organisation (Lukens et al. 2004; Gaeta et al. 2007; Szadkowski et al. 2010), with associated changes in the distribution of epigenetic marks (Gaeta et al. 2007; Xu et al. 2009). An ongoing question has been how much this genic diversity and plasticity gives rise to the phenotypic plasticity observed in many cruciferous crops.

It is becoming apparent that epigenetic marks such as modified histones and DNA methylation can modulate plant phenotype by regulation of gene transcription affecting developmental plasticity and interactions with the environment. There is increasing evidence that epigenetic variation plays an important role in plant adaptation to environment, and this has the potential to be harnessed for crop breeding (King et al. 2010). The range of target sites and level of DNA methylation in plants is greater than found in many animal systems, and this inherent variation is likely to affect many important traits required for crops to adapt to changing environments. Among the crop traits that appear to be under epigenetic control are interactions with abiotic and biotic environmental stresses, vegetative variation, seed development, cold-adaption, fruit ripening and hybrid stability (Eichten et al. 2011; Haun et al. 2007; Manning et al. 2006; King et al. 2010). Within *Brassica* oilseed crops, variation in DNA methylation marks across the genome has been shown to be associated with selection for energy use efficiency (Hauben et al. 2009). Dominance relationships amongst S-alleles of the *B. rapa* pollen self-incompatibility locus are also affected by tissue-specific monoallelic *de novo* DNA methylation (Shiba et al. 2006). Novel variation may be induced in *Brassica* species by stochastic DNA hypomethylation of the genome (King 1995; Amoah et al. 2012).

The role of epigenetic variation in modulating gene function and phenotypic response to the environment has started to be unravelled in Arabidopis (Becker et al. 2011; Stokes et al. 2002; Vaughn et al. 2007). Recent studies indicate similar or greater plasticity in crop species (Amoah et al. 2012; Karan et al. 2012; King et al. 2010). High resolution genomic maps of epigenetic marks have been developed for Arabidopsis (Cokus et al. 2008; Lister et al. 2008), but have yet to be compiled for *Brassica* genomes. However, the distribution of DNA methylation marks in *Brassica* has been surveyed at increasing resolution from marker-based studies (Lukens et al. 2006; Salmon et al. 2008; Long et al. 2011). A recent survey of the stability and distribution of DNA methylation in *B. napus*, based on methylation sensitive AFLP (MSAP) and retrotransposon epimarkers, indicated a surprising degree of stability, with the majority of marker alleles heritable and conserved in different growing environments and distinct developmental stages (Long et al. 2011).

Ribosomal DNAs (rDNAs) represent a significant component of plant genomes located at nucleolar organiser regions. The behaviour of the tandem arrays of thousands of rDNA gene copies is of particular interest in crops, since their capacity for parallel transcription may provide a constraint on cell proliferation and protein processing at key stages of development, such as seed filling. An epigenetic phenomenon, nucleolar dominance describes nucleolus formation around rRNA genes inherited from only one progenitor of an interspecific hybrid or allopolyploid. Thus in *B. napus*, A genome rRNA transcripts are detected, but C genome transcripts are not (Chen and Pikaard 1997a, b). Analysis of rDNAs associated with particular donor genomes in both natural and newly synthesised amphidiploids indicate that dominance affecting transcription of rRNA genes is genome specific, and independent of maternal effect, ploidy or rRNA gene dosage (Chen and Pikaard 1997a). Hypomethylation with 5-aza-2′-deoxycytidine leads to high levels of transcription for the previously silent C genome rRNA genes (Chen and Pikaard 1997b).

6.13 Organelle Genomes

Mitochrondria and chloroplasts play central roles in the metabolism and physiology of all plants, and contribute to the performance of crop species. There has been relatively little effort focused on characterising variation of organellar genomes within the Brassicaceae. Until recently, most systematic information has been accumulated at the level of haplotype variation. For mitochondria this may have a profound effect on the presence and efficiency of cytoplasmic male sterility (CMS), with variation in the copy number of different mitotypes contributing to the CMS phenotype (Chen et al. 2011). Recent comparative analysis of the complete mitochondrial genomes from *B. rapa* (A), *B. oleracea* (C) with the three amphidiploid species revealed variation in genome size from 219,747 to 360,271 bp (Chang et al. 2011). These authors found that genes of known function were conserved in all mitotypes, but that there was variation in the open reading frames (ORFs) of genes with unknown functions. Overall, there was evidence of genome compaction and inheritance in the course of *Brassica* mitotype evolution.

The contribution of variation in chloroplast haplotype to photosynthetic efficiency or overall growth parameters has not been deeply studied in cruciferous crops, although haplotyping has provided a framework for understanding maternal relationships between different sub-taxa of *B. oleracea* (Allender et al. 2007) and between those of the A and C diploid genome species and domesticated amphidiploid *B. napus* (Allender and King 2010). There is considerable scope for comparative genome sequence analysis of *Brassica* organelle genomes.

6.14   Data Integration and Informatics

The ability to navigate between trait loci and underlying functional genetic units can be a limiting factor in the rapid application of genomic information to improvements in crop breeding and agronomic practice. A central requirement is to define ordered arrays of whole genome information and provide the experimental evidence that indicates how these correspond to chromosome architecture and genetic linkage maps. Online resources exist, based on collation of *Brassica* reference linkage maps (http://www.brassica.info/resource/maps/published-data.php; Wang et al. 2011b). More comprehensive integrated databases and curation tools have been developed that allow navigation between annotated genome sequence, genetic and QTL maps (Love et al. 2012).

An important role of annotated genomic information anchored to genetic and physical maps that are true representations of linear chromosomes, is to facilitate the manipulation of crop traits. In particular, where the function of a single or collection of gene alleles is known, genomic knowledge can assist in locating these in the relevant genomic context. This is turn enables analysis and manipulation of regulatory structures, interpretation of evolutionary selection pressures, and phenomena such as linkage drag, gene conversion and reciprocal or non-reciprocal translocation between homoeologous chromosomes. In more immediate practical terms, this knowledge may then be used to develop locus-specific molecular genetic markers for use in Marker Assisted Selection, or to introduce novel alleles, either via site-selected mutagenesis or transgenic modification.

The inference of genomic regulatory networks from eQTL, associative transcriptomics and mQTL will in future provide a framework for integrating knowledge at the level of gene expression with higher level behaviours of crop physiology, development and responses to external environments. This clearly requires development of more sophisticated data repositories and bioinformatic tools.

6.15 Prospects

Over the past decade there have been progressive and significant advances in our understanding of the genome composition and organisation of crop brassicas. These have been driven by improvements in technological and analytical capability. The genomes are characterised by complexity at different levels of organisation, with early indications of the consequences for evolution, plasticity and cultivation of these crops. As with many crop species, identification and functional characterisation of candidate genes has benefited from the reference Arabidopsis. However, the closer relationship with crop brassicas has driven the development of innovative analytical techniques and platform resources based on genomic information. This provides a sound basis on which to investigate and manipulate this important crop genus.

Notwithstanding the evolutionary proximity of Arabidopsis and crucifer crops, there remains considerable scope to carry out systematic studies that characterise functional genes, and to understand in more detail the plasticity afforded by presence of multiple close paralogous genes. It can be anticipated that the role of epigenetic variation in accounting for the considerable plasticity observed within crucifer crops will be unravelled, now that genomic tools are available and being refined for resolution of epigenetic marks. More generally, crucifer crops provide an excellent framework for establishing a deep understanding of genome and phenotypic plasticity, and this information can be used to underpin efficient selective and hybrid breeding. Developing this understanding in *Brassica* and related genera is likely to provide scope to generate novel crops to meet global demands of food security in times of rapid climate change.

Access to increased genomic knowledge will also provide the tools to enable the resolution of potential progenitors of the domesticated amphidiploid crop brassicas. Although the *Brassica* A and C genomes remain essentially intact in amphidiploid *B. napus* (Parkin et al. 2005; Wang et al. 2011), this species is not found in natural populations. The exact locations of the original hybridisation events remain uncertain (Allender and King 2010), although the major centres of natural diversity for both *B. rapa* and *B. oleracea* are located in Europe and Asia Minor. These regions are therefore the starting point for any investigation into the origins of stable crop *B. napus*, and comparative sequence analysis of transcriptome and other representations of the genome across the genepool will provide a valuable resource for these investigations.

The integration of heterogeneous sources of data, information and knowledge will pose considerable challenges in terms of bioinformatic analysis and visualisation. The deeper insights of value to crop production and breeding are likely to arise from integration of information from the different realms of genomics, transcriptomics, metabolomics and various other aspects of biochemistry and physiological signalling and beyond to embrace systematic phenomic characterietion.

As a footnote, the Brassicales Map Alignment Project (BMAP) was initiated in 2010 (http://www.brassica.info/resource/sequencing/bmap.php) with the aim to generate high quality contiguous genome sequence datasets for representative taxa across the Brassiaceae, now extended to the Brassicales. As this project progresses it will provide a rich resource to understand genome evolution and functional diversity, and hopefully contribute to the development of a wider range of cruciferous crops.

References

Alix K, Ryder C, Moore J, King G, Heslop-Harrison JS (2005) The genomic organization of retrotransposons in *Brassica oleracea*. Plant Mol Biol 59:839–851

Alix K, Joets J, Ryder CD, Moore J, Barker GC, Bailey JP, King GJ, Heslop-Harrison JS (2008) The CACTA transposon Bot1 played a major role in *Brassica* genome divergence and gene proliferation. Plant J 56:1030–1044

Allender CJ, Allainguillaume J, Lynn JR, King GJ (2007) Chloroplast SSRs reveal uneven distribution of genetic diversity in C genome n=9 *Brassica* species. Theor Appl Genet 114:609–618

Allender C, King GJ (2010) Clarifying the origins of the amphiploid species *Brassica napus* L. using chloroplast and nuclear molecular markers. BMC Plant Biol 10:54

Amoah S, Kurup S, Rodriguez Lopez CM, Welham SJ, Powers SJ, Hopkins CJ et al (2012) A hypomethylated population of *Brassica rapa* for forward and reverse epi-genetics. BMC Plant Biol 12:193

The Arabidopsis Genome Initiative (AGI) (2000) Analysis of the genome of the flowering plant Arabidopsis thaliana. Nature 408:816–820

Ayele M, Haas BJ, Kumar N, Wu H, Xiao Y, Van Aken S et al (2005) Whole genome shotgun sequencing of *Brassica oleracea* and its application to gene discovery and annotation in *Arabidopsis*. Genome Res 15:487–495

Arumuganathan K, Earle ED (1991) Nuclear DNA content of some important plant species. Plant Mol Biol Rep 9:211–215

Bancroft I, Barnes S, Li J, Meng J, Qiou D, Schmidt R et al (2006) Establishment of an integrated marker system for oilseed rape breeding (IMSORB). Acta Hortic 706:195–202

Bancroft I, Morgan C, Fraser F, Higgins J, Wells R, Clissold L et al (2011) Dissecting the genome of the polyploid crop oilseed rape by transcriptome sequencing. Nat Biotechnol 29:762–766

Barker GC, Larson TR, Graham IA, Lynn JR, King GJ (2007) Novel insights into seed fatty acid synthesis and modification pathways from genetic diversity and QTL analysis of the *Brassica* C genome. Plant Physiol 144:1827–1842

Barker MS, Vogel H, Schranz ME (2009) Paleopolyploidy in the Brassicales: analyses of the Cleome transcriptome elucidate the history of genome duplications in Arabidopsis and other Brassicales. Genome Biol Evol 1:391–399

Basunanda P, Radoev M, Ecke W, Friedt W, Becker H, Snowdon RJ (2010) Comparative mapping of quantitative trait loci involved in heterosis for seedling and yield traits in oilseed rape *Brassica napus* L. Theor Appl Genet 120:271–281

Becker C, Hagmann J, Müller J, Koenig D, Stegle O, Borgwardt K et al (2011) Spontaneous epigenetic variation in the *Arabidopsis thaliana* methylome. Nature 480:245–249

Beilstein MA, Nagalingum NS, Clements MD, Manchester SR, Mathews S (2010) Dated molecular phylogenies indicate a Miocene origin for Arabidopsis thaliana. Proc Natl Acad Sci USA 107:18724–18728

Bohuon EJ, Ramsay LD, Craft JA, Arthur AE, Marshall DF, Lydiate DJ et al (1998) The association of flowering time quantitative trait loci with duplicated regions and candidate loci in *Brassica oleracea*. Genetics 150:393–401

Brown JS (2001) Fit of form and function, diversity of life, and procession of life as an evolutionary game. Chapter 4. In: Orzack SH, Sober E (ed) Adaptationism and optimality. Cambridge, England University Press

Burns MJ, Barnes SR, Bowman JG, Clarke MH, Werner CP, Kearsey MJ (2003) QTL analysis of an intervarietal set of substitution lines in *Brassica napus*: (i) Seed oil content and fatty acid composition. Heredity 90:39–48

Cárdenas PD, Gajardo HA, Huebert T, Parkin IAP, Iniguez-Luy FL, Federico ML (2012) Retention of triplicated phytoene synthase *PSY* genes in *Brassica napus* L. and its diploid progenitors during the evolution of the *Brassiceae*. Theor Appl Genet 124:1215–1228

Chang S, Yang T, Du T, Huang Y, Chen J, Yan J et al (2011) Mitochondrial genome sequencing helps show the evolutionary mechanism of mitochondrial genome formation in *Brassica*. BMC Genomics 12:497

Chapman NH, Bonnet J, Grivet L, Lynn J, Graham N, Smith R et al (2012) High resolution mapping of a fruit firmness-related QTL in tomato reveals epistatic interactions associated with a complex combinatorial locus. Plant Physiol 159:1644–1657

Chen JZ, Pikaard CS (1997a) Transcriptional analysis of nucleolar dominance in polyploid plants: biased expression/silencing of progenitor rRNA genes is developmentally regulated in *Brassica*. Proc Natl Acad Sci USA 94:3442–3447

Chen JZ, Pikaard CS (1997b) Epigenetic silencing of RNA polymerase I transcription: a role for DNA methylation and histone modification in nucleolar dominance. Genes Dev 11:2124–2136

Chen J, Guan R, Chang S, Du T, Zhang H, Xing H (2011) Substoichiometrically different mitotypes coexist in mitochondrial genomes of *Brassica napus* L. PLoS One 6(3):e17662

Choi SR, Teakle GR, Plaha P, Kim JH, Allender CJ, Beynon E et al (2007) The reference genetic linkage map for the multinational *Brassica rapa* genome sequencing project. Theor Appl Genet 115:777–792

Cokus SJ, Feng S, Zhang X, Chen Z, Merriman B, Haudenschild CD et al (2008) Shotgun bisulphite sequencing of the Arabidopsis genome reveals DNA methylation patterning. Nature 452:215–219

Colinas J, Birnbaum K, Benfey PN (2002) Using cauliflower to find conserved non-coding regions in Arabidopsis. Plant Physiol 129:451–454

Darwin C (1875) The variation of animals & plants under domestication, 2nd edn. John Murray, London, pp 398–403

Edwards D, Batley K, Parkin I, Kole C (2012) Genetics, genomics and breeding of oilseed Brassicas. Science Publishers, Enfield, USA

Eichten SR, Swanson-Wagner RA, Schnable JC, Waters AJ, Hermanson PJ, Liu S et al (2011) Heritable epigenetic variation among maize inbreds. PLoS Genet 7(11):e1002372

Eshed Y, Zamir D (1994) A genomic library of Lycopersicon Pennellii in Lycopersicon-Esculentum – a tool for fine mapping of genes. Euphytica 79:175–179

Feng J, Long Y, Shi L et al (2012) Characterization of metabolite quantitative trait loci and metabolic networks that control glucosinolate concentration in the seeds and leaves of *Brassica napus*. New Phytol 193:96–108

Gaeta RT, Pires JC, Iniguez-Luy F, Leon E, Osborn TC (2007) Genomic changes in resynthesized *Brassica napus* and their effect on gene expression and phenotype. Plant Cell 19:3403–3417

Gao M, Li G, McCombie WR, Quiros CF (2005) Comparative analysis of a transposon-rich *Brassica oleracea* BAC clone with its corresponding sequence in *A.thaliana*. Theor Appl Genet 111:949–955

Geleta M, Heneen W, Stoute A, Muttucumaru N, Scott R, King GJ et al (2012) Assigning Brassica microsatellite markers to the nine C-genome chromosomes using *Brassica rapa* var. trilocularis-B. *oleracea* var. alboglabra monosomic alien addition lines. Theor Appl Genet 125:455–466

Gómez-Campo C (1999) Biology of *Brassica* coenospecies. Elsevier Science, Amsterdam

Haberer G, Mader MT, Kosarev P, Spannag M, Yang L, Mayer KFX (2006) Large-scale cis-element detection by analysis of correlated expression and sequence conservation between Arabidopsis and *Brassica oleracea*. Plant Physiol 142:1589–1602

Hammond JP, Mayes S, Bowen HC, Graham NS, Hayden RM, Love CG et al (2011) Regulatory hotspots are associated with plant gene expression under varying soil phosphorus P supply in *Brassica rapa*. Plant Physiol 156:1230–1241

Harper AL, Trick M, Higgins J, Fraser F, Clissold L, Wells R et al (2012) Associative transcriptomics of traits in the polyploid crop species *Brassica napus*. Nat Biotechnol 30:798–802

Hauben M, Haesendonckx B, Standaert E, Van Der Kelen K, Azmi A, Akpo H et al (2009) Energy use efficiency is characterized by an epigenetic component that can be directed through artificial selection to increase yield. Proc Natl Acad Sci USA 106:20109

Haun WJ, Laoueille-Duprat S, O'Connell MJ, Spillane C, Grossniklaus U, Phillips AR et al (2007) Genomic imprinting, methylation and molecular evolution of maize Enhancer of zeste Mez homologs. Plant J 49:325–337

Hayward A, Mason A, Dalton Morgan J, Zander M, Edwards D et al (2012) SNP discovery and applications in *Brassica napus*. J Plant Biotechnol 39:1–12

Higgins JA, Magusin A, Trick M, Fraser F, Bancroft I (2012) Use of mRNA-Seq to discriminate contributions to the transcriptome from the constituent genomes of the polyploid crop species *Brassica napus*. BMC Genomics 13:247. doi:10.1186/1471-2164-13-247

Hong CP, Plaha P, Koo D-H, Yang T-J, Choi SR, Lee YK et al (2006) A survey of the *Brassica rapa* genome through BAC-end sequence analysis, and comparative analysis with *Arabidopsis thaliana*. Mol Cells 22:300–307

Howell EC, Armstrong SJ, Barker GC, Jones GH, King GJ, Ryder CD et al (2005) Physical organisation of the major duplication on *Brassica oleracea* chromosome O6 revealed through fluorescence in situ hybridisation with Arabidopsis and *Brassica* BACs. Genome 48:1093–1103

Howell EC, Barker GC, Jones GH, Kearsey MJ, King GJ, Kop EP et al (2002) Integration of the cytogenetic and genetic linkage maps of *Brassica oleracea*. Genetics 161:1225–1234

Howell EC, Kearsey MJ, Jones GH, King GJ, Armstrong SJ (2008) A and C genome distinction and chromosome identification in *Brassica napus* by sequential FISH and GISH. Genetics 180:1849–1857

Irwin JA, Lister C, Soumpourou E, Zhang Y, Howell EC, Teakle G et al (2012) Functional alleles of the flowering time regulator FRIGIDA in the *Brassica oleracea* genome. BMC Plant Biol 12:21

Karan R, DeLeon T, Biradar H, Subudhi PK (2012) Salt stress induced variation in DNA methylation pattern and its influence on gene expression in contrasting rice genotypes. PLoS One 7:e40203

Kim JS, Chung TY, King GJ, Jin M, Yang TJ, Jin YM et al (2006) A sequence-tagged linkage map of *Brassica rapa*. Genetics 174:29–39

King G (1995) Morphological development in *Brassica oleracea* is modulated by in vivo treatment with 5-azacytidine. J Hortic Sci 70:333–342

King GJ (2007) Utilization of Arabidopsis and *Brassica* genomic resources to underpin genetic analysis and improvement of *Brassica* crops. In: Varshney RK, Koebner RMD (eds) Model plants: crop improvement. CRC, Boca Ratan, pp 33–69

King GJ, Amoah S, Kurup S (2010) Exploring and exploiting epigenetic variation in crop plants. Genome 53:856–868

Kliebenstein D (2009) Quantitative genomics: analyzing intraspecific variation using global gene expression polymorphisms or eQTLs. Annu Rev Plant Biol 60:93–114

Koo DH, Plaha P, Lim YP, Hur YK, Bang JW (2004) A high-resolution karyotype of *Brassica rapa ssp.* pekinensis revealed by pachytene analysis and multicolor fluorescence in situ hybridization. Theor Appl Genet 109:1346–1352

Koo DH, Hong CP, Batley J, Chung YS, Edwards D, Bang JW et al (2011) Rapid divergence of repetitive DNAs in *Brassica* relatives. Genomics 97:173–185

Lagercrantz U, Lydiate DJ (1995) RFLP mapping in *Brassica nigra* indicates differing recombination rates in male and female meiosis. Genome 38:255–264

Lim KB, de Jong H, Yang TJ, Park JY, Kwon SJ, Kim JS et al (2005) Characterization of rDNAs and tandem repeats in the heterochromatin of *Brassica rapa*. Mol Cells 19:436–444

Lim KB, Yang TJ, Hwang YJ, Kim JS, Park JY, Kwon SJ et al (2007) Characterization of the centromere and pericentromere retrotransposons in *Brassica rapa* and their distribution in related *Brassica* species. Plant J 49:173–183

Lister R, O'Malley RC, Tonti-Filippini J, Gregory BD, Berry CC, Millar AH et al (2008) Highly integrated single-base resolution maps of the epigenome in Arabidopsis. Cell 133:523–536

Long Y, Xia W, Li R, Wang J, Shao M, Feng J et al (2011) Epigenetic QTL mapping in *Brassica napus*. Genetics 189:1093–1102

Love C, Andongabo A, Wang J, Carion P, Rawlings C, King GJ (2012) InterstoreDB: a generic intergration resource for genetic and genomic data. J Integr Plant Biol 54:345–355

Love CG, Graham NS, Ó Lochlainn S, Bowen HC, May ST, White PJ et al (2010) A *Brassica* exon array for whole-transcript gene expression profiling. PLOS One 5(9):e12812

Lukens LN, Pires JC, Leon E, Vogelzang R, Oslach L, Osborn TC (2006) Patterns of sequence loss and cytosine methylation within a population of newly resynthesized *Brassica napus* allopolyploids. Plant Physiol 140:336–348

Lukens LN, Quijada PA, Udall J, Pires JC, Schranz ME, Osborn TC (2004) Genome redundancy and plasticity within ancient and recent *Brassica* crop species. Biol J Linn Soc 82:665–674

Lysak MA, Koch MA, Pecinka A, Schubert I (2005) Chromosome triplication found across the tribe *Brassiceae*. Genome Res 15:516–525

Manning K, Tor M, Poole M, Hong Y, Thompson A, King GJ et al (2006) A naturally occurring epigenetic mutation in an SBP-box gene inhibits tomato fruit ripening. Nat Genet 38:948–952

Mun JH, Kwon SJ, Yang TJ, Kim HS, Choi BS, Baek S et al (2008) The first generation of a BAC-based physical map of *Brassica rapa*. BMC Genomics 9:280

Mun JH, Kwon SJ, Yang TJ, Seol YJ, Jin M, Kim JA et al (2009) Genome-wide comparative analysis of the *Brassica rapa* gene space reveals genome shrinkage and differential loss of duplicated genes after whole genome triplication. Genome Biol 10:R111

Mun JH, Kwon SJ, Seol YJ, Kim JA, Jin M, Kim JS et al (2010) Sequence and structure of *Brassica rapa* chromosome A3. Genome Biol 11:R94

Nagaharu U (1935) Genomic analysis in *Brassica* with special reference to the experimental formation of *B. napus* and peculiar mode of fertilisation. Jpn J Bot 7:389–452

Østergaard L, King GJ (2008) Standardized gene nomenclature for the *Brassica* genus. Plant Methods 4:10

Panjabi P, Jaqanath A, Bisht NC, Padmaja KL, Sharma C, Gupta V et al (2008) Comparative mapping of *Brassica juncea* and *Arabidopsis thaliana* using intron polymorphism IP markers: homoeologous relationships, diversification and evolution of the A, B and C *Brassica* genomes. BMC Genomics 3:113–132

Park JY, Koo DH, Hong CP, Lee SJ, Jeon JW, Lee SH et al (2005) Physical mapping and microsynteny of *Brassica rapa ssp.* pekinensis genome corresponding to a 222 kb gene-rich region of Arabidopsis chromosome 4 duplicated on chromosome 5. Mole Genet Genomics 274:579–588

Parkin IAP, Clarke WE, Sidebottom C, Zhang W, Robinson SJ, Links MG et al (2010) Towards unambiguous transcript mapping in the allotetraploid *Brassica napus*. Genome 53:929–938

Parkin IAP, Gulden SM, Sharpe AG, Lukens L, Trick M, Osborn TC et al (2005) Segmental structure of the *Brassica napus* genome based on comparative analysis with *Arabidopsis thaliana*. Genetics 171:765–781

Parkin IA, Sharpe AG, Keith DJ, Lydiate DJ (1995) Identification of the A and C genomes of amphidiploid *Brassica napus* (oilseed rape). Genome 38:1122–1131

Prakash S, Bhat SR, Quiros CF, Kirti PB, Chopra VL (2009) *Brassica* and its close allies: cytogenetics and evolution. Plant Breed Rev 31:21–187

Quiros CF, Farnham M (2011) The genetics of *Brassica oleracea*. In: Bancroft I, Schmidt R (eds) Genetics and genomics of the Brassicaceae, Plant genetics and genomics series. Springer, New York

Rae AM, Howell EC, Kearsey MJ (1999) More QTL for flowering time revealed by substitution lines in *Brassica oleracea*. Heredity 83:586–596

Ramchiary N, Lim YP (2011) Genetics of *Brassica rapa* L. In: Schmidt R, Bancroft I (eds) Genetics and genomics of the Brassicaceae, vol 9, Plant genetics and genomics: crops and models. Springer, New York, pp 215–260

Rana D, van den Boogaart T, Bent E, O'Neill CM, Hynes L, Macpherson L et al (2004) Conservation of the microstructure of genome segments in *Brassica napus* and its diploid relatives. Plant J 40:725–733

Ryder CD, Smith LB, Teakle GR, King GJ (2001) Contrasting genome organisation: two regions of the *Brassica oleracea* genome compared with collinear regions of the *Arabidopsis thaliana* genome. Genome 44:808–817

Salmon A, Clotault J, Jenczewski E, Chable V, Manzanares-Dauleux MJ (2008) *Brassica oleracea* displays a high level of DNA methylation polymorphism. Plant Sci 174:61–70

Schranz ME, Lysak MA, Mitchell-Olds T (2005) The ABC's of comparative genomics in the Brassicaceae: building blocks of crucifer genomes. Trends Plant Sci 11:1360–1385

Schranz ME, Song BH, Windsor AJ, Mitchell-Olds T (2007) Comparative genomics in the Brassicaceae: a family-wide perspective. Curr Opin Plant Biol 10:168–175

Sharpe AG, Parkin IA, Keith DJ, Lydiate DJ (1995) Frequent nonreciprocal translocations in the amphidiploid genome of oilseed rape *Brassica napus*. Genome 38:1112–1121

Shi J, Li R, Qiu D, Jiang C, Long Y, Morgan C et al (2009) Unraveling the complex trait of crop yield with quantitative trait loci mapping in *Brassica napus*. Genetics 182:851–861

Shiba H, Kakizaki T, Iwano M, Tarutani Y, Watanabe M, Isogai A et al (2006) Dominance relationships between self-incompatibility alleles controlled by DNA methylation. Nat Genet 38:297–299

Smooker AM, Wells R, Morgan C, Beaudoin F, Cho K, Fraser F et al (2010) The identification and mapping of candidate genes and QTL involved in the fatty acid desaturation pathway in *Brassica napus*. Theor Appl Genet 122:1075–1090

Snowdon R (2007) Cytogenetics and genome analysis in *Brassica* crops. Chromosome Res 15:85–95

Stephenson P, Baker D, Girin T, Perez A, Amoah S, King GJ et al (2010) A rich TILLING resource for studying gene function in *Brassica rapa*. BMC Plant Biol 10:62

Stokes TL, Kunkel BN, Richards EJ (2002) Epigenetic variation in *Arabidopsis* disease resistance. Genes Dev 16:171–182

Suwabe K, Morgan C, Bancroft I (2008) Integration of *Brassica* A genome genetic linkage map between *Brassica napus* and *B. rapa*. Genome 51:169–176

Szadkowski E, Eber F, Huteau V, Lodé M, Huneau C, Belcram H et al (2010) The first meiosis of resynthesized *Brassica napus*, a genome blender. New Phytol 186:102–112

The Brassica rapa Genome Consortium (2011) The genome of the mesopolyploid crop species *Brassica rapa*. Nat Genet 43:1035–1039

Town CD, Cheung F, Maiti R, Crabtree J, Haas BJ, Wortman JR et al (2006) Comparative genomics of *Brassica oleracea* and Arabidopsis thaliana reveals gene loss, fragmentation and dispersal following polyploidy. Plant Cell 18:1348–1359

Trick M, Cheung F, Drou N, Fraser F, Lobenhofer EK, Hurban P et al (2009a) A newly-developed community microarray resource for transcriptome profiling in*Brassica* species enables the confirmation of *Brassica*-specific expressed sequences. BMC Plant Biol 9:50

Trick M, Kwon SJ, Choi SR, Fraser F, Soumpourou W, Drou N et al (2009b) Complexity of genome evolution by segmental rearrangement in *Brassica rapa* revealed by sequence-level analysis. BMC Genomics 10:539

Trick M, Long Y, Meng J, Bancroft I (2009c) Single nucleotide polymorphism SNP discovery in the polyploid *Brassica napus* using Solexa transcriptome sequencing. Plant Biotechnol J 7:334–346

Vaughn MW, Tanurdžić M, Lippman Z, Jiang H, Carrasquillo R et al (2007) Epigenetic natural variation in *Arabidopsis thaliana*. PLoS Biol 5(7):e174

Wang J, Hopkins C, Hou J, Zou X, Wand C, Long Y et al (2012) Promoter variation and transcript divergence in brassicaceae lineages of flowering locus T. PLoS One 7(10):e47127

Wang J, Lydiate DJ, Parkin IAP, Falentin C, Delourme R, Carion P et al (2011a) Integration of linkage maps for the Amphidiploid *Brassica napus*, and comparative mapping with Arabidopsis and *Brassica rapa*. BMC Genomics 12:101

Wang X, Torres MJ, Pierce G, Lemke C, Nelson LK, Yuksel B et al (2011b) A physical map of *Brassica oleracea* shows complexity of chromosomal changes following recursive paleopoly-ploidizations. BMC Genomics 12:470

Wang N, Wang Y, Tian F, King GJ, Zhang C, Long Y et al (2008) A functional genomics resource for *Brassica napus*: development of an EMS mutagenized population and discovery of FAE1 point mutations by TILLING. New Phytol 180:751–765

Wang J, Wu B, Liu J, Jiang C, Long Y, Shi L et al (2009) The evolution of *Brassica napus* FLOWERING LOCUS T paralogues in the context of inverted chromosomal duplication blocks. BMC Evol Biol 9:271

Xiong Z, Pires JC (2011) Karyotype and identification of all homoeologous chromosomes of allo-polyploid *Brassica napus* and its diploid progenitors. Genetics 187:37–49

Xu Y, Zhong L, Wu X, Fang X, Wang J (2009) Rapid alterations of gene expression and cytosine methylation in newly synthesized *Brassica napus* allopolyploids. Planta 229:471–483

Xu J, Qian X, Wang X, Li R, Cheng X, Yang T et al (2010) Construction of an integrated genetic linkage map for the A genome of *Brassica napus* using SSR markers derived from sequenced BACs in *B. rapa*. BMC Genomics 11:594

Yang B, Qiu D, Quiros CF (2010) Variation of five major glucosinolate genes in *Brassica rapa* in relation to *Brassica oleracea* and Arabidopsis thaliana. Span J Agric Res 8:662–671

Zhang XY, Wessler SR (2004) Genome-wide comparative analysis of the transposable elements in the related species Arabidopsis thaliana and *Brassica oleracea*. Proc Natl Acad Sci USA 101:5589–5594

Chapter 7
Genetic Engineering of Lipid Biosynthesis in Seeds

Stacy D. Singer, Michael S. Greer, Elzbieta Mietkiewska, Xue Pan, and Randall J. Weselake

Abstract The demand for vegetable oils, including those derived from crucifer (Brassicaceae) species, has been increasing rapidly over recent years for use in both food and industrial applications. In order to meet these demands, biotechnological approaches will almost certainly be a necessity to generate crops with improved lipid traits. In addition to the clear need to increase the seed oil content of crucifer species, there has also been growing interest in generating transgenic lines that display improved compositions of fatty acids and non-acyl lipids, including carotenoids and tocochromanols, for enhanced nutritional or industrial applicability. Fortunately, knowledge concerning oilseed metabolism and lipid biosynthesis are accumulating at a rapid pace, which is enabling attempts to genetically engineer crucifer species with enhanced oil content and quality. This chapter outlines the various attempts and successes in this field to date.

Keywords Brassicaceae • Oil content • Fatty acid composition • Carotenoids • Tocochromanols • Plant biotechnology

Abbreviations

ACCase	Acetyl-CoA carboxylase
ACP	Acyl-carrier protein
CoA	Coenzyme A
DAG	sn-1,2-diacylglycerol

S.D. Singer, Ph.D. • M.S. Greer, M.Sc. • E. Mietkiewska, Ph.D. • X. Pan, M.Sc.
R.J. Weselake, B.Sc., M.Sc., Ph.D. (✉)
Agricultural, Food Nutritional Science, University of Alberta,
410 Agriculture/Forestry Centre, Edmonton, Alberta T6G 2P5, Canada
e-mail: randall.weselake@ualberta.ca

S.K. Gupta (ed.), *Biotechnology of Crucifers*, DOI 10.1007/978-1-4614-7795-2_7,
© Springer Science+Business Media, LLC 2013

DGAT	Diacylglycerol acyltransferase
DHA	Docosahexaenoic acid
ER	Endoplasmic reticulum
EPA	Eicosapentaenoic acid
FA	Fatty acid
FAD	Fatty acid desaturase
FAE	Fatty acid elongation complex
Fat	Fatty-acyl ACP thioesterase
γ-TMT	γ-tocopherol methyltransferase
GGDP	Geranylgeranyl diphosphate
GGR	Geranylgeranyl reductase
GLA	γ-linolenic acid
GPAT	sn-glycerol-3-phoshate acyltransferase
G3P	sn-glycerol-3-phosphate
HEAR	High erucic acid rapeseed
HPPD	p-hydroxyphenylpyruvate dioxygenase
HPT	Homogentisate phytyltransferase
KAS	β-ketoacyl-ACP synthase
LPAAT	Lysophosphatidic acid acyltransferase
LPCAT	Lysophosphatidylcholine acyltransferase
mFA	Modified fatty acid
MGGBQ	2-methyl-6-geranylgeranyl-1,4-benzoquinol
MPBQ	2-methyl-6-phytyl-1,4-benzoquinol
mtPDC	Mitochondrial pyruvate dehydrogenase complex
PA	Phosphatidic acid
PC	Phosphatidylcholine
PDAT	Phospholipid: diacylglycerol acyltransferase
PDCT	Phosphatidylcholine: diacylglycerol cholinephosphotransferase
PDHK	Pyruvate dehydrogenase kinase
PKS	Polyketide synthase
PSY	Phytoene synthase
PUFA	Polyunsaturated fatty acid
SDA	Stearidonic acid
SMCFA	Short- and medium-chain saturated fatty acids
TAG	Triacylglycerol
TC	Tocopherol cyclase
VLC-PUFA	Very long-chain polyunsaturated fatty acid

7.1   Introduction

Members of the crucifer (Brassicaceae) family are highly valuable to the plant lipid
field of research and agricultural industry. For example, *Arabidopsis thaliana* provides a model system for the study of lipid biosynthesis and as a platform for

investigating the effects of modifying enzymes related to lipid metabolism. In terms of agricultural value, crucifer species are also of major economic importance globally, with oilseed rape (mostly *Brassica napus*) being second in oilseed production only to soybean and palm. Although the majority of oilseed rape is grown in Europe (~40 %), the top producers, including China, Canada and India, produce over half of the world's annual supply (Food and Agriculture Organization of the United Nations 2012). Indeed, Canada is the world's leading exporter (Food and Agriculture Organization of the United Nations 2012), where the rapid growth in rapeseed value has recently led the plant to supplant wheat as the most seeded crop (Statistics Canada 2012). In part, this high global value of rapeseed derives from successful breeding programs that have resulted in the development of numerous varieties that are well-suited to either industrial applications or human consumption.

The Canadian variety of *B. napus*, canola, has been branded as a healthy oil as a result of its high oleic acid content and low levels of glucosinolates and erucic acid. In contrast, the predominant varieties grown in Europe include those bearing high levels of erucic acid to primarily serve industrial needs for erucamide, which is used in numerous applications in the oleochemical industry. In addition to the appeal of generating improved rapeseed varieties, interest has also arisen in developing other crucifer species to meet similar industrial and food market requirements. For example, the seed oil of *Crambe abyssinica* contains high erucic acid levels and could be used similarly to high erucic acid (HEAR) *B. napus* seed oil. Medical uses have also been suggested for *Lunaria annua*, which produces 23 % nervonic acid; a fatty acid (FA) that is used in the treatment of multiple sclerosis (Warwick 2011). Moreover, several crucifer oilseed varieties have been proposed as future biofuel sources; a market that is expected to grow by 50–60 % globally in the next 20 years (Canola Council of Canada 2011; Gahukar 2012).

The manipulation of lipid biosynthetic pathways through genetic engineering has provided novel opportunities to bolster our understanding of plant lipid metabolism, and also produce higher value oilseed varieties (Warwick 2011). The economic impact of these breakthroughs could be quite significant in crucifer species due to their broad range of oil profiles, agronomic traits and potential applications. For instance, the Canadian seed crushing and oil extraction industry would increase its profits by an estimated 90 million dollars US per annum if provided with canola germplasm producing 1 % more oil than present in currently available varieties (Canola Council of Canada 2011). Beyond increasing seed oil yield, tailoring the lipid composition of Brassicaceae oil seeds to better suit industrial or nutritional needs would also greatly increase the economic value of these crops. Toward this latter goal, both modifying the concentration of major oil components, such as saturated and unsaturated FAs, as well as the alteration of minor, but valuable, components such as carotenoids or tocopherols, have been gaining momentum in recent years. Thus far, a range of genetic engineering strategies have been employed to alter lipid biosynthesis in crucifer species. This chapter first reviews the basic biosynthetic pathways involved in lipid metabolism, with a particular emphasis on crucifer seed oil biosynthesis, and then highlights achievements made in engineering these pathways to better produce seeds with higher, or modified, lipid contents.

7.2 Lipid Biosynthesis in Seeds

7.2.1 Triacylglycerol Biosynthesis

Numerous high qualitity review articles have recently been published describing the process of triacylcerol biosynthesis in plants (Bates and Browse 2012; Baud and Lepiniec 2010; Napier and Graham 2010; Snyder et al. 2009; Wallis and Browse 2010; Weselake et al. 2009). Our description of seed lipid biosynthesis primarily relies upon these reviews. In seeds, the biosynthesis of storage triacylglycerol (TAG) begins with the production of FAs, which are generated in the stroma of plastids (Fig. 7.1). Here, acetyl-CoAs derived primarily from pyruvate through the catalytic action of plastidial pyruvate dehydrogenase (PDH) are committed to FA synthesis by their carboxylation to produce malonyl-CoA. Bicarbonate is consumed in this ATP-dependant reaction, which is catalyzed by acetyl-coenzyme A carboxylase (ACCase). These malonyl-acyl carrier protein (ACP) subunits are then incorporated into a growing acyl-ACP chain through condensation catalyzed by β-ketoacyl-ACP synthases (KASs). Following this condensation, the β-ketoacyl-ACP is then reduced, dehydrated, and reduced again by the action of β-ketoacyl-ACP reductase, β-hydroxyacyl-ACP dehydratase (Brown et al. 2009) and enoyl-ACP reductase (de Boer et al. 1998), respectively. These four reactions are repeated sequentially to extend the acyl-ACP chain by two carbons up to a final length of C18. KAS I catalyzes the acetylation of acyl-ACPs with chain lengths between C4 and C14, and KAS II elongates C16 to C18. KAS III catalyzes the condensation of acetyl-CoA and acyl-ACP to produce the initial acetoacetyl-ACP (C4) moiety, which is elongated by the action of other KAS enzymes. Subsequently, a soluble plastidial stearoyl-ACP desaturase can catalyze the introduction of a double bond at the $\Delta\text{-}^9$ position to produce oleoyl-ACP prior to cleavage of the acyl-ACP thioester bond promoted by the fatty-acyl ACP thioesterase, FatA. Cleavage of the palmitoyl and stearoyl-ACP thioester bonds is catalyzed by FatB. The liberated 'free' FAs are then exported from the stroma and quickly converted into acyl-CoAs by the action of acyl-CoA synthase. The endoplasmic reticulum (ER)-associated fatty acid elongation (FAE) complex can then catalyze the extension of carbon chain length of the acyl-CoAs in a cycle of reactions highly similar to *de novo* FA synthesis. Homomeric acetyl-CoA carboxylase provides the malonyl-CoA necessary to extend acyl-CoAs outside of the plastid.

In an acyl-CoA-dependant pathway, TAG is then produced by the sequential acylation of a glycerol backbone by enzymes localized to the ER (Fig. 7.1). sn-Glycerol-3-phosphate (G3P), generated from dihydroxyacetone phosphate (DHAP) by the catalytic action of glyceraldehyde 3-phosphate dehydrogenase (G3PDH), is incorporated into the ER membrane by *sn*-glycerol-3-phoshate acyltransferase (GPAT), esterifying an acyl moiety to the *sn*-1 position of the glycerol backbone (Zheng et al. 2003). The *sn*-2 position of this backbone is then acylated by the catalytic action of lysophosphatidic acid acyltransferase (LPAAT) to produce phosphatidic acid (PA) (Maisonneuve et al. 2010), which is then dephosphorylated by the

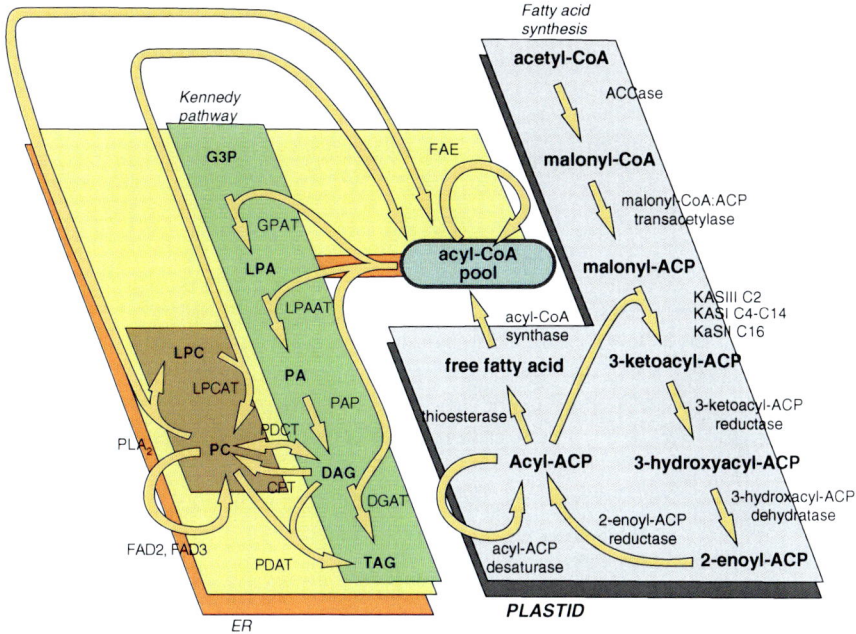

Fig. 7.1 Biosynthesis of triacylglycerol. Fatty acid synthesis is initiated in the plastid, where acetyl-CoA and malonyl-CoA (derived from the catalytic action of acetyl-Co A carboxylase [ACCase]) are condensed by the catalytic action of 3-ketoacyl-ACP synthase (KAS) III to produce acetoacyl-acyl carrier protein (ACP). This acyl-ACP intermediate is then catalytically reduced by 3-ketoacyl-ACP reductase, dehydrated by 3-hydroxyacyl-ACP dehydratase, and reduced again by 2-enoyl-ACP reductase to yield an acyl-ACP molecule containing two additional carbon atoms. KAS I then catalyzes the initiation of six more cycles of acyl-ACP extension to yield palmitoyl-ACP. Stearoyl-ACP can be produced by one further cycle initiated by KAS II, and can be desaturated to produce oleoyl-ACP. Thioesterases then catalyze the cleavage of the acyl chains from ACP to produce free fatty acids, which are exported to the cytosol and immediately converted into acyl-CoAs on the outside of the plastid. Further elongation of acyl-CoAs can occur in the ER via the catalytic action of fatty acid elongase (FAE). In the Kennedy pathway, these acyl-CoAs are used to sequentially acylate a glycerol backbone. sn-glycerol-3-phosphate acyltransferase (GPAT) first catalyzes the acylation of sn-glycerol-3-phosphate (G3P) at the sn-1 position to produce lyso-phosphatidic acid (LPA). LPA is then acylated at the sn-2 position by the catalytic action of lyso-phosphatidic acyltransferase (LPAAT) to produce phosphatidic acid (PA), which is dephosphorylated via the catalytic action of phosphatidic acid phosphatase (PAP) to produce sn-1, 2-diacylglycerol (DAG). DAG can then be converted to triacylglycerol (TAG) by the catalytic action of diacylglyc-erol acyltransferase (DGAT). The DAG skeleton can move into the phospholipid pool via the for-ward reaction of cholinephosphotransferase (CPT) or via reversible phosphocholine headgroup exchange catalyzed by phosphatidylcholine: diacylglycerol cholinephosphotransferase (PDCT). Acyl chains in phosphatidylcholine (PC) can be desaturated by the catalytic action of fatty acid desaturase (FAD) 2 and FAD3 before being returned to the DAG pool to produce triacylglycerol (TAG). The acyl chains at the sn-2 position of PC may be exchanged with those in the acyl-CoA pool through the catalytic action of phospholipase A_2 (PLA$_2$), acyl-CoA synthetase (ACS) and the forward reaction of lysophosphatidylcholine acyltransferase (LPCAT) to reacylate lysophosphati-dylcholine (LPC). LPCAT may also catalyze acyl exchange at the sn-2 position of PC with the acyl-CoA pool via the combined forward/reverse reactions of LPCAT. Phospholipid: diacylglyc-erol acyltransferase (PDAT) catalyzes the acyl-CoA-independent transfer of an acyl moiety from PC to DAG to generate TAG (Adapted from Snyder et al. (2009) and Baud et al. (2010))

catalytic action of PA phosphatase (PAP or lipins) to produce sn-1,2-diacylglycerol (DAG) (Mietkiewska et al. 2010). The acylation of DAG can then be catalyzed by diacylglycerol acyltransferase (DGAT) (Taylor et al. 2009b) or converted into phospholipids by the catalytic action of CPD-choline: sn-1,2-diacylglycerol choline phosphotransferase (CPT). The conversion of DAG into phosphatidylcholine (PC) provides an opportunity for FA desaturases FAD2 and FAD3 to sequentially act upon the PC carbon chains to catalyze the introduction of double bonds at the Δ^{12} and Δ^{15} positions (Yang et al. 2012). These acyl chains can then be incorporated into TAG through the sequential action of phosphatidylcholine: diacylglycerol choline-phosphotransferase (PDCT), which catalzyses the exchange of headgroups between PC and DAG, and DGAT, which acts upon the newly formed DAG (Bates and Browse 2012). The acyl chains of PC may also be removed by phospholipases, and replaced with new chains by the catalytic action of lysophosphatidylcholine acyl-transferase (LPCAT) in an acyl-editing process known as the Land's cycle (Zheng et al. 2012).

TAGs can also be produced independently of the acyl-CoA pool by the action of phospholipid: diacylglycerol acyltransferase (PDAT), which catalyzes the transfer of acyl moieties from the sn-2 position of PC to the sn-3 position of DAG, generating lysophosphatidylcholine and TAG. Similarly, diacylglycerol transacylase (DGTA) functions to catalyze the transfer of the sn-2 acyl moiety from DAG to the sn-3 position of a second DAG, generating lysophosphatidic acid and TAG (Stobart et al. 1997). A gene encoding DGTA, however, has yet to be identified and its activity has been suggested to be a side reaction of PDAT, not that of an independent enzyme.

7.2.2   Non-acyl Lipid Biosynthesis

While the relative content of major lipid components can significantly affect the perceived health attributes and economic value of Brassicaceae oilseeds, the same is true for minor lipid components such as non-acyl lipids, and in particular the carotenoids and tocochromanols (vitamin E). In the case of carotenoids, proteomic analyses have indicated that the bulk of enzymes required to catalyze the synthesis of carotenoids are present in thylakoid membrane-associated plastoglobules (Ytterberg et al. 2006). Geranylgeranyl diphosphate (GGDP) synthase promotes the condensation of one dimethylallyl diphosphate (DMAPP) and three isopentenyl diphosphate (IPP) molecules, both products of the 2-C-methyl-D-erythritol-4-phosphate (MEP) pathway (Phillips et al. 2008), to produce the 20 carbon GGDP (Okada 2000). This GGDP serves as a substrate for the first committed step in carotenoid biosynthesis, whereby phytoene synthase (PSY) catalyzes the condensation of two GGDP molecules to form a linear 40 carbon tetraterpene molecule (15-cis-phytoene) (Scolnik and Bartley 1994). Four double bonds are then introduced by the action of phytoene desaturase (PDS) and ζ-carotene desaturase (ZDS) to produce 9,9'-di-cis ζ-carotene (Bartley et al. 1999). In the absence of light and chlorophyll, which can promote auto-isomerization, the isomerization of 9,9'-di-*cis* ζ-carotene is catalyzed by carotenoid isomerase (CRTISO) to yield *trans*-lycopene (Park et al. 2002). Cyclases then cyclize both ends of the *trans*-lycopene. The first cyclization can be catalyzed by β-cylase (LYCB) or ε-cyclase (LYCE), leading to the eventual production of either β-carotene or α-carotene, respectively (Cunningham et al. 1996; Pogson et al. 1996). Irrespective of the first reaction, LYCB cyclizes the opposing end of the α- or β-carotene molecule (Cunningham et al. 1996). β-carotene hydrolases can then catalyze the introduction of hydroxyl groups into both cyclic groups on β-carotene to produce zeaxanthin. Similarly, β- and ε-hydrolases can catalyze the introduction of two hydroxyl moieties at the ends of α-carotene to produce lutein (Meyer and Kinney 2010). Although not naturally occurring in the majority of plants, a small number of organisms can then go on to synthesize a large number of ketocarotenoids through the action of β-carotene ketolases, which oxygenate the C4 position of β rings of numerous carotenoid species (Zhu et al. 2009; Fig. 7.2).

GGDP, when not committed to carotenoid biosynthesis, can also serve as a precursor for tocochromanol production. Tocochromanols are amphipathic lipids possessing a hydroxylated and methylated aromatic (with five carbon epoxy ring) head group linked to a diterpene (C20) side chain. Tocochromanols with saturated side chains are classified as tocopherols, while tocotrienols are tocochromanols with three double bonds in their side chains (Meyer and Kinney 2010). Brassicaceae can be genetically engineered to accumulate tocotrienols, but are not known to do so naturally. Both tocopherols and tocotrienols can be divided into four sub-groups (α, β, γ, δ), which specify the number and position of the methyl groups on the head group's aromatic ring (Falk and Munne-Bosch 2010). This head group is derived from homogentisate, which itself is the product of *p*-hydroxyphenylpyruvate dioxygenase (HPPD) action (Collakova and DellaPenna 2003). The isoprenyl side chain is derived from reduced GGDP (catalyzed by geranylgeranyl reductase [GGR]), which is linked to the homogentisate by the action of homogentisate phytyltransferase (HPT, VTE2) in the first committed reaction in tocochromanol biosynthesis (Venkatesh et al. 2006). The product of VTE2 activity is 2-methyl-6-phytyl-1,4-benzoquinol (MPBQ), which can then be methylated by the action of MPBQ methyltransferase (VTE3) at the benzyl's 3 position leading to the production of α- and γ-tocopherol. In the absence of this methylation, MPBQ becomes a substrate for β- and δ-tocopherol biosynthesis (van Eenennaam et al. 2003). Tocopherol cyclase (TC, VTE1) catalyzes the formation of a cyclic-five carbon ether- structure adjacent to the aromatic ring, in both MPBQ and 2-methyl-6-geranylgeranyl-1,4-benzoquinol (MGGBQ), to produce δ -tocopherol and γ-tocopherol, respectively (Porfirova et al. 2002). The fifth aromatic carbon of both tocopherols can then be methylated in a reaction catalyzed by γ-tocopherol methyltransferase (γ-TMT, VTE4) (Bergmüller et al. 2003). The biosynthesis of tocotrienols mimics that of tocopherols with the exception that the GGR-catalyzed reduction of GGDP is omitted and GGDP is instead condensed with homogentisate directly to produce MGGBQ (Cahoon et al. 2003; Fig. 7.2).

tyrosine, shikimate

glyceraldehyde + pyruvate

15-cis-phytoene

PDS

9,9' dicis-ζ-carotene

p-hydroxyphenyl

3 isopentenyl-PP + dimethylallyl

PSY

ZDS

HPPD

GGDPS

7,9,7' ,9' -tetra-cis-lycopene

GGR

homogentisate

GGDP

CRTISO

PDP

all trans-lycopene

VTE2

HGGT

LYCE   LYCB

2-methyl-6-phytyl-BQ

MGGBQ

δ-carotene   Υ-carotene

VTE3

VTE3

LYCB

VTE1

VTE1

2,3-dimethyl-6-phytyl-BQ

DMGGBQ

α-carotene   β-carotene

VTE1

VTE1

β-carotene hydroxylase

δ-tocopherol   Υ-tocopherol

δ-tocotrienol   Υ-tocotrienol

zeinoxanthin   β-cryptoxanthin

VTE4

VTE4

ε-carotene hydroxylase

β-carotene hydroxylase

β-tocopherol   α-tocopherol

β-tocotrienol   α-tocotrienol

lutein   zeaxanthin

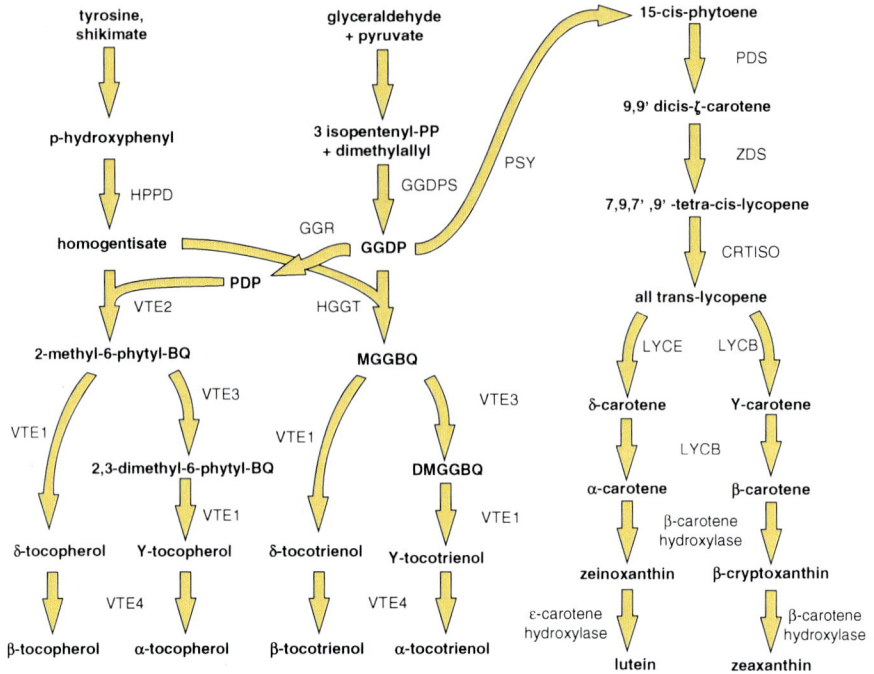

Fig. 7.2 Biosynthesis of tocochromanols and carotenes. The catalytic action of geranylgeranyl reductase converts GGDP into phytyl diphosphate, which can then be condensed with homogentisate through the catalytic action of homogentisate phytyltransferase (VTE2) to produce 2-methyl-6-phytyl-1,4-benzylquinol (MPBQ). This product can then either be cyclized by the catalytic action of tocopherol cyclase (VTE1), leading to δ- and β-tocopherol production, or methylated by the catalytic action of MPBQ methyltransferase (VTE3) leading to γ- and α-tocopherol production. Alternatively, GGDP can be directly condensed with homogentisate to produce MGGBQ, leading to the production of all four tocotrienols. PSY can also devote GGDP to carotene production by catalyzing its conversion into 15-*cis*-phytoene. PDS and ZDS catalyze the introduction of four double bonds, which are converted to *trans* form by CRTISO. Carotene biosynthesis then bifurcates to produce either δ- or γ-carotene depending on which enzyme first cyclizes lycopene. Carotenes α- and ß- are then produced by a second cyclization catalyzed by LYCB. Hydroxylases can use α-carotenes to catalyze the production of zeinoxanthin and lutein, or they can convert β-carotene to β-cryptoxanthin and zeaxanthin. BQ benzoquinol, CRTISO carotenoid isomerase, DMGGBQ 2,3-dimethyl-6-geranylgeranyl-1,4-benzoquinol, GGDP geranylgeranyl diphosphate, GGPDS GGDP synthase, GGR geranylgeranyl reductase, HGGT homogentisate geranylgeranyl transferase, HPPD p-hydroxyphenylpyruvate dioxygenase, LYCB lycopene β-cyclase, LYCE lycopene ε-cyclase, MGGBQ 2-methyl-6-geranylgeranyl-1,4-benzoquinol, PDP phytyl diphosphate, PDS phytoene desaturase, PSY phytoene synthase, VTE vitamin E, ZDS ζ-carotene desaturase

7.3   Increasing Seed Lipid Content

The consumption of vegetable oils has increased by more than 50 % during the past decade; this phenomenon has been driven by both an elevation in global affluence and the mounting use of plant-based industrial oils such as biodiesel. This strong demand for vegetable oils, along with the growing limitation of arable land, has led to a considerable rise in the price of plant-derived oils over recent years (reviewed by Lu et al. 2011). Therefore, increasing seed oil content, and thus the yield of oil per unit area of arable land, is of the utmost importance in the struggle to supply enough oil for both food and non-food applications at affordable prices (Weselake et al. 2009). In the past, the direct manipulation of seed oil content using genetic engineering approaches has been limited in many cases by a lack of knowledge concerning the regulation of lipid metabolism, and therefore conventional breeding techniques have provided the main source of crop plants exhibiting improved seed oil content to date (Ohlrogge and Jaworski 1997). Although our understanding of seed lipid accumulation is still somewhat inadequate, extensive research in recent years has led to an unprecedented understanding of the genes and pathways involved in this process, especially in the model crucifer, Arabidopsis (Baud and Lepiniec 2009). As a result, a growing number of studies involving the genetic manipulation of specific genes with various roles in plant development have led to the production of transgenic lines displaying significant increases in seed oil content (see Tables 7.1 and 7.2).

7.3.1   Enhancing the Production of Triacylgycerol Building Blocks

One strategy that has been utilized in an attempt to elevate seed oil content using genetic engineering has been to enhance the production of TAG building blocks, including FAs and glycerol backbones. In the case of FAs, their *de novo* synthesis takes place in plastids, the first key step of which is catalyzed by ACCase (Ohlrogge and Jaworski 1997). The majority of plants contain two forms of ACCase that are differentially compartmentalized, with prokaryotic, or heteromeric, ACCase located within the plastids and eukaryotic, or homomeric, ACCase located within the cytosol (Konishi et al. 1996). Somewhat surprisingly, it has been found that the seed-specific over-expression of *BCCP2*, which encodes the biotin binding subunit of plastidial heteromeric ACCase, in Arabidopsis resulted in decreased seed oil content compared to untransformed lines (Thelen and Ohlrogge 2002). Conversely, expression of a plastid-targeted homomeric ACCase from Arabidopsis in *B. napus* seeds resulted in a small, heritable, relative increase in oil content of 5 % (Roesler et al. 1997). This discrepancy between the two studies has been suggested to be attributable to the possibility that the homomeric ACCase utilized in the latter study was able to escape any negative feedback regulation exerted over the heteromeric isoforms normally found in plastids (Baud and Lepeniec 2010).

More pronounced increases in TAG accumulation have also been obtained via a proposed augmentation of acetyl moieties for FA synthesis through the manipulation of pyruvate dehydrogenase kinase (PDHK), which is a negative regulator of the mitochondrial pyruvate dehydrogenase complex (mtPDC). This complex acts as a link between glycolytic carbon metabolism and the tricarboxylic acid cycle, and

Table 7.1 List of genes whose over-expression results in enhanced seed oil content

Gene	Encodes	Promoter	Donor	Host	Reference
ACC1	Plastid-targeted homomeric ACCase	napin	Arabidopsis	B. napus	Roesler et al. (1997)
GPD1	G3PDH	napin	S. cerevisiae	B. napus	Vigeolas et al. (2007)
plsB	GPAT	35S	E. coli	Arabidopsis	Jain et al. (2000)
GPAT	GPAT without plastidial targeting	35S	C. tinctorius	Arabidopsis	Jain et al. (2000)
SLC1-1	LPAAT	35S	S. cerevisiae	B. napus	Zou et al. (1997)
SLC1-1		35S	S. cerevisiae	Arabidopsis	Zou et al. (1997)
CaLPAAT		35S	C. abyssinica	B. napus	Xu et al. (2010)
BAT1.5		napin	B. napus	Arabidopsis	Maisonneuve
BAT1.13		napin	B. napus	Arabidopsis	et al. (2010)
AtDGAT1	DGAT1	napin	Arabidopsis	Arabidopsis	Jako et al. (2001)
				B. napus	Taylor et al. (2009b)
TmDGAT1		napin	T. majus	Arabidopsis B. napus	Xu et al. (2008)
BnDGAT1		napin	B. napus	B. napus	Weselake et al. (2008), Taylor et al. (2009b)
AHb2	Hemoglobin-2	USP	Arabidopsis	Arabidopsis	Vigeolas et al. (2011)
BnLEC1	LEC1	Truncated napin	B. napus	Arabidopsis	Tan et al. (2011)
BnL1L	LEC1-LIKE				
AtWRI1	WRI1	35S	Arabidopsis	Arabidopsis	Cernac and Benning (2004)
BnWRI1		35S	B. napus	Arabidopsis	Liu et al. (2010)
GmDOF4	DOF4	35S	G. max	Arabidopsis	Wang et al. (2007b)
GmDOF11	DOF11				
STM	STM	35S	B. napus	B. napus	Elhiti et al. (2012)
BnGRF2a	GRF2	35S napin	B. napus	Arabidopsis	Liu et al. (2012)

thus plays a role in cell respiration (Zou et al. 1999). Down-regulation of *AtPDHK* via seed-specific antisense expression in Arabidopsis resulted in elevated mtPDC activity and enhanced levels of respiration. In addition to generating energy and reducing equivalents, the respiration process also yields numerous building blocks for a number of crucial biosynthetic processes, such as FA synthesis. Correspondingly, the transgenic plants exhibited relative increases of up to 50 % in seed oil content, with no changes in FA composition, along with an augmentation in seed weight (Marillia et al. 2003).

Table 7.2 List of genes whose down-regulation results in enhanced seed oil content

Gene	Encodes	Host	Modification	Reference
AtPDHK	Mitochondrial pyruvate dehydrogenase kinase	Arabidopsis	Antisense	Marillia et al. (2003)
G6PD5 G6PD6	Glucose-6-phosphate dehydrogenase	Arabidopsis	T-DNA double mutants	Wakao et al. (2008)
AAP2	Amino acid permease 2	Arabidopsis	T-DNA mutant	Zhang et al. (2010)
PEPC	Phosphoenolpyruvate carboxylase	B. napus	Antisense	Chen et al. (1999)
At1g01050 At3g53620	Pyrophosphatase	Arabidopsis	Antisense	Meyer et al. (2012)
GL2	Glabra2	Arabidopsis	T-DNA mutant	Shen et al. (2006) Shi et al. (2012)

In a related approach, an attempt was made to enhance seed levels of G3P, which provides the glycerol backbones necessary for TAG biosynthesis and has been suggested to co-limit the rate of TAG synthesis during the oil accumulation phase in seeds (Vigeolas and Geigenberger 2004). To generate this increase, a yeast-cytosolic *GLYCEROL-3-PHOSPHATE DEHYDROGENASE* (*GPD1*), which encodes an enzyme that catalyzes the conversion of glycolysis-derived dihydroxyacetone phosphate to G3P, was expressed in developing *B. napus* seeds. Transgenic lines displayed a 40 % relative elevation in seed oil content under greenhouse conditions, with no significant changes in FA composition. Furthermore, there was no obvious reduction in either protein or transient starch levels in these lines, and other morphological characteristics, such as plant growth and silique number, were also normal (Vigeolas et al. 2007).

A similar outcome was attained through the disruption of two cytosolic forms of glucose-6-phosphate dehydrogenase, which catalyze the production of D-glucono-1,5-lactone 6-phosphate and NADPH from D-glucose-6-phosphate and $NADP^+$. While it has been proposed that this enzyme may play a role in the provision of reducing power for FA biosynthesis in maturing seeds, it was found that double Arabidopsis mutants with reduced enzyme activity led to an increase in carbon substrates for oil synthesis rather than a decrease in NADPH supply for FA biosynthesis. In line with this, the mutants also exhibited an approximately 5 % relative average increase in seed oil content and elevated seed weight compared to wild-type controls with no alteration in the carbon to nitrogen ratio or FA composition (Wakao et al. 2008)

7.3.2 Modification of Carbon Partitioning

Seed storage compounds generally consist of proteins, carbohydrates (mainly starch), and/or lipids most often in the form of TAG, with proportions and tissue localizations varying enormously depending on the plant species. For example, in Arabidopsis and other oilseeds, the starch deposition pathway is far less prominent than the lipid biosynthetic pathway, as only a transient accumulation of starch is observed during seed development in these species (reviewed by Weselake et al. 2009). However, regardless of their relative proportions, these three pathways must compete for carbon derived from plant photosynthesis. Thus, one approach that has been taken to enhance seed oil in crucifers has been to divert the flow of carbon from competing pathways and channel it towards TAG biosynthesis.

In an effort to redirect carbon from protein synthesis to that of oil in seeds, Arabidopsis lines bearing T-DNA disruptions within the *AMINO ACID PERMEASE2* (*AAP2*) gene, which encodes an enzyme involved in phloem loading and amino acid distribution to the embryo, were analyzed for modifications in oil content. As anticipated, mutant lines displayed a significant reduction in protein levels, and while the total carbon levels remained unchanged, FA levels were elevated by up to 14 % on a relative basis, mainly as a result of increased C18:1, C18:2 and C20:1. As there was no difference in seed weight between the mutants and wild-type seeds, it seemed that other seed storage compounds were likely elevated in the mutants. Therefore, it was suggested that the lower total N levels within mutant seeds caused more carbon to be channeled into oil synthesis (Zhang et al. 2010). Similarly, the antisense expression of *PEPC*, which encodes an enzyme that catalyzes the conversion of phosphoenolpyruvate to oxaloacetate, in *B. napus* resulted in a 6.4–18 % relative elevation in seed oil content and a concomitant reduction in protein levels. The enhanced seed oil phenotype likely resulted due to an increase in the availability of phosphoenolpyruvate, which is an early substrate of FA synthesis, thus providing additional FA content for TAG biosynthesis (Chen et al. 1999). Recently, it was also found that seed-preferred RNAi-mediated silencing of Arabidopsis At1g01050 and At3g53620, which encode soluble, cytosolic pyrophosphatases, led to small, absolute, seed oil increases of 1–4 %, mainly at the expense of seed storage protein (Meyer et al. 2012). The mechanism behind this finding, however, remains enigmatic.

7.3.3   Over-Expression of Genes Encoding Acyltransferases

Many efforts to enhance seed oil content have involved the up-regulation of genes encoding ER-bound acyltransferases involved in the Kennedy pathway and hence, TAG assembly. For example, cDNAs encoding either plastidial *GPAT* from *Carthamus tinctorius* with its plastidial targeting sequence removed or a *GPAT* (*plsB*) from *Escherichia coli*, which encode enzymes that catalyze the first step in the Kennedy pathway, were expressed in Arabidopsis developing seeds. The resulting transgenic lines exhibited relative seed oil content enhancement of 22 % and 15 %, respectively, as well as increased seed weight (Jain et al. 2000). While similar findings have been obtained with the second enzyme in the pathway (LPAAT), the results have been somewhat inconsistent, depending on the source of the transgenic LPAAT. One of the first attempts at utilizing an LPAAT to elevate seed oil content involved the constitutive expression of a variant of the *SPHINGOLIPID COMPENSATION1* (*SLC1-1*) *LPAAT* from *Saccharomyces cerevisiae* in *B. napus* and Arabidopsis (Zou et al. 1997). Relative increases in seed oil content of up to 22 % in transgenic *B. napus* and up to 49 % in transgenic Arabidopsis, respectively, were noted under greenhouse conditions while subsequent studies with *SLC1-1* expression in *B. napus* exhibited up to 13.5 % enhancement in oil content under field conditions (Taylor et al. 2002), with no concomitant reduction in seed storage protein content. However, the transgenic lines did display alterations in FA composition, with enhanced synthesis and incorporation of long chain FAs into the *sn*-2 position of TAG. Similarly, constitutive expression of an *LPAAT* from *C. abyssinica* (*CaLPAAT*) in *B. napus* resulted in increased oil content in transgenic seeds (Xu et al. 2010). Likewise, the seed-specific expression of cDNAs encoding two microsomal LPAAT isozymes from *B. napus* (*BAT1.5* and *BAT1.13*) in Arabidopsis resulted in greater seed mass and up to 27 % relative enhancements in seed TAG content, with no other discernible morphological or developmental abnormalities (Maisonneuve et al. 2010). In contrast to *SLC1-1*, heterologous expression of *BAT1.5* and *BAT1.13*, as well as *CaLPAAT*, did not yield any differences in FA composition of TAG present in the seeds of the transgenic plants compared to untransformed plants (Maisonneuve et al. 2010; Xu et al. 2010). Conversely, the seed-specific expression of *LPAAT* cDNAs from *Limnanthes douglasii* (Brough et al. 1996) or *L. alba* (Lassner et al. 1995) in *B. napus* yielded altered FA composition at the *sn*-2 position in TAG, but did not result in any changes in seed oil content.

To date, the most extensively studied enzyme in terms of increasing seed oil content has been DGAT, which is the final enzyme of the Kennedy pathway, and its activity level has been suggested to limit the flow of carbon into TAG (reviewed by Chen et al. 2010). Most plants have at least two forms of DGAT (DGAT1 and -2), which share no homology and appear to have non-redundant functions in lipid metabolism (Shockey et al. 2006). Since it seems that DGAT2 enzymes may exhibit high specificities for particular FAs and have little effect on overall TAG content, the majority of attempts to promote elevated oil content thus far have centered on DGAT1. In an initial study, seed-specific over-expression of *AtDGAT1* resulted in increases in Arabidopsis seed oil content of up to 28 % in homozygous lines (Jako et al. 2001). Similarly, expression of a *Tropaeolum majus DGAT1* in *B. napus* or Arabidopsis resulted in relative enhancements of seed oil content of 11 % and 30 %, respectively (Xu et al. 2008). Interestingly, this group was also able to achieve an 80 % increase in DGAT catalytic activity through the site-directed mutagenesis of a single residue within TmDGAT1, as well as a subsequent elevation in relative oil content of 50 % in transgenic Arabidopsis expressing the mutated gene. These results emphasize the potential for engineering DGAT enzymes, and possibly other acyltransferases, to possess enhanced catalytic activity. Since then, lower, but still significant, increases in seed oil content have been obtained through the expression of Arabidopsis or *B. napus DGAT1* in the developing seeds of *B. napus* under both greenhouse and field conditions (Taylor et al. 2009b; Weselake et al. 2008). It has also been shown that expression of *AtDGAT1* in *B. napus* substantially reduced the penalty in seed oil content normally induced by drought, which suggests that the transgenic expression of *DGAT1* may be most effective at enhancing oil content under conditions where the plant is unlikely to reach its full capacity due to environmental stress (Weselake et al. 2008).

7.3.4 Up-regulation of Nonsymbiotic Hemoglobin-2

Non-symbiotic hemoglobins are a group of oxygen-binding proteins that have been found in various plant species and can be divided into two classes (Hb1 and Hb2) based on phylogenetic characteristics, oxygen-binding properties, and gene expression patterns (Trevaskis et al. 1997; Hoy and Hargrove 2008). Since seeds exhibit a relatively low internal oxygen concentration (Vigeolas et al. 2003), it seems plausible that increasing this concentration may promote mitochondrial respiration and thus, storage metabolism. Indeed, seed-specific over-expression of the Arabidopsis class 2 nonsymbiotic hemoglobin, Arabidopsis hemoglobin-2 (*AHb2*), but not a class 1 gene, resulted in a 40 % relative elevation in the total TAG content, and a slight decrease in protein levels, in seeds. These alterations may have been attributable to a putative improvement in oxygen availability within the transgenic seeds. The FA composition within seeds was also altered, with increases in polyunsaturated linoleic ($18:2^{cis}\Delta^{9,12}$) and α-linolenic ($8:3^{cis}\Delta^{9,12,15}$) acid (Vigeolas et al. 2011).

7.3.5 Altering the Expression of Transcription Factors

While all of the aforementioned studies have indicated that seed oil content can indeed be regulated, at least to some extent, by the alteration of single key enzymes, it is now known that seed oil accumulation is controlled by a complex, coordinated regulatory network that is not only relevant to lipid metabolism, but also carbon metabolism in general. Since transcription factors can affect a large number of genes within a pathway simultaneously, the secondary bottlenecks often encountered in transgenic plants bearing modifications to a single enzyme are less likely to be problematic (Courchesne et al. 2009). Therefore, the modulation of transcription factors to effect changes in lipid biosynthesis has provoked widespread interest in recent years.

During seed development, four transcription factors (LEC1, LEC2, FUSCA3 and ABI3) control virtually all aspects of the maturation process (Ohto et al. 2007). While the over-expression of cDNAs encoding these transcription factors can result in increased seed oil content, their expession also causes a variety of developmental abnormalities and even lethality (Stone et al. 2001; Santos-Mendoza et al. 2005; Wang et al. 2007a; Mu et al. 2008), which has made their use for the genetic amelioration of oil content challenging. In a novel approach to minimize developmental abnormalities, Tan et al. (2011) utilized truncated seed-specific napin promoters with reduced levels of activity to drive the expression of two *B. napus LEAFY COTYLEDON 1 (LEC1)* homologues, respectively, in Arabidopsis. The resulting transgenic seeds exhibited up-regulation of a subset of genes involved in FA biosynthesis and glycolysis, and produced elevated relative oil contents of 2–20 % with slight reductions in protein levels and no other detrimental effects.

Similarly, the WRINKLED1 (WRI1) transcription factor, which has been suggested to play a pivotal role in the regulation of glycolysis and certain aspects of FA biosynthesis during seed development and functions downstream of the master regulators mentioned above, has recently garnered much attention for its potential ability to alter seed oil content in various plant species. Indeed, constitutive over-expression of *AtWRI1*, as well as *B. napus WRI1* orthologs, in Arabidopsis resulted in approximately 10–20 % relative increases in seed oil content (Cernac and Benning 2004; Liu et al. 2010). This *WRI1*-mediated enhancement of seed oil content, however, may not be a general rule, as expression of *AtWRI1* driven by the seed-specific *OLEOSIN* promoter yielded elevated transcript levels for several *WRI1* target genes, but did not promote any significant change in oil accumulation (Baud et al. 2009). While this discrepancy may be attributable to the different promoters utilized to drive *WRI1* expression, it is also possible that it may have resulted from the use of different ecotypes of Arabidopsis (col-0 vs. Ws), which can exhibit variations in storage lipid synthesis (Ruuska et al. 2002).

Two other transcription factors from *Glycine max* termed DNA-BINDING WITH ONE FINGER4 and 11 (GmDOF4 and 11), both derived from a large gene family with very diverse roles in plants (for example Mena et al. 1998; Plesch et al. 2001; Skirycz et al. 2006), have also been found to play a role in the regulation of seed oil content. Expression of these cDNAs in Arabidopsis using the *35S* promoter resulted in increased seed weight, as well as a 5.3–9.1 % relative elevation in oil content with a decrease in linoleic acid and increases in α-linolenic acid and *cis*-11-eicosenoic acid ($20:1^{cis}\Delta^{11}$) in *GmDOF4*-expressing seeds, and a 4.7–8.1 % relative elevation in oil content with no changes in FA composition in GmDOF11-expressing seeds. In both cases, these changes were associated with the up-regulation of genes associated with FA biosynthesis and the down-regulation of storage protein genes (Wang et al. 2007b).

The modulation of several transcription factors with roles that are apparently not directly related to FA or TAG biosynthesis have also recently been found to increase oil content in transgenic crucifers. For example, the *GLABRA2* gene from Arabidopsis, which encodes a class IV homeodomain-leucine zipper transcription factor that is required for the correct differentiation of several epidermal types, has been found to regulate seed oil levels. Loss-of-function *gl2* mutants displayed up to an approximately 10 % relative elevation in oil content compared to wild-type plants, along with abnormalities in trichome morphology, root hair density and extrusion of seed coat mucilage (Shen et al. 2006; Shi et al. 2012). While the exact mechanism behind these results remains largely unknown, it was suggested that seed mucilage and oil may compete for photosynthate within developing seeds, and in the absence of mucilage formation in the mutants, the unused sucrose was instead utilized for oil biosynthesis (Shi et al. 2012). Unexpectedly, in a related study, both the expression of sense and antisense copies of a *GL2* homologue from *B. napus* (*BnaC.GL2.b*) in Arabidopsis resulted in small relative seed oil enhancements of 3.5–5.0 % and 2.5–6.1 %, respectively (Chai et al. 2010). In both types of transgenic plants there were decreased levels of endogenous Arabidopsis *GL2* expression, which suggests that over-expression of *GL2* may have somehow affected the function of the endogenous gene (Chai et al. 2010).

Furthermore, the *SHOOTMERISTEMLESS* (*STM*) gene, which encodes a class I KNOTTED1-LIKE homeodomain protein and is required for the formation and maintenance of the shoot apical meristem, has also recently been shown to have an effect on seed oil content (Elhiti et al. 2012). Constitutive over-expression of the *B. napus STM* gene caused a small (1.5–2 %) absolute enhancement of seed oil content without altering protein levels, which almost certainly resulted from the accompanying up-regulation of several key transcription factors (such as *BnLEC1*, *BnLEC2* and *BnWRI1*), along with numerous enzymes involved in sucrose metabolism, glycolysis and FA biosynthesis. In addition to changes in oil content in the transgenic seeds, there were also alterations in FA composition, with a decrease in oleic acid ($18:1^{cis}\Delta^9$) and slight increases in linoleic and α-linolenic acids (Elhiti et al. 2012). Similarly, both constitutive and seed-specific expression of a cDNA encoding the *B. napus* GROWTH-REGULATING FACTOR 2-LIKE transcription factor (BnGRF2a) in Arabidopsis resulted in approximately 50 % and 40 % increases in seed oil content, respectively. This was suggested to be largely due to the up-regulation of various genes associated with cell proliferation, oil synthesis and photosynthesis (Liu et al. 2012).

7.4 Modification of Fatty Acid Composition

Since FA composition can determine both the quality and the potential use of plant oils, considerable focus has also been placed on improving this trait. For example, oils bearing enhanced omega-3 FAs are desirable for their nutritional benefits, while those exhibiting increased unusual FAs, such as hydroxylated FAs, are of value for industrial uses. As a result of this recent interest in the production of high-value plant oils with modified FA profiles, success stories are accumulating in this field through the use of genetic engineering, which is enabling the generation of Brassicaceae seed oil with improved processing characteristics, food and industrial applications.

7.4.1 Saturated Fatty Acids

Plant oils with a high content of short, medium and long chain saturated FAs are used in a number of food products, such as chocolate, candy coating, and non-dairy creams, as well as industrial applications such as detergents and cosmetics (Scarth and Tang 2006; Dehesh 2001). Although coconut and palm oils are the main commercial sources of short and medium chain saturated fatty acids (SMCFAs) (Carlsson et al. 2011), the richest sources of SMCFAs such as caprylic (8:0) and capric (10:0) acids, which are in particular demand in the medical field, are species from the *Cuphea* genus. Unfortunately, although efforts have been made to domesticate *Cuphea* species, several obstacles were encountered related to seed dormancy and shattering (Knapp 1993). Conversely, the seed oil of members of the Brassicaceae contains only traces of SMCFAs with small amounts of stearic acid (18:0); however, the engineering of *Brassica* crop species could potentially yield an economical source of these oils (Dehesh 1996; Voelker et al. 1996).

Fatty acid chain length is determined during their *de novo* synthesis in plastids through the catalytic action of chain length-specific acyl-ACP thioesterases (Dehesh et al. 1998; Dyer et al. 2008; Voelker et al. 1997). Voelker et al. (1996) carried out one of the first attempts in which a medium chain-specific fatty-acyl ACP thioesterase (FatB) from the California bay tree (*Umbellularia californica*) was utilized to increase the content of lauric acid (12:0) in *B. napus*, resulting in the production of high levels of lauric acid (up to 50 %) in seed oil under field conditions. This work demonstrated that specialized thioesterases have the ability to redirect the synthesis of common FAs in *B. napus* to those of medium chain-length FAs. Unfortunately, subsequent attempts to induce the production of other SMCFAs, such as capric or caprylic acids, in *B. napus* were less successful, resulting in medium chain FA proportions of only 8 % and 30 % in the seed oil, respectively (Wiberg et al. 2000).

Vegetable oils with a high content of stearic acid, such as those derived from cocoa and mangosteen (Hawkins and Kridl 1998), are useful for the production of food products such as margarine and shortening. As follows, an acyl-ACP thioesterase specific for the stearoyl moiety (FatA) from mangosteen (*Garcinia mangostana*) has also been used to enhance the accumulation of this FA in *B. napus* seeds up to 22 % of the total FA content compared to 2.5 % in the wild-type control (Hawkins and Kridl 1998). In addition, the level of stearic acid is also controlled by the catalytic activity of stearoyl-ACP (Δ^9) desaturase, and its antisense down-regulation in *B. napus* resulted in an increase of stearic acid up to 40 % (Knutzon et al. 1992).

In most cases, accumulation of SMCFAs in transgenic seeds over-expressing *FatB* was lower compared to the levels observed in plants that naturally produce them (Thelen and Ohlrogge 2002). One possible explanation for this could be a relatively small pool of short/medium chain acyl-ACP, which does not allow the accumulation of high levels of SMCFAs in seeds. These results indicate that in addition to thioesterases, other enzymes such as KASs could also be involved in the regulation of chain length, as well as the amount of FAs that accumulate in the seeds (Dehesh 2001). Over the years, a number of medium chain-specific KAS IV genes were isolated from *Cuphea* species and were subsequently expressed in combination with chain-specific *FatB* genes in an attempt to increase the content of SMCFAs to even higher levels (Dehesh et al. 1998; Stoll et al. 2005; Voelker and Kinney 2001). For instance, concomitant over-expression of a medium chain-specific *KAS IV* and *FatB* from *C. hookeriana* resulted in a 30–40 % increase in the content of medium chain FAs compared to lines expressing only *FatB* (Dehesh et al. 1998).

Similarly, it was found in the same study that a *C. hookeriana KAS IV* could increase the accumulation of lauric acid through its co-expression with the *U. californica FatB* (Dehesh 2001). Another factor that limits the production of SMCFAs in Brassicaceae species is a lack of medium chain-specific acyltransferases. In particular, *B. napus* LPAAT has a preference for oleoyl-CoA and lacks activity with lauryl-CoA (Frentzen 1998). To overcome this limitation, a laurate-specific *LPAAT* from coconut (*Cocos nucifera*) was co-expressed with the *UcFATB*. The resulting transgenic seeds accumulated significantly higher levels of lauric acid, making up to 65 % of the total FA pool compared to 48 % in seeds expressing only the *FatB* (Knutzon et al. 1999).

Unfortunately, the production of saturated FAs in transgenic Brassicaceae seeds was often accompanied by various morphological problems, such as reduced germination of high stearic acid *GmFATA B. napus* seeds (Hawkins and Kridl 1998) and induction of β-oxidation in high lauric acid *UcFATB* seeds (Eccleston and Ohlrogge 1998), which has implications for their use as future crops. In any case, it is anticipated that higher levels of SMCFAs in transgenic plants could be obtained via the co-expression of medium chain-specific FatB, KAS IV, GPAT, LPAAT and DGAT enzymes (Carlsson et al. 2011).

7.4.2 Monounsaturated Fatty Acids

Monounsaturated FAs can also be used in a number of food and chemical applications (reviewed by Metzger and Bornscheuer 2006) and can be divided into two groups. The first group contains FAs found in Brassicaceae seeds, such as oleic, erucic ($22:1^{cis}\Delta^{13}$) and small traces of nervonic acid ($24:1^{cis}\Delta^{15}$) acids (Scarth and Tang 2006; Taylor et al. 2010). The second group is represented by unusual monoenic FAs such as petroselinic ($18:1^{cis}\Delta^6$), $16:1^{cis}\Delta^6$ and palmitoleic ($16:1^{cis}\Delta^9$) acids present in, for example, *Coriandrum sativum*, *Thunbergia alata* and *Doxantha unguis-cati* L. seeds, respectively (Cahoon et al. 1992, 1994, 1998). Unfortunately, these latter species are not suitable for large-scale agricultural production and therefore the utilization of these plants as resources for the aforementioned FAs is very limited (Suh et al. 2002). Considering the demand for all of these FAs, there is an ongoing interest in increasing their content using metabolic engineering approaches.

The content of oleic acid, which increases the oxidative stability of oil and has various applications in soap, detergents, lubricants and cosmetics (Dyer et al. 2008; Metzger and Bornscheuer 2006), is largely regulated by the catalytic activity of a microsomal Δ^{12} desaturase encoded by the *FAD2* gene (Miquel and Browse 1992; Drexler et al. 2003). To date, considerable progress has been made towards increasing oleic acid levels by down-regulating the activity of *FAD2* using seed-specific co-suppression, antisense expression or RNA interference approaches (Kinney et al. 1994; Stoutjesdijk et al. 2000; Peng et al. 2010). Indeed, these transgenic approaches resulted in oleic acid contents of up to 85 % of total FAs in *B. napus* seeds. In addition, further increases in oleic acid levels may be possible through the down-regulation of *FatB*, which would lower the amount of saturated FAs released from plastids (Carlsson et al. 2011). A possible next step in the generation of transgenic plants bearing even higher oleic acid content could be the down-regulation of two additional genes: *PDCT* and *PDAT* (Lu et al. 2009; Zhang et al. 2009).

Since oils high in erucic acid have over 200 applications, efforts have been ongoing to increase the levels of this FA in Brassicaceae species. High erucic acid rapeseed cultivars (HEAR) contain erucic acid predominantly in the *sn*-1 and *sn*-3 positions of the glycerol backbone, presumably due to the poor affinity of *B. napus* LPAAT for very long chain FAs, which limits its accumulation to a maximum of 66 % (Dyer et al. 2008; Lassner et al. 1995; Weier et al. 1997). HEAR seed oil with greater than 80 % erucic acid would be desirable to reduce production costs (Mietkiewska et al. 2004). The first attempts to enhance erucic acid production in *B. napus* seeds focused on the expression of novel *LPAAT* genes from *L. alba* and *L. douglasii*, which resulted in increased trierucin accumulation (Brough et al. 1996; Frentzen 1998), but did not enhance the overall level of erucic acid in HEAR oil, possibly as a result of a limited pool of erucoyl-CoA synthesized by the FAE complex from its precursor, oleic acid (Scarth and Tang 2006; Taylor et al. 2011). Using a different strategy, several *FAE* genes encoding β-ketoacyl-CoA synthases with different acyl-CoA specificities were isolated and used to enhance the accumulation of erucic acid in the seeds of Brassicaceae species. For example, Katavic et al. (2001) over-expressed Arabidopsis *FAE1* in a HEAR cultivar of *B. napus*, resulting in erucic acid levels of up to 48–53 % of the total FA pool compared to 43 % in the wild-type control. Even more pronounced increases in the content of erucic acid were achieved in *Brassica carinata* by expressing a *C. abyssinica FAE*, which resulted in erucic acid accumulation up to 52 %; a net increase of 40 % compared to wild-type plants (Mietkiewska et al. 2007). This same group also demonstrated that by combining RNAi-mediated silencing of *FAD2*, which yielded an increased pool of oleic acid, with over-expression of the *C. abyssinica FAE*, it was possible to bring the content of erucic acid up to 58 %; a 45 % relative increase over wild-type levels (Mietkiewska et al. 2008; Taylor et al. 2010). These results indicate that there may be a limit to the amount of erucic acid that can be produced in transgenic seeds through the alteration of one or two genes; therefore, it is likely that the manipulation of multiple factors may be required to generate further increases (Nath et al. 2009). Intriguingly, such a gene-stacking strategy was applied to increase erucic acid levels up to 72 % by crossing transgenic *B. napus* lines that over-expressed the *LdLPAAT* and *BnFAE* cDNAs with a high oleic acid mutant line. A similar strategy was also recently described to increase erucic acid content in *C. abyssinica* seeds (Li et al. 2012).

Substantial progress has also been made to increase the content of nervonic acid in Brassicacea oils, which has important pharmaceutical and nutritional applications in the treatment of numerous neurological disorders (Sargent et al. 1994; Taylor et al. 2009a). For example, over-expression of *FAE* genes from *L. annua* and *Cardamine graeca* in *B. carinata* caused the levels of nervonic acid to increase from 2.8 % in the wild-type control to 30 % and 42.5 % in *B. carinata*, respectively (Guo et al. 2008; Taylor et al. 2009a, 2010, 2011).

Plant oils with a high content of unusual monoenic FAs, such petroselinic, and palmitoleic acids, are often used in the production of biodegradable lubricants, surfactants and plastic precursors (Carlsson et al. 2011; Thelen and Ohlrogge 2002; Scarth and Tang 2006). Unfortunately, the first attempts to engineer plants for the production of these FAs resulted in only limited success. In these instances, when specific Δ^4 16:0-ACP, Δ^6 16:0-ACP and Δ^9 16:0-ACP desaturases involved in the synthesis of petroselinic, 16:1$^{cis}\Delta^6$ and palmitoleic acids in *C. sativum*, *T. alata* and *D. unguis-cati*, respectively, were expressed in Arabidopsis or *B. napus*, only a limited accumulation (1–15 %) of the corresponding FAs was observed (Bondaruk et al. 2007; Cahoon et al. 1992, 1994, 1998; Suh et al. 2002). A major breakthrough in the production of palmitoleic acid in transgenic plants, however, was achieved recently through the use of an engineered Δ^9 desatuarse, which exhibited more than 100-fold higher activity in converting C16:0 to C16:1$^{cis}\Delta^9$ than that of its natural form (Carlsson et al. 2011; Nguyen et al. 2010). Seed-specific over-expression of this engineered desaturase together with a pair of fungal cytosolic 16:0Δ^9 desaturases increased the level of ω-7 FAs up to 71 % in Arabidopsis plants carrying a mutation in the *KAS II* gene (Carlsson et al. 2011; Nguyen et al. 2010). Although it has yet to be tested, it has also been suggested that a multiple set of genes, including those encoding ferrodoxin, KAS and thioesterase may be required for higher accumulation of these monoenic FAs in transgenic plants (Dyer et al. 2008).

7.4.3 Polyunsaturated Fatty Acids

There is an increasing interest in engineering oilseed crops with enhanced polyunsaturated fatty acid (PUFA) content due to the fact that clinical evidence has indicated that PUFAs, and especially omega-3 very long chain-PUFAs (VLC-PUFAs; e.g. eicosapentaenoic acid [EPA, 20:5$^{cis}\Delta^{5,8,11,14,17}$] and docosahexaenoic acid [DHA, 22:6$^{cis}\Delta^{4,7,10,13,16,19}$]), have multiple beneficial effects on human health (Ruiz-López et al. 2012). All higher plants contain the enzymes to produce the C18 PUFAs, linoleic acid and α-linolenic acid, while a few plant species have the ability to further metabolize α-linolenic acid to stearidonic acid (SDA, C18:4$^{cis}\Delta^{6,9,12,15}$) and linoleic to γ-linolenic acid (GLA, 18:3$^{cis}\Delta^{6,9,12}$). To date, few attempts have been made to enhance linoleic or α-linolenic acid contents in plants; however, considerable progress has been made towards the engineering of high-yielding crops to produce Brassicaceae seed oils with elevated levels of GLA and SDA, respectively. For example, Liu et al. (2001) demonstrated that co-expression of a Δ^6 and a Δ^{12} desaturase isolated from *Mortierella alpina* in *B.napus* resulted in a significant accumulation of GLA (up to 43 % of total seed FA). Similar levels of GLA were generated in the seeds of *B. juncea* through the expression of a Δ^6 desaturase from the fungus *Pythium irregulare* (Hong et al. 2002). Furthermore, hybridization between transgenic *B. napus* lines containing the *M. alpina* Δ^6 and Δ^{12} desaturase genes and independently produced lines containing the *B. napus* Δ^{15} desaturase gene generated up to 23 % SDA (Ursin 2003).

Unfortunately, plants do not possess the enzymes required to convert C18 PUFAs into VLC-PUFAs (Napier 2007; Napier and Graham 2010; Ruiz-López et al. 2012). Currently, marine fish are the primary dietary source of omega-3 VLC-PUFAs, such as EPA and DHA. However, due to the diminishing marine stock and increased demand for VLC-PUFAs, especially omega-3 VLC-PUFAs, the possibility of engineering plants as an alternative sustainable source of EPA and DHA has been thoroughly investigated. The main strategy employed for the production of VLC-PUFAs in plants has been the reconstitution of the VLC-PUFA biosynthetic pathway. Two different pathways have been proposed for VLC-PUFA synthesis: the anaerobic polyketide synthase (PKS)-like pathway, which exists in thraustochytrids and some marine bacteria (Metz et al. 2001; Napier 2002; Okuyama et al. 2007) and the aerobic pathway, which involves an alternating process of FA desaturation and elongation, and is used in most eukaryotes to generate VLC-PUFAs. As of yet, there has been only a single successful example of engineering the anaerobic PKS-like pathway for VLC-PUFAs in plants, which yielded up to 0.8 % DHA and 1.7 % docosapentaenoic acid of total FAs in Arabidopsis seeds (Metz et al. 2006). Efforts are ongoing, however, to further optimize the engineering of this pathway in oilseeds.

Conversely, far more success has been achieved through the use of the aerobic pathway of VLC-PUFA biosynthesis. The desaturases in this pathway comprise a unique class of oxygenases termed 'front-end desaturases' that catalyze double bond formation between the pre-existing double bond and carboxyl end of PUFAs (see Meesapyodsuk and Qiu 2012 for a detailed review). There are two main distinct aerobic pathways leading to the synthesis of VLC-PUFAs: the "Δ^6" and "Δ^8" pathways. The conventional "Δ^6" pathway exists in most VLC-PUFA-synthesizing eukaryotes and involves sequential Δ^6 desaturation, elongation, and Δ^5 desaturation to produce arachidonic acid and EPA. The "Δ^8" pathway has been reported in some protists and is initiated by Δ^9 elongation, followed by Δ^8 desaturation and Δ^5 desaturation to generate arachidonic acid ($20:4^{cis}\Delta^{5,8,11,14}$) and EPA. EPA generated from both pathways is subsequently elongated and desaturated to yield DHA. In the past few years, all primary genes required for aerobic VLC-PUFA synthesis have been cloned (Sayanova and Napier 2004). The most impressive demonstration of re-engineering the complete aerobic pathway for EPA production in Brassicaceae species was reported by Cheng et al. (2010), which accumulated an average of 20 % of EPA in transgenic zero-erucic acid B. carinata seeds. This study also suggested that the choice of host species had substantial influence on VLC-PUFA level, with zero-erucic acid B. carinata appearing to be a better host than B. juncea. Recently, several further studies also demonstrated the practical feasibility of the metabolic engineering of Brassicaceae species for DHA production in seed oil. For example, Robert et al. (2005) described the successful reconstitution of the DHA biosynthetic pathway in Arabidopsis, which produced up to 0.5 % DHA in the seeds of T_1 plants. Improved results were obtained by introducing a nine-gene construct into the B. juncea breeding line 1424, whereby Wu et al. (2005) were able to obtain up to 1.5 % DHA in transgenic seeds.

Despite the substantial progress that has been made towards the engineering of plants for enhanced VLC-PUFA contents, the amount of EPA and DHA produced is still far less than that present in fish oil. One of the major bottlenecks for producing high levels of VLC-PUFAs in transgenic plants is the well-documented "substrate-dichotomy" (Napier 2007), which occurs as a result of the different substrate requirements of desaturase and elongase, namely glycerolipid-linked substrates for desaturases and acyl-CoA for elongases. Many different strategies have been proposed to prevent acyl chain exchange between the phospholipid and acyl-CoA pool for a series of desaturation and elongation steps. The most obvious of these methods is to incorporate acyl-CoA-dependent desaturases into the engineering pathway. However, while several groups (Hoffmann et al. 2008; Petrie et al. 2010; Robert et al. 2005) have attempted this strategy, the seed levels of VLC-PUFAs (EPA and DHA) remained extremely low. The reasons behind these low levels of desired FAs remains to be further investigated. Therefore, although considerable preliminary success has been achieved towards engineering plants for increased contents of VLC-PUFAs, further research will be necessary to generate transgenic seed oil that is substantially equivalent to marine oil.

7.4.4 Unusual Fatty Acids

Fatty acids with additional functional groups, including hydroxylated, epoxidized and conjugated FAs, as well as acetylenic and cyclopropene FAs, are valuable for a wide range of industrial applications, such as lubricants and drying oils. Even though some of these exotic FAs can accumulate upwards of 90 % in the seed oils of various native plants, including ricinoleic acid in castor oil and vernolic acid in *Bernardia pulchella* oil, these plants either have a fairly low seed yield or have a very restricted growing area (Carlsson et al. 2011). Therefore, substantial research has been carried out to genetically engineer high yielding oilseeds to produce these unusual FAs. In most cases, the enzymes responsible for the synthesis of these FAs are related to, and probably evolved from, the FAD2 oleoyl-PC Δ^{12}-desaturase, which catalyzes the conversion of oleic acid to linoleic acid (Cahoon et al. 1999; Dyer et al. 2002; Lee et al. 1998; van de Loo et al. 1995). Thus, in a similar fashion to desaturation, FA modification catalyzed by these FAD2-related enzymes occurs using PC as the substrate.

   To date, many FAD2-related enzymes have been identified in plants that naturally produce modified fatty acids (mFAs), such as the hydroxylase from castor (van de Loo et al. 1995), the epoxygenase from *Vernonia galamensis* (Kinney et al. 2002) as well as the conjugase from *Momordica charantia* (Cahoon et al. 1999), and these enzymes have proven to be very useful for engineering plants for novel industrial oils (Cahoon et al. 2006; Lu et al. 2006b; Smith et al. 2003). However, the amount of mFAs that accumulate in these transgenic seeds in which the expression of a single *FAD2*-like gene has been over-expressed is relatively low (<20 %) compared to the levels found in natural plant sources (Broun and Somerville 1997; Cahoon et al. 1999; Lee et al. 1998). The major bottleneck for producing high amounts of mFAs in transgenic seed oil seems to be channeling these FAs from their site of synthesis on PC to storage TAG (Bates and Browse 2011; Dyer et al. 2008).

One popular hypothesis is that, in natural mFA-producing plants, enzymes involved in the TAG biosynthetic pathway have co-evolved with the FAD2-related enzymes, which enable them to effectively incorporate mFAs into TAG (Lu et al. 2006b). Thus, co-expressing cDNAs encoding these co-evolved enzymes with the FAD2-related enzymes could facilitate the flow of mFAs from PC and thus relieve the bottleneck. This hypothesis has been validated by several experiments. For example, Li et al. (2010) reported that transgenic expression of *V. galamensis* *DGAT1* or *DGAT2* enhanced the level of the epoxidized vernolic acid in mature transgenic soybean seeds co-expressing a Δ^{12}-epoxygenase from *Stokesia laevis*. In addition, combining the expression of the acyltransferases, *RcDGAT2* or *PDAT1A*, along with *FAH12* in Arabidopsis showed significantly increased levels (approximately 10 % of total FA content) of hydroxy FA accumulation in seed oil (Burgal et al. 2008; van Erp et al. 2011). Besides the acyltransferases involved in the final step of TAG synthesis, enzymes including LPCAT, LPAAT, specific phospholipases, GPAT and the recently discovered PDCT, may also be important for further increases in the production of mFAs in transgenic plants (Carlsson et al. 2011). Another aspect that requires consideration is that expressing single *FAD2*-like genes in transgenic plants had detrimental effects on oil content (Li et al. 2010; van Erp et al. 2011). This negative effect could be at least partially overcome by co-expression with co-evolved genes, because these enzymes may prevent a futile cycle of mFA synthesis and breakdown.

Recently, it has also been suggested that the acyl-ACP-dependent feedback inhibition of FA synthesis may be a major limiting factor for engineering plants for mFAs (Snapp and Lu 2012); however, this theory has yet to be proven and will require further study. Taken together, these examples indicate that understanding specific mFA metabolic pathways in mFA-producing plants and the utilization of gene-stacking techniques for mFA production will likely allow vital optimization of this metabolic engineering strategy.

7.5 Engineering Non-acyl Lipid Biosynthesis in Seeds

In addition to the acyl lipids that make up seed oil storage reserves, plants also produce a wide variety of non-acyl lipids, including sterols, chlorophyll, carotenoids and tocochromanols. To date, the antioxidant carotenoids and tocochromanols have garnered the most interest for biotechnological modification in plant seeds due to their essential roles in the diets of animals, including humans (reviewed in Meyer and Kinney 2010). Interestingly, while genes responsible for virtually all aspects of both pathways have been elucidated from genetic and genomics-based approaches in model species, efforts to enhance the content or modify the composition of each type of compound have been met with mixed success.

7.5.1 Carotenoids

Carotenoids make up a large group of over 750 tetraterpene lipids (C40) that are often highly coloured and exert important physiological functions in a wide range of organisms, including plants, as well as some algae, bacteria and fungi. In plants, these molecules accumulate in specialized plastid-derived chromoplasts or the thylakoid membrane system of chloroplasts where they are thought to have an antioxidant function and in some cases are integral components of the photosynthetic apparatus (Bartley and Scolnik 1995). Unlike photosynthetic organisms, animals are not able to synthesize carotenoids *de novo*, and as such, must acquire them through their diets. β-carotene is possibly the most well-known of these molecules for its dietary requirement in animals, as it serves as a precursor to vitamin A. However, in addition to their role in vitamin A production, carotenoids have also attracted much attention for their putative nutraceutical properties, such as their possible effects in the prevention of various types of cancer (Zhang et al. 2012; Nishino et al. 2009; Levy et al. 2011), cardiovascular disease (Fassett and Coombes 2011), and age-related macular degeneration (Carpentier et al. 2009).

While seed carotenoid biosynthesis is essential for the generation of abscisic acid and its associated regulation of seed dormancy, it is a minor constituent of most seeds, accumulating to approximately ten-fold lower levels than other non-acyl lipids. These levels are also approximately a factor of ten lower than the carotenoid content of photosynthetic leaves (Howitt and Pogson 2006). Due to the prevalence and tragic outcomes of vitamin A deficiencies in humans, various attempts have been made to enhance pro-vitamin A carotenoid levels in seeds, many of which have centered upon the over-expression of *PHYTOENE SYNTHASE* (*PSY*), which catalyzes the first committed step in carotenoid biosynthesis and is considered to be a regulatory point within the pathway. The first of these attempts, whereby a plastid-targeted *PSY* gene from *Pantoea ananatis* (formerly known as *Erwinia uredovora*; *crtB*) was expressed in a seed-specific manner in *B. napus*, was exceptionally successful (Shewmaker et al. 1999). The resulting transgenic embryos were visibly orange and exhibited up to a 50-fold increase in carotenoid content compared to untransformed controls. This increase was predominantly due to increases in α- and β-carotenes, with no substantial change in lutein, which is the predominant carotenoid in most seeds. This vast increase in carotenoids appeared to occur at the expense of tocopherols and chlorophyll, as tocopherol levels decreased three-fold in mature transgenic seeds and transient chlorophyll was reduced by six-fold in developing seeds. Interestingly, and perhaps somewhat disappointingly, the fatty acyl composition in these seeds was also altered, exhibiting a higher percentage of oleic acid and decreased proportions of linoleic and α-linolenic acid, which could be considered undesirable in terms of nutrition (Shewmaker et al. 1999). Similar, although slightly less spectacular, results were obtained in a subsequent study in which the seed-specific over-expression of an endogenous Arabidopsis *PSY* led to a significant increase in total carotenoid content, with an average relative 43-fold increase in the amount of β-carotene (Lindgren et al. 2003). However, there were several notable differences from the previous study, including the fact that many other carotenoids, including lutein, were also increased in the transgenic seeds, as were levels of chlorophyll, which suggests that slight functional differences may exist in bacterial and plant-derived *PSY* genes.

As there was also a small buildup of phytoene in the lines expressing the bacterial *PSY* gene (Shewmaker et al. 1999), it was proposed that the desaturation and isomerization steps that occur early in the pathway may have been experiencing a bottleneck to some degree. Indeed, a follow-up study in which the same *PSY* gene was co-expressed with bacterial *PHYTOENE DESATURASE* (*crtI*) and *LYCOPENE β-CYCLASE* (*crtY*) genes eliminated this phytoene excess and increased the ratio of β- to α-carotene from 2:1 to 3:1 without exhibiting any further increase in total carotenoid levels in *B. napus* seeds (Ravanello et al. 2003). Along the same lines, constitutive RNAi-mediated down-regulation of the *B. napus LYCOPENE ε-CYCLASE* (*ε-CYC*), which catalyzes the formation of α-carotene and lutein from lycopene, resulted in enhanced β-carotene, zeaxanthin, violaxanthin and lutein levels in seeds and an overall decrease in FA content (Yu et al. 2008).

Enhancement of seed carotenoid content has also been elicited through the engineering of genetic elements that are not directly involved in the carotenoid biosynthetic pathway. For example, constitutive silencing of the *B. napus DE-ETIOLATED-1* (*DET1*) gene, which is a negative regulator of light-mediated responses in plants and plays a pleiotropic role in plant development (Schäfer and Bowler 2002), resulted in 28–230 % increases in carotenoids (Wei et al. 2009). Interestingly, seed-specific silencing of the same gene resulted in a lesser degree of carotenoid enhancement, with increases of only 34–76 %. While the transgenic plants had normal plant morphology and productivity, poor germination rates were noted in the transgenic seeds, which would adversely affect their use in an agricultural context. Similarly, constitutive, but not seed-specific, expression of the Arabidopsis *AtmiR156b* gene, which regulates a network of genes through the targeting of numerous transcription factors (Schwab et al. 2005; Gandikota et al. 2007), in *B. napus* resulted in 1.3 to 2-fold increases in carotenoid content in seeds, with a 4.5-fold elevation in β-carotene levels (Wei et al. 2010). At present, the mechanisms underlying the roles of these genes in carotenoid biosynthesis are not fully understood.

In recent years, attempts have also been made to increase the production of particular carotenoids in seeds, such as the ketocarotenoids, which have distinct value in human health and as a feed ingredient (Giuliano et al. 2008). Astaxanthin in particular has become a main focus of this type of research as it is a high-value antioxidant that is expensive to synthesize chemically and is only produced by a limited number of organisms, including marine bacteria, as well as certain freshwater algae and basidiomycete yeast (Johnson and An 1991; Boussiba 2000; Del Campo et al. 2004; Johnson 2003). The majority of higher plants are not able to produce ketocarotenoids due to their lack of a gene encoding β-carotene ketolase (4,4'-oxygenase) (Cunningham and Gantt 2005). In one instance, Stålberg et al. (2003) expressed a plastid-targeted *β-CAROTENE KETOLASE* (*bkt2*) gene from the alga *Haematococcus pluvialis* in Arabidopsis in a seed-specific manner, which resulted in the production of several ketocarotenoids, including small amounts of astaxanthin. Interestingly, the relative amount of ketocarotenoids were able to be increased through the co-expression of this cassette with one bearing the endogenous Arabidopsis *PSY*, which yielded seeds with a 4.6-fold increase in total pigment and a 13-fold increase in the levels of three major ketocarotenoids (4-keto-lutein, canthaxanthin and adonirubin) compared to seeds of transgenic plants bearing only the *bkt2* construct (Stålberg et al. 2003).

More recently, an attempt was made to genetically engineer *B. napus* seeds to yield enhanced levels and composition of carotenoids through the expression of seven key plastid-targeted bacterial genes found previously to be involved in ketocarotenoid formation (Fujisawa et al. 2009). These included genes encoding PSY, β-carotene ketolase, β-carotene hydroxylase, isopentenyl pyrophosphate isomerase, geranylgeranyl pyrophosphate synthase, lycopene β-cyclase and phytoene desaturase/carotene isomerase. While ketocarotenoids were generated in the transgenic seeds, only very low levels of astaxanthin were produced and in every case, the levels of total carotenoids in transgenic seeds were similar to or less than values reported for seeds transformed with *PSY* alone (Shewmaker et al. 1999), which suggests that any increase in total carotenoids was mainly due to the catalytic activity of PSY (Fujisawa et al. 2009).

Therefore, while great success has been achieved in engineering seeds that possess an increase in native pro-vitamin A-related carotenoids, it seems that the generation of non-native carotenoids, such as particular ketocarotenoids, is going to be a far more challenging task. It has become evident that not only ketolase activity, but also an increased flux of substrate into the carotenoid biosynthetic pathway will be required, and it is possible that modification of the native carotenoid profile may also be necessary to limit the accumulation of non-desired intermediates, possibly via RNAi-mediated silencing (reviewed by Meyer and Kinney 2010). In addition, it is possible that the ability to engineer novel, plastid-derived compartments within the seed may also be required for the efficient biosynthesis and storage of carotenoids. For example, the Arabidopsis *or* mutant effected the constitutive formation of chromoplasts, which led to carotenoid accumulation without any concomitant up-regulation of carotenoid biosynthetic genes (Lu et al. 2006a), and has the potential to be of extreme value in the production of ketocarotenoids in sufficient abundance.

7.5.2   Tocochromanols

Tocochromanols are a group of non-acyl lipids found in all photosynthetic organisms. In terms of human nutrition, the tocochromanols all constitute forms of vitamin E, which is an essential nutrient in the diet of mammals. Indeed, vitamin E intakes in excess of the recommended daily allowance have been associated with improved immune function, decreased risk of cardiovascular disease and some cancers, as well as slowing the progression of several degenerative conditions associated with aging such as arthritis and cataracts. Unfortunately, obtaining these therapeutic levels of vitamin E from an average diet is incredibly difficult (reviewed by Meyer and Kinney 2010).

As is the case for the carotenoid biosynthetic pathway, genes for virtually every step in tocochromanol biosynthesis have been isolated, which has facilitated efforts to manipulate this pathway. Attempts to enhance vitamin E levels in seeds have been carried out through either efforts to increase the flux of substrate through the biosynthetic pathway in order to quantitatively elevate total tocopherol levels, or by shifting the composition of the tocopherol pool towards elevated amounts of tocopherols with the highest vitamin E activities (Ajjawi and Shintani 2004).

Increasing total amounts of tocochromanols in seeds would be of potential use for improving the antioxidant capacity of vegetable oils, which could in turn enhance their oxidative stability for food processing and their use as high temperature lubricants, and could also increase the vitamin E activity of oils. To this end, over-expression of many of the genes directly involved in tocochromanol biosynthesis, including *TYRA*, *HPPD*, *HPT* (*VTE2*), and *TC* (*VTE1*), has been carried out in transgenic Brassicaceae seeds; unfortunately, the results have been modest at best (Karunanandaa et al. 2005; Tsegaye et al. 2002; Raclaru et al. 2006; Savidge et al. 2002; Collakova and DellaPenna 2003; Kumar et al. 2005). In addition, over-expression of *TC* has also been linked to compositional changes, with increases in the proportion of δ-tocopherols (Kumar et al. 2005).

The fact that only small, but significant, increases in tocochromanol content were observed in transgenic plants expressing a number of single genes involved in the tocochromanol biosynthetic pathway indicates that the metabolic flux to this particular pathway is not controlled by a single enzyme-catalyzed reaction. Therefore, several attempts have been made to enhance the expression of multiple regulatory steps simultaneously, with slightly improved results. For example, co-expression of *HPPD*, *HPT* and *TC* in *B. napus* seeds resulted in an average increase of tocopherol content of 1.2-fold, with an enhancement in the proportion of δ-tocopherol due to the over-expression of the *TC* gene (Raclaru et al. 2006). Similarly, co-expression of *TYRA*, *HPPD* and *HPT* in Arabidopsis and *B. napus* seeds yielded up to five-fold increases in tocochromanol levels, which were suggested to be almost exclusively due to increased tocotrienols (Karunanandaa et al. 2005). Unfortunately, many of the latter lines also exhibited impairments in germination (Karunanandaa et al. 2005), which would likely preclude their use in crop improvement.

The inability to engineer large increases in tocopherol content in oilseeds indicates that there remains a lack of understanding concerning this particular pathway in plants. Efforts are currently underway to identify additional genes that are not directly involved in the tocochromanol biosynthetic pathway, that exert control over the tocopherol content and profile of seeds. Indeed, it has been suggested that there may be a limit to the amount that these compounds can be increased without a concomitant enhancement of the flux through the MEP pathway (Savidge et al. 2002) and it has been shown that GGDP may be limiting for tocopherol biosynthesis (Furuya et al. 1987). This information will undoubtedly allow further success in the engineering of enhanced tocopherol content in transgenic seeds in the future.

Of all the tocochromanols, α-tocopherol content has been found to be particularly crucial from a nutritional perspective as it possesses the highest vitamin E activity. While α-tocopherol tends to be the most abundant tocochromanol in photosynthetic tissues, seed oil generally contains 10–20 times higher levels of tocochromanols than leaves and are thus the major dietary source of vitamin E, but α-tocopherol is very often a minor component (reviewed by DellaPenna and Pogson 2006). Fortunately, efforts to modify oilseed tocopherol composition to yield higher proportions α-tocopherol from γ-tocopherol have yielded excellent results. The fact that γ-tocopherol methyltransferase (γ-TMT), which is a methyltransferase involved in the conversion of γ- and δ-tocochromanols to their α- and β- counterparts, was thought to be a limiting factor in α-tocopherol production suggested that conversion of the large pool of γ-tocopherol in seeds to the more nutritionally desirable α-tocopherol may be possible through its targeted over-expression. As expected, the seed-specific over-expression of the Arabidopsis γ-TMT ortholog (VTE4) resulted in the nearly complete conversion of γ-tocopherol to α-tocopherol and a resulting nine-fold enhancement in vitamin E activity (Shintani and DellaPenna 1998) without any significant change in total tocochromanol content. More recently, a similar approach has been taken in the crop plant, B. juncea, whereby Arabidopsis VTE4 was constitutively expressed, resulting in a six-fold increase in the α- to γ-tocopherol ratio in seeds, with the total tocopherol content remaining largely unchanged compared to controls (Yusuf and Sarin 2007).

Furthermore, since the modest increase in seed tocopherol levels noted in transgenic Arabidopsis over-expressing the HPT1 (VTE2) gene was found to be primarily due to an increase in γ-tocopherol (Collakova and DellaPenna 2003; Savidge et al. 2002), an attempt was made to simultaneously over-express VTE2 and VTE4 to yield higher levels of α-tocopherol. As anticipated, this co-expression yielded an additive effect, increasing total tocochromanols while at the same time converting virtually the entire pool of γ-tocopherol to α-tocopherol and resulting in a nearly 12-fold enhancement in vitamin E activity with no decreases in total chlorophyll or carotenoid levels (Collakova and DellaPenna 2003).

7.6   Conclusion and Future Directions

In recent years, there has been a growing interest in the utilization of genetic engineering to generate Brassicaceae crop species bearing modified lipid compositions, novel lipids, or increased lipid content. These oils have the potential to be incredibly useful for a variety of industrial and food applications, and may provide a more sustainable means of generating these products. However, despite the great advances that have been achieved in our understanding of the various FA and lipid biosynthetic pathways, there remain obvious gaps in our knowledge, which continue to hamper advancements in the development of transgenic plants bearing modified seed oil phenotypes (Weselake et al. 2009; Baud and Lepeniec 2010). To complicate matters even further, the processes of acyl- and non-acyl lipid synthesis are both highly complex, involving carbon metabolism and often various environmental factors in addition to the pathways in which their synthesis occurs. Therefore, the genetic manipulation of plants to enhance these qualities will almost certainly necessitate intricate strategies that rely upon the simultaneous over-expression and/ or down-regulation of a number of target genes (reviewed by Baud and Lepeniec 2010). It seems likely that this type of approach will provide the means to yield transgenic seeds with higher levels of particular FAs and lipids, which would increase their potential usefulness as commercial crops in the future.

Acknowledgements RJW is grateful to Alberta Enterprise and Advanced Education, Alberta Innovates Bio Solutions, AVAC Ltd., the Canada Foundation for Innovation, the Canada Research Chairs Program and the Natural Sciences and Engineerng Research Council of Canada for supporting his research on seed oil formation.

References

Ajjawi I, Shintani D (2004) Engineered plants with elevated vitamin E: a nutraceutical success story. Trends Biotechnol 22:104–107

Bartley GE, Scolnik PA (1995) Plant carotenoids: pigments for photoprotection, visual attraction, and human health. Plant Cell 7:1027–1038

Bartley GE, Scolnik PA, Beyer P (1999) Two *Arabidopsis thaliana* carotene desaturases, phytoene desaturase and ζ-carotene desaturase, expressed in *Escherichia coli*, catalyze a poly-*cis* pathway to yield pro-lycopene. Eur J Biochem 259:396–403

Bates PD, Browse J (2011) The pathway of triacylglycerol synthesis through phosphatidylcholine in Arabidopsis produces a bottleneck for the accumulation of unusual fatty acids in transgenic seeds. Plant J 68:387–399

Bates PD, Browse J (2012) The significance of different diacylglycerol synthesis pathways on plant oil composition and bioengineering. Front Plant Sci 3:1–11

Baud S, Lepeniec L (2010) Physiological and developmental regulation of seed oil production. Prog Lipid Res 49:235–249

Baud S, Lepiniec L (2009) Regulation of de novo fatty acid synthesis in maturing oilseeds of Arabidopsis. Plant Physiol Biochem 47:448–455

Baud S, Wuilleme S, To A, Rochat C, Lepiniec L (2009) Role of *WRINKLED1* in the transcriptional regulation of glycolytic and fatty acid biosynthetic genes in Arabidopsis. Plant J 60:933–947

Bergmüller E, Porfirova S, Dormann P (2003) Characterization of an Arabidopsis mutant deficient in γ-tocopherol methyltransferase. Plant Mol Biol 52:1181–1190

Bondaruk M, Johnson S, Degafu A, Boora P, Bilodeau P, Morris J et al (2007) Expression of a cDNA encoding palmitoyl-acyl carrier protein desaturase from cat's claw (*Doxantha unguiscati* L.) in *Arabidopsis thaliana* and *Brassica napus* seeds leads to accumulation of unusual unsaturated fatty acids and increased stearic acid content in the seed oil. Plant Breeding 126:186–194

Boussiba S (2000) Carotenogenesis in the green alga *Haematococcus pluvialis*: cellular physiology and stress response. Physiol Plantarum 108:111–117

Brough CL, Coventry JM, Christie WW, Kroon JTM, Brown AP, Barsby TL et al (1996) Towards the genetic engineering of triacylglycerols of defined fatty acid composition: major changes in erucic acid content at the *sn*-2 position affected by the introduction of a 1-acyl-*sn*-glycerol-3-phosphate acyltransferase from *Limnanthes douglasii* into oil seed rape. Mol Breeding 2:133–142

Broun P, Somerville C (1997) Accumulation of ricinoleic, lesquerolic, and densipolic acids in seeds of transgenic Arabidopsis plants that express a fatty acyl hydroxylase cDNA from castor bean. Plant Physiol 113:933–942

Brown A, Affelck V, Kroon J, Slabas A (2009) Proof of function of a putative 3-hydroxyacyl-acyl carrier protein dehydratase from higher plants by mass spectrometry of product formation. FEBS Lett 583:363–368

Burgal J, Shockey J, Lu C, Dyer J, Larson T, Graham I et al (2008) Metabolic engineering of hydroxy fatty acid production in plants: RcDGAT2 drives dramatic increases in ricinoleate levels in seed oil. Plant Biotechnol J 6:819–831

Cahoon EB, Shanklin J, Ohlrogge JB (1992) Expression of a coriander desaturase results in petroselinic acid production in transgenic tobacco. Proc Natl Acad Sci USA 89:11184–11188

Cahoon EB, Cranmer AM, Shanklin J, Ohlrogge JB (1994) Δ^6 hexadecenoic acid is synthesized by the activity of a soluble Δ^6 palmitoyl-acyl carrier protein desaturase in *Thunbergia alata* endosperm. J Biol Chem 269:27519–27526

Cahoon EB, Shah S, Shanklin J, Browse J (1998) A determinant of substrate specificity predicted from the acyl-acyl carrier protein desaturase of developing cat's claw seed. Plant Physiol 117:593–598

Cahoon EB, Carlson TJ, Ripp KG, Schweiger BJ, Cook GA, Hall SE et al (1999) Biosynthetic origin of conjugated double bonds: production of fatty acid components of high-value drying oils in transgenic soybean embryos. Proc Natl Acad Sci USA 96:12935–12940

Cahoon EB, Hall SE, Ripp KG, Ganzke TS, Hitz WD, Coughlan SJ (2003) Metabolic redesign of vitamin E biosynthesis in plants for tocotrienol production and increased antioxidant content. Nat Biotechnol 21:1082–1087

Cahoon EB, Dietrich CR, Meyer K, Damude HG, Dyer JM, Kinney AJ (2006) Conjugated fatty acids accumulate to high levels in phospholipids of metabolically engineered soybean and Arabidopsis seeds. Phytochemistry 67:1166–1176

Canola-council.org [INTERNET] Canola Council of Canada (2011) [cited 10 Sep 2012]. Available from http://www.canola-council.org/

Canola-council.org [INTERNET] Canola Council of Canada, 2011 Annual Report (2011) [cited 10 Sep 2012]. Available from http://www.canola-council.org/media/503315/2011_annual_report_web.pdf

Carlsson AS, Yilmaz JL, Green AG, Stymne S, Hofvander P (2011) Replacing fossil oil with fresh oil – with what and for what? Eur J Lipid Sci Technol 113:812–831

Carpentier S, Knaus M, Suh M (2009) Associations between lutein, zeaxanthin, and age-related macular degeneration: an overview. Crit Rev Food Sci Nutr 49:313–326

Cernac A, Benning C (2004) WRINKLED1 encodes an AP2/EREB domain protein involved in the control of storage compound biosynthesis in Arabidopsis. Plant J 40:575–585

Chai G, Bai Z, Wei F, King GJ, Wang C, Shi L, Dong C, Chen H, Liu S (2010) Brassica *GLABRA2* genes: analysis of function related to seed oil content and development of functional markers. Theor Appl Genet 120:1597–1610

Chen JQ, Lang CX, Hu ZH, Liu ZH, Huang RZ (1999) Antisense PEP gene regulates to ratio of protein and lipid content in *Brassica napus* seeds. J Agric Biotechnol 7:316–320

Chen JM, Qi WC, Wang SY, Guan RZ, Zhang HS (2010) Correlation of Kennedy pathway efficiency with seed oil content of canola (*Brassica napus* L.) lines. Can J Plant Sci 91:251–259

Cheng B, Wu G, Vrinten P, Falk K, Bauer J, Qiu X (2010) Towards the production of high levels of eicosapentaenoic acid in transgenic plants: the effects of different host species, genes and promoters. Transgenic Res 19:221–229

Collakova E, DellaPenna D (2003) Homogentisate phytyltransferase activity is limiting for tocopherol biosynthesis in Arabidopsis. Plant Physiol 131:632–642

Courchesne NM, Parisien A, Wang B, Lan CQ (2009) Enhancement of lipid production using biochemical, genetic and transcription factor engineering approaches. J Biotechnol 141:31–41

Cunningham FX Jr, Gantt E (2005) A study in scarlet: enzymes of ketocarotenoid biosynthesis in the flowers of *Adonis aesrivalis*. Plant J 41:478–492

Cunningham FX Jr, Pogson B, Sun Z, McDonald KA, DellaPenna D, Gantt E (1996) Functional analysis of the β and ε lycopene cyclase enzymes of Arabidopsis reveals a mechanism for control of cyclic carotenoid formation. Plant Cell 8:1613–1626

de Boer G-J, Kater MM, Fawcett T, Slabas AR, Nijkamp HJJ, Stuitje AR (1998) The NADH-specific enoyl-acyl carrier protein reductase: characterization of a housekeeping gene involved in storage lipid synthesis in seeds of Arabidopsis and other plant species. Plant Physiol Biochem 36:473–486

Dehesh K (2001) How can we genetically engineer oilseed crops to produce high levels of medium-chain fatty acids? Eur J Lipid Sci Technol 103:688–697

Dehesh K, Jones A, Knutzon DS, Voelker TA (1996) Production of high levels of 8:0 and 10:0 fatty acids in transgenic canola by overexpression of ChFatB2, a thioesterase cDNA from *Cuphea hookeriana*. Plant J 9:167–172

Dehesh K, Edwards P, Fillatti J, Slabaugh M, Byrne J (1998) KAS IV: a 3-ketoacyl-ACP synthase from *Cuphea sp.* is a medium chain specific condensing enzyme. Plant J 15:383–390

Del Campo JA, Rodriguez H, Moreno J, Vargas MA, Rivas J, Guerrero MG (2004) Accumulation of astaxanthin and lutein in *Chlorella zofingiensis* (Chlorophyta). Appl Microbiol Biotechnol 64:848–854

DellaPenna D, Pogson BJ (2006) Vitamin synthesis in plants: tocopherols and carotenoids. Annu Rev Plant Biol 57:711–738

Drexler H, Spiekermann P, Meyer A, Domergue F, Zank T, Sperling P et al (2003) Metabolic engineering of fatty acids for breeding of new oilseed crops: strategies, problems and first results. J Plant Physiol 160:779–802

Dyer JM, Chapital DC, Kuan J-W, Mullen RT, Turner C, McKeon TA et al (2002) Molecular analysis of a bifunctional fatty acid conjugase/desaturase from tung. Implications for the evolution of plant fatty acid diversity. Plant Physiol 130:2027–2038

Dyer JM, Stymne S, Green AG, Carlsson AS (2008) High-value oils from plants. Plant J 54:640–655

Eccleston VS, Ohlrogge JB (1998) Expression of lauroyl-acyl carrier protein thioesterase in *Brassica napus* seeds induces pathways for both fatty acid oxidation and biosynthesis and implies a set point for triacylglycerol accumulation. Plant Cell 10:613–622

Elhiti M, Yang C, Chan A, Durnin DC, Belmonte MF, Ayele BT, Tahir M, Stasolla C (2012) Altered seed oil and glucosinolate levels in transgenic plants overexpressing the *Brassica napus SHOOTMERISTEMLESS* gene. J Exp Bot 63:4447–4461

Falk J, Munne-Bosch S (2010) Tocochromanol functions in plants: antioxidation and beyond. J Exp Biol 61:1549–1566

FAOSTAT [INTERNET]. Food and Agriculture Organization of the United Nations (2012) [cited 10 Sep 2012]. Available from http://www.fao.org/index_en.htm

Fassett RG, Coombes JS (2011) Astaxanthin: a potential therapeutic agent in cardiovascular disease. Mar Drugs 9:447–465

Frentzen M (1998) Acyltransferases from basic science to modified seed oils. Fett-Lipid 100:161–166

Fujisawa M, Takita E, Harada H, Sakurai N, Suzuki H, Ohyama K, Shibata D, Misawa N (2009) Pathway engineering of *Brassica napus* seeds using multiple key enzyme genes involved in ketocarotenoid formation. J Exp Bot 60:1319–1332

Furuya T, Yoshikawa T, Kimura T, Kaneko H (1987) Production of tocopherols by cell culture of safflower. Phytochemistry 26:2741–2747

Gahukar RT (2012) New sources of feed stocks for biofuels production: Indian perspectives. JPTAF 3:24–28

Gandikota M, Birkenbihl RP, Hohmann S, Cardon GH, Saedler H, Huijser P (2007) The miRNA156/157 recognition element in the 3'UTR of the Arabidopsis SBP box gene SPL3 prevents early flowering by translational inhibition in seedlings. Plant J 49:683–693

Giuliano G, Tavazza R, Diretto G, Beyer P, Taylor MA (2008) Metabolic engineering of carotenoid biosynthesis in plants. Trends Biotechnol 26:139–145

Guo Y, Mietkiewska E, Francis T, Katavic V, Brost JM, Giblin EM, Barton DL, Taylor DC (2008) Increase of nervonic acid content in transformed yeast and transgenic plants by introduction of a *Lunaria annua* L. *3-Ketoacyl-CoA Synthase (KCS)* gene. Plant Mol Biol 69:565–575

Hawkins DJ, Kridl J (1998) Characterization of acyl-ACP thioesterases of mangosteen (*Garcinia mangostana*) seed and high levels of stearate production in transgenic canola. Plant J 13:743–752

Hoffmann M, Wagner M, Abbadi A, Fulda M, Feussner I (2008) Metabolic engineering of ω3-very long chain polyunsaturated fatty acid production by an exclusively acyl-CoA-dependent pathway. J Biol Chem 283:22352–22362

Hong H, Datla N, Reed DW, Covello PS, MacKenzie SL, Qiu X (2002) High-level production of γ-linolenic acid in *Brassica juncea* using a Δ6 desaturase from Pythium irregulare. Plant Physiol 129:354–362

Howitt CA, Pogson BJ (2006) Carotenoid accumulation and function in seeds and non-green tissues. Plant Cell Environ 29:435–445

Hoy JA, Hargrove MS (2008) The structure and function of plant hemoglobins. Plant Physiol Biochem 46:371–379

Jain RK, Coffey M, Lai K, Kumar A, MacKenzie SL (2000) Enhancement of seed oil content by expression of glycerol-3-phosphate acyltransferase genes. Biochem Soc Trans 28:958–961

Jako C, Kumar A, Wei Y, Zou J, Barton DL, Giblin EM, Covello PS, Taylor DC (2001) Seed-specific over-expression of an Arabidopsis cDNA encoding a diacylglycerol acyltransferase enhances seed oil content and seed weight. Plant Physiol 126:861–874

Johnson EA (2003) *Phaffia rhodozyma*: colorful odyssey. Int Microbiol 6:169–174

Johnson EA, An GH (1991) Astaxanthin from microbial sources. Crit Rev Biotechnol 11:297–26

Karunanandaa B, Qi Q, Hao M, Baszis SR, Jensen PK, Wong Y-HH, Jiang J, Venkatramesh M, Gruys KJ, Moshiri F, Post-Beittenmiller D, Weiss JD, Valentin HE (2005) Metabolically engineered oilseed crops with enhanced seed tocopherol. Metab Eng 7:384–400

Katavic V, Friesen W, Barton DL, Gossen KK, Giblin EM, Luciw T et al (2001) Improving erucic acid content in rapeseed through biotechnology: what can the Arabidopsis *FAE1* and the yeast *SLC1-1* genes contribute. Crop Sci 41:739–747

Kinney AJ, Cahoon EB, Hitz WD (2002) Manipulating desaturase activities in transgenic crop plants. Biochem Soc Trans 30:1099–1103

Kinney AJ, Cahoon EB, Damude HG, Hitz WD, Kolar CW, Liu ZB, inventors; EI Dupont De Nemours and Co., assignee (2004) Production of very long chain polyunsaturated fatty acids in oilseed plants. WO 2004071467

Knapp SJ (1993) Breakthroughs towards the domestification of *Cuphea*. In: Janick J, Simon JE (eds) New crops. Wiley, New York, pp 372–379

Knutzon DS, Thompson GA, Radke SE, Johnson WB, Knauf VC, Kridl JC (1992) Modification of *Brassica* seed oil by antisense expression of a stearoyl-acyl carrier protein desaturase gene. Proc Natl Acad Sci USA 89:2624–2628

Knutzon DS, Hayes TR, Wyrick A, Xiong H, Davies HM, Voleker TA (1999) Lysophosphatidic acid acyltransferase from coconut endosperm mediates the insertion of laurate at the sn-2 position of triacylglycerols in lauric rapeseed oil and can increase total laurate levels. Plant Physiol 120:739–746

Konishi T, Shinohara K, Yamada K, Sasaki Y (1996) Acetyl-CoA carboxylase in higher plants: most plants other than gramineae have both the prokaryotic and the eukaryotic forms of this enzyme. Plant Cell Physiol 37:117–122

Kumar R, Raclaru M, Schüßeler T, Gruber J, Sadre R, Lühs W et al (2005) Characterisation of plant tocopherol cyclases and their overexpression in transgenic *Brassica napus* seeds. FEBS Lett 579:1357–1364

Lassner MW, Levering CK, Davies HM, Knutzon DS (1995) Lysophosphatidic acid acyltransferase from meadowfoam mediates insertion of erucic acid at the sn-2 position of triacylglycerol in transgenic rapeseed oil. Plant Physiol 109:1389–1394

Lee M, Lenman M, Banaś A, Bafor M, Singh S, Schweizer M et al (1998) Identification of non-heme diiron proteins that catalyze triple bond and epoxy group formation. Science 280:915–918

Levy J, Salfisch S, Atzmon A, Hirsch K, Khanin M, Linnewiel K, Morag Y et al (2011) The role of tomato lycopene in cancer prevention. In: Mutanen M, Pajari A-M (eds) Vegetables, whole grains, and their derivatives in cancer prevention. Springer, New York, pp 47–66

Li R, Yu K, Hatanaka T, Hildebrand DF (2010) Vernonia DGATs increase accumulation of epoxy fatty acids in oil. Plant Biotechnol J 8:184–195

Li X, van Loo EN, Gruber J, Fan J, Guan R, Frentzen M et al (2012) Development of ultra-high erucic acid oil in the industrial oil crop *Crambe abyssinica*. Plant Biotechnol J 10:862–870

Lindgren LO, Stålberg KG, Höglund A-S (2003) Seed-specific overexpression of an endogenous Arabidopsis phytoene synthase gene results in delayed germination and increased levels of carotenoids, chlorophyll, and abscisic acid. Plant Physiol 132:779–785

Liu JW, Huang YS, DeMichele S, Bergana M, Bobik E, Hastilow C et al (2001) Evaluation of the seed oils from a canola plant genetically transformed to produce high level of γ-linolenic acid. In: Huang YS, Ziboh A (eds) γ-linolenic acid: recent advances in biotechnology and clinical applications. AOCS, Champaign, pp 61–71

Liu J, Hua W, Zhan G, Wei F, Wang X, Liu G et al (2010) Increasing seed mass and oil content in transgenic Arabidopsis by the overexpression of *wri1*-like gene from *Brassica napus*. Plant Physiol Biochem 48:9–15

Liu J, Hua W, Yang H-L, Zhan G-M, Li R-J, Deng L-B et al (2012) The *BnGRF2* gene (*GRF2-like* gene from *Brassica napus*) enhances seed oil production through regulating cell number and plant photosynthesis. J Exp Bot 63:3727–3740

Lu C, Fulda M, Wallis JG, Browse J (2006a) A high-throughput screen for genes from castor that boost hydroxy fatty acid accumulation in seed oils of transgenic Arabidopsis. Plant J 45:847–856

Lu S, Van Eck J, Zhou X, Lopez AB, O'Halloran DM, Cosman KM et al (2006b) The cauliflower *Or* gene encodes a DnaJ cysteine-rich domain-containing protein that mediates high levels of β-carotene accumulation. Plant Cell 18:3594–3605

Lu CF, Xin ZG, Ren ZH, Miquel M, Browse J (2009) An enzyme regulating triacylglycerol composition is encoded by the ROD1 gene of Arabidopsis. Proc Natl Acad Sci U S A 106:18837–18842

Lu C, Napier JA, Clemente TE, Cahoon EB (2011) New frontiers in oilseed biotechnology: meeting the global demand for vegetable oils for food, feed, biofuel, and industrial applications. Curr Opin Biotechnol 22:252–259

Maisonneuve S, Bessoule J-J, Lessire R, Delseny M, Roscoe TJ (2010) Expression of rapeseed microsomal lysophosphatidic acid acyltransferase isozymes enhances seed oil content in Arabidopsis. Plant Physiol 152:670–684

Marillia E-F, Micallef BJ, Bicallef M, Weninger A, Pedersen KK, Zou J, Taylor DC (2003) Biochemical and physiological studies of *Arabidopsis thaliana* transgenic lines with repressed expression of the mitochondrial pyruvate dehydrogenase kinase. J Exp Bot 54:259–270

Meesapyodsuk D, Qiu X (2012) The front-end desaturase: structure, function, evolution and biotechnological use. Lipids 47:227–237

Mena M, Vicente-Carbajosa J, Schmidt RJ, Carbonero P (1998) An endosperm-specific DOF protein from barley, highly conserved in wheat, binds to and activates transcription from the prolamin-box of a native B-hordein promoter in barley endosperm. Plant J 16:53–62

Metz JG, Roessler P, Facciotti D, Levering C, Dittrich F, Lassner M et al (2001) Production of polyunsaturated fatty acids by polyketide synthases in both prokaryotes and eukaryotes. Science 293:290–293

Metz JG, Flatt JH, Kuner, JM, inventors (2006) The genes for the enzymes of the polyunsaturated fatty acid polyketide synthase of *Schizochytrium* and their use in the manufacture of polyunsaturated fatty acids. WO 2006135866

Metzger JO, Bornscheuer U (2006) Lipids as renewable resources: current state of chemical and biotechnological conversion and diversification. Appl Microbiol Biotechnol 71:13–22

Meyer K, Kinney AJ (2010) Biosynthesis and biotechnology of seed lipids including sterols, carotenoids and tocochromanols. In: Wada H, Murata N (eds) Lipids in photosynthesis. Springer, New York, pp 407–444

Meyer K, Stecca KL, Ewell-Hicks K, Allen SM, Everard JD (2012) Oil and protein accumulation in developing seeds is influenced by the expression of a cytosolic pyrophosphatase in Arabidopsis. Plant Physiol 159:1221–1234

Mietkiewska E, Giblin EM, Wang S, Barton DL, Dirpaul J, Brost JM et al (2004) Seed-specific heterologous expression of a nasturtium *FAE* gene in Arabidopsis results in a dramatic increase in the proportion of erucic acid. Plant Physiol 136:266–275

Mietkiewska E, Brost JM, Giblin EM, Barton DL, Taylor DC (2007) Cloning and functional characterization of the *Fatty Acid Elongase 1* (*FAE1*) gene from high erucic *Crambe abyssinica* cv. Prophet. Plant Biotechnol J 5:636–645

Mietkiewska E, Hoffman TL, Brost JM, Giblin EM, Barton DL, Francis T et al (2008) Hairpin-RNA mediated silencing of endogenous FAD2 gene combined with heterologous expression of *Crambe abyssinica FAE* gene causes an increase in the level of erucic acid in transgenic *Brassica carinata* seeds. Mol Breeding 22:619–627

Mietkiewska E, Siloto RMP, Dewald J, Shah S, Brindley DN, Weselake RJ (2010) Lipins from plants are phosphatidate phosphatases that restore lipid synthesis in a pah1D mutant strain of *Saccharomyces cerevisiae*. FEBS J 278:764–775

Miquel M, Browse J (1992) Arabidopsis mutants deficient in polyunsaturated fatty acid synthesis. Biochemical and genetic characterization of a plant oleoyl-phosphatidylcholine desaturase. J Biol Chem 267:1502–1509

Mu J, Tan H, Zheng Q, Fu F, Liang Y, Zhang J et al (2008) LEAFY COTYLEDON1 is a key regulator of fatty acid biosynthesis in Arabidopsis. Plant Physiol 148:1042–1054

Napier JA (2002) Plumbing the depths of PUFA biosynthesis: a novel polyketide synthase-like pathway from marine organisms. Trends Plant Sci 7:51–54

Napier JA (2007) The production of unusual fatty acids in transgenic plants. Annu Rev Plant Biol 58:295–319

Napier JA, Graham IA (2010) Tailoring plant lipid composition: designer oilseeds come of age. Curr Opin Plant Biol 13:330–337

Nath UK, Wilmer JA, Wallington EJ, Becker HC, Möllers C (2009) Increasing erucic acid content through combination of endogenous low polyunsaturated fatty acids alleles with *Ld-LPAAT+Bn-fae1* transgenes in rapeseed (*Brassica napus* L.). Theor Appl Genet 118:765–773

Nguyen HT, Mishra G, Whittle E, Pidkowich MS, Bevan SA, Merlo AO et al (2010) Metabolic engineering of seeds can achieve levels of ω-7 fatty acids comparable with the highest levels found in natural plant sources. Plant Physiol 154:1897–1904

Nishino H, Murakoshi M, Tokuda J, Satomi Y (2009) Cancer prevention by carotenoids. Arch Biochem Biophys 483:165–168

Ohlrogge JB, Jaworski JG (1997) Regulation of fatty acid synthesis. Annu Rev Plant Phys 48:109–136

Ohto M, Stone SL, Harada JJ (2007) Genetic control of seed development and seed mass. In: Bradford K, Nonogaki H (eds) Seed development, dormancy and germination. Blackwell, Oxford, pp 1–24

Okada K, Saito T, Nakagawa T, Kawamukai M, Kamiya Y (2000) Five geranylgeranyl diphosphate synthases expressed in different organs are localized into three subcellular compartments in Arabidopsis. Plant Physiol 122:1045–1056

Okuyama H, Orikasa Y, Nishida T, Watanabe K, Morita N (2007) Bacterial genes responsible for the biosynthesis of eicosapentaenoic and docosahexaenoic acids and their heterologous expression. Appl Environ Microbiol 73:665–670

Park H, Kreunen SS, Cuttriss AJ, DellaPenna D, Pogson BJ (2002) Identification of the carotenoid isomerase provides insight into carotenoid biosynthesis, prolamellar body formation, and photomorphogenesis. Plant Cell 14:321–332

Peng Q, Hu Y, Wei R, Zhang Y, Guan C, Ruan Y, Liu C (2010) Simultaneous silencing of *FAD2* and *FAE1* genes affects both oleic acid and erucic acid contents in *Brassica napus* seeds. Plant Cell Rep 29:317–325

Petrie JR, Shrestha P, Mansour MP, Nichols PD, Liu Q, Singh SP (2010) Metabolic engineering of omega-3 long-chain polyunsaturated fatty acids in plants using an acyl-CoA Δ6-desaturase with ω3-preference from the marine microalga micromonas pusilla. Metab Eng 12:233–240

Phillips MA, Leon P, Boronat A, Rdoriguez-Concepcion M (2008) The plastidial MEP pathway: unified nomenclature and resources. Trends Plant Sci 13:619–623

Plesch G, Ehrhardt T, Mueller-Roeber B (2001) Involvement of TAAAG elements suggests a role for Dof transcription factors in guard cell-specific gene expression. Plant J 28:455–464

Pogson B, McDonald KA, Truong M, Britton G, DellaPenna D (1996) Arabidopsis carotenoid mutants demonstrate that lutein is not essential for photosynthesis in higher plants. Plant Cell 8:1627–1639

Porfirova S, Bergmüller E, Tropf S, Lemke R, Dormann P (2002) Isolation of an Arabidopsis mutant lacking vitamin E and identification of a cyclase essential for all tocopherol biosynthesis. Proc Natl Acad Sci U S A 99:1181–1190

Raclaru M, Gruber J, Kumar R, Sadre R, Lühs W, Zarhloul MK et al (2006) Increase of the tocochromanol content in transgenic *Brassica napus* seeds by overexpression of key enzymes involved in prenylquinone biosynthesis. Mol Breeding 18:93–107

Ravanello MP, Ke D, Alvarez J, Huang B, Shewmaker CK (2003) Coordinate expression of multiple bacterial carotenoid genes in canola leading to altered carotenoid production. Metab Eng 5:255–263

Robert SS, Singh SP, Zhou X, Petrie JR, Blackburn SI, Mansour PM et al (2005) Metabolic engineering of Arabidopsis to produce nutritionally important DHA in seed oil. Funct Plant Biol 32:473–479

Roesler K, Shintani D, Savage L, Boddupalli S, Ohlrogge J (1997) Targeting of the Arabidopsis homomeric acetyl-coenzyme a carboxylase to plastids of rapeseeds. Plant Physiol 113:75–81

Ruiz-López N, Sayanova O, Napier JA, Haslam RP (2012) Metabolic engineering of the omega-3 long chain polyunsaturated fatty acid biosynthetic pathway into transgenic plants. J Exp Bot 63:2397–2410

Ruuska SA, Girke T, Benning C, Ohlrogge JB (2002) Contrapuntal networks of gene expression during Arabidopsis seed filling. Plant Cell 14:1191–1206

Santos-Mendoza M, Dubreucq B, Miquel M, Caboche M, Lepiniec L (2005) LEAFY COTYLEDON 2 activation is sufficient to trigger the accumulation of oil and seed specific mRNAs in Arabidopsis leaves. FEBS Lett 579:4666–4670

Sargent JR, Coupland K, Wilson R (1994) Nervonic acid and demyelinating disease. Med Hypotheses 42:237–242

Savidge B, Weiss JD, Wong Y-HH, Lassner MW, Mitsky TA, Shewmaker CK, Post-Beittenmiller D, Valentin HE (2002) Isolation and characterization of homogentisate phytyltransferase genes from *Synechocystis* sp. PCC 6803 and Arabidopsis. Plant Physiol 129:321–332

Sayanova OV, Napier JA (2004) Eicosapentaenoic acid: biosynthetic routes and the potential for synthesis in transgenic plants. Phytochemistry 65:147–158

Scarth R, Tang J (2006) Mofification of *Brassica* oil using conventional and transgenic approaches. Crop Sci 46:1225–1236

Schäfer E, Bowler C (2002) Phytochrome-mediated photoperception and signal transduction in higher plants. EMBO Rep 3:1042–1048

Schwab R, Palatnik JF, Riester M, Schommer C, Schmid M, Weigel D (2005) Specific effects of microRNAs on the plant transcriptome. Dev Cell 8:517–527

Scolnik PA, Bartley GE (1994) Nucleotide sequence of an Arabidopsis cDNA for phytoene synthase. Plant Physiol 104:1471–1472

Shen B, Sinkevicius KW, Selinger DA, Tarczynski MC (2006) The homeobox gene *GLABRA2* affects seed oil content in Arabidopsis. Plant Mol Biol 60:377–387

Shewmaker CK, Sheehy JA, Daley M, Colburn S, Ke DY (1999) Seed-specific overexpression of phytoene synthase: increase in carotenoids and other metabolic effects. Plant J 20:401–412

Shi L, Katavic V, Yu Y, Kunst L, Haughn G (2012) Arabidopsis *glabra2* mutant seeds deficient in mucilage biosynthesis produce more oil. Plant J 69:37–46

Shintani DK, DellaPenna D (1998) Elevating the vitamin E content of plants through metabolic engineering. Science 282:2098–2100

Shockey JM, Gidda SK, Chapital DC, Kuan J-C, Dhanoa PK, Bland JM et al (2006) Tung tree DGAT1 and DGAT2 have nonredundant functions in triacylglycerol biosynthesis and are localized to different subdomains of the endoplasmic reticulum. Plant Cell 18:2294–2313

Skirycz A, Reichelt M, Burow M et al (2006) DOF transcription factor AtDof1.1 (OBP2) is part of a regulatory network controlling glucosinolate biosynthesis in Arabidopsis. Plant J 47:10–24

Smith MA, Moon H, Chowrira G, Kunst L (2003) Heterologous expression of a fatty acid hydroxylase gene in developing seeds of *Arabidopsis thaliana*. Planta 217:507–516

Snapp AR, Lu C (2012) Engineering industrial fatty acids in oilseeds. Front Biol. doi:10.1007/s11515-012-1228-9

Snyder C, Yurchenko O, Siloto R, Chen X, Liu Q, Mietkiewska E et al (2009) Acyltransferase action in the modification of seed oil biosynthesis. New Biotechnol 26:11–16

Stålberg K, Lindgren O, Ek B, Höglund A-S (2003) Synthesis of ketocarotenoids in the seed of *Arabidopsis thaliana*. Plant J 36:771–779

Statistics Canada.gc.ca [INTERNET] Statistics Canada, Stocks of principal field crops (2012) 31 July 2012 [Date modified 7 Sep 2012; [cited 2012 Sep 10]. Available from http://www.statcan.gc.ca/daily-quotidien/120907/dq120907d-eng.htm

Stobart K, Mancha M, Lenman M, Dahlqvist A, Stymne S (1997) Triacylglycerols are synthesized and utilized by transacylation reactions in microsomal preparations of developing safflower (*Carthamus tinctorius* L.) seeds. Planta 203:58–66

Stoll C, Luhs W, Zarhloul MK, Friedt W (2005) Genetic modification of saturated fatty acids in oilseed rape (*Brassica napus*). Eur J Lipid Sci Technol 107:244–248

Stone SL, Kwong LW, Yee KM, Pelletier G, Lepiniec L, Fisher RL et al (2001) *LEAFY COTYLEDON2* encodes a B3 domain transcription factor that induces embryo development. Proc Natl Acad Sci USA 98:11806–11811

Stoutjesdijk PA, Hurlstone CJ, Singh SP, Green AG (2000) High-oleic acid Australian *Brassica napus* and *B. juncea* varieties produced by co-suppression of endogenous Δ12-desaturases. Biochem Soci Trans 28:938–940

Suh MC, Schultz DJ, Ohlrogge JB (2002) What limits production of unusual monoenoic fatty acids in transgenic plants? Planta 215:584–595

Tan H, Yang X, Zhang F, Zheng X, Qu C, Mu J et al (2011) Enhanced seed oil production in canola by conditional expression of *Brassica napus LEAFY COTYLEDON1* and *LEC1-LIKE* in developing seeds. Plant Physiol 156:1577–1588

Taylor DC, Katavic V, Zou J-T, MacKenzie SL, Keller WA, An J et al (2002) Field testing of transgenic rapeseed cv. Hero transformed with a yeast *sn-2* acyltransferase results in increased oil content, erucic acid content and seed yield. Mol Breeding 4:317–322

Taylor DC, Francis T, Guo Y, Brost JM, Katavic V, Mietkiewska E et al (2009a) Molecular cloning and characterization of a *KCS* gene from *Cardamine graeca* and its heterologous expression in *Brassica* oilseeds to engineer high nervonic acid oils for potential medical and industrial use. Plant Biotechnol J 7:925–938

Taylor DC, Zhang Y, Kumar A, Francis T, Giblin EM, Barton DL et al (2009b) Molecular modification of triacylglycerol accumulation by over-expression of DGAT1 to produce canola with increased seed oil content under field conditions. Botany 87:533–543

Taylor DC, Falk KC, Palmer CD, Hammerlindl J, Babic V, Mietkiewska E et al (2010) *Brassica carinata* – a new molecular farming platform for delivering bio-industrial oil feedstocks: case studies of genetic modifications for improvement in seed very long-chain fatty acids and oil content. Biofuel Bioprod Bior 4:538–561

Taylor DC, Smith MA, Fobert P, Mietkiewska E, Weselake RJ (2011) Plant systems. Metabolic engineering of higher plants to produce bio-industrial oils. In: Moo-Young M (ed) Comprehensive biotechnology, vol 4, 2nd edn. Elsevier, Amsterdam/Boston, pp 67–85

Thelen JJ, Ohlrogge JB (2002) Both antisense and sense expression of biotin carboxyl carrier protein isoform 2 inactivates the plastid acetyl-coenzyme A carboxylase in *Arabidopsis thaliana*. Plant J 32:419–431

Trevaskis B, Watts RA, Andersson CR, Llewellyn DJ, Hargrove MS, Olson JS, Dennis ES, Peacock WJ (1997) Two hemoglobin genes in *Arabidopsis thaliana*: the evolutionary origins of leghemoglobins. Proc Natl Acad Sci U S A 94:12230–12234

Tsegaye Y, Shintani DK, DellaPenna D (2002) Overexpression of the enzyme *p-hydroxyphenolpyruvate* dioxygenase in Arabidopsis and its relation to tocopherol biosynthesis. Plant Physiol Biochem 40:913–920

Ursin VM (2003) Modification of plant lipids for human health: development of functional land-based omega-3 fatty acids. J Nutr 133:4271–4274

van de Loo FJ, Broun P, Turner S, Somerville C (1995) An oleate 12-hydroxylase from *Ricinus communis* L. is a fatty acyl desaturase homolog. Proc Natl Acad Sci U S A 92:6743–6747

van Eenennaam AL, Lincoln K, Durrett TP, Valentin HE, Shewmaker CK, Thorne GM et al (2003) Engineering vitamin E content: from Arabidopsis mutant to soy oil. Plant Cell 15:3007–3019

van Erp H, Bates PD, Burgal J, Shockey J, Browse J (2011) Castor phospholipid: diacylglycerol acyltransferase facilitates efficient metabolism of hydroxy fatty acids in transgenic Arabidopsis. Plant Physiol 155:683–693

Venkatesh TV, Karunanandaa B, Free DL, Rottnek JM, Baszis VJE (2006) Identification and characterization of an Arabidopsis homogentisate phytyltransferase paralog. Planta 223:1134–1144

Vigeolas H, Geigenberger P (2004) Increased levels of glycerol-3-phosphate lead to a stimulation of flux into triacylglycerol synthesis after supplying glycerol to developing seeds of *Brassica napus* L. in planta. Planta 219:827–835

Vigeolas H, van Dongen JT, Waldeck P, Hühn D, Geigenberger P (2003) Lipid storage metabolism is limited by the prevailing low oxygen concentrations within developing seeds of oilseed rape. Plant Physiol 133:2048–2060

Vigeolas H, Waldeck P, Zank T, Geigenberger P (2007) Increasing seed oil content in oil-seed rape (*Brassica napus* L.) by over-expression of a yeast glycerol-3-phosphate dehydrogenase under the control of a seed-specific promoter. Plant Biotechnol J 5:431–441

Vigeolas H, Hühn D, Geigenberger P (2011) Nonsymbiotic hemoglobin-2 leads to an elevated energy state and to a combined increase in polyunsaturated fatty acids and total oil content when overexpressed in developing seeds of transgenic Arabidopsis plants. Plant Physiol 155:1435–1444

Voelker TA, Kinney AJ (2001) Variations in the biosynthesis of seed storage lipids. Annu Rev Plant Physiol 52:335–361

Voelker TA, Hayes TR, Cranmer AC, Turner JC, Davies HM (1996) Genetic engineering of a quantitative trait: metabolic and genetic parameters influencing the accumulation of laurate in rapeseed. Plant J 9:229–241

Voelker TA, Jones A, Cranmer AM, Davies HM, Knutzon DS (1997) Broad-range and binary-range acyl-acyl-carrier protein thioesterases suggest an alternative mechanism for medium-chain production in seeds. Plant Physiol 114:669–677

Vogel G, Browse J (1996) Cholinephosphotransferase and diacylglycerol acyltransferase. Plant Physiol 110:923–931

Wakao S, Andre C, Benning C (2008) Functional analyses of cytosolic glucose-6-phosphate dehydrogenases and their contribution to seed oil accumulation in Arabidopsis. Plant Physiol 146:277–288

Wallis JG, Browse J (2010) Lipid biochemists salute the genome. Plant J 61:1092–1106

Wang H, Guo J, Lambert KN, Lin Y (2007a) Developmental control of Arabidopsis seed oil biosynthesis. Planta 226:773–783

Wang HW, Zhang B, Hao YJ, Huang J, Tian AG, Liao Y et al (2007b) The soybean Dof-type transcription factor genes, GmDof4 and GmDof11, enhance lipid content in the seeds of transgenic Arabidopsis plants. Plant J 52:716–729

Warwick SI (2011) Brassicaceae in agriculture. In: Schmidt R, Bancroft I (eds) Genetics and genomics of the Brassicaceae, plant genetics and genomics: crops and models. Springer Science, New York, pp 33–65

Wei S, Li X, Gruber MY, Li R, Zhou R, Zebarjadi A, Hannoufa A (2009) RNAi-mediated suppres-
     sion of *DET1* alters the levels of carotenoids and sinapate esters in seeds of *Brassica napus*.
     J Agric Food Chem 57:5326–5333

Wei S, Yu B, Gruber MY, Khachatourians GG, Hegedus DD, Hannoufa A (2010) Enhanced seed
     carotenoid levels and branching in transgenic *Brassica napus* expressing the Arabidopsis
     miR156b gene. J Agric Food Chem 58:9572–9578

Weier D, Hanke C, Eickelkamp A, Lühs W, Dettendorfer J, Schaffert E et al (1997) Trierucoylglycerol
     biosynthesis in transgenic plants of rapeseed (*Brassica napus* L.). FETT-Lipid 99:160–165

Weselake RJ, Shah S, Tang M, Quant PA, Snyder CL, Furukawa-Stoffer TL et al (2008) Metabolic
     control analysis is helpful for informed genetic manipulation of oilseed rape (*Brassica napus*)
     to increase seed oil content. J Exp Bot 59:3543–3549

Weselake RJ, Taylor DC, Rahman MH, Shah S, Laroche A, McVetty PBE, Harwood JL (2009)
     Increasing the flow of carbon into seed oil. Biotechnol Adv 27:866–878

Wiberg E, Edwards P, Byrne J, Stymne S, Dehesh K (2000) The distribution of caprylate, caprate
     and laurate in lipids from developing and mature seeds of transgenic *Brassica napus* L. Planta
     212:33–40

Wu G, Truksa M, Datla N, Vrinten P, Bauer J, Zank T et al (2005) Stepwise engineering to produce
     high yields of very long-chain polyunsaturated fatty acids in plants. Nat Biotechnol
     23:1013–1017

Xu J, Francis T, Mietkiewska E, Gibline EM, Barton DL, Zhang Y et al (2008) Cloning and char-
     acterization of an acyl-CoA-dependent *diacylglycerol acyltransferase 1* (*DGAT1*) gene from
     Tropaeolum majus, and a study of the functional motifs of the DGAT protein using site-directed
     mutagenesis to modify enzyme activity and oil content. Plant Biotechnol J 6:799–818

Xu K, Yang Y, Li X (2010) Ectopic expression of *Crambe abyssinica* lysophosphatidic acid acyl-
     transferase in transgenic rapeseed increases its oil content. Afr J Biotechnol 9:3904–3910

Yang Q, Fan C, Guo Z, Qin J, Wu J, Li Q et al (2012) Identification of FAD2 and FAD3 genes in
     Brassica napus genome and development of allele-specific markers for high oleic and low lino-
     lenic acid contents. Theor Appl Genet 125:715–729

Ytterberg A, Peltier J-B, van Wijk KJ (2006) Protein profiling of plastoglobules in chloroplasts and
     chromoplasts. A surprising site for differential accumulation of metabolic enzymes. Plant
     Physiol 140:984–997

Yu B, Lydiate DJ, Young LW, Schäfer UA, Hannoufa A (2008) Enhancing the carotenoid content
     of *Brassica napus* seeds by downregulating lycopene epsilon cyclase. Transgenic Res
     17:573–585

Yusuf MA, Sarin NB (2007) Antioxidant value addition in human diets: genetic transformation of
     Brassica juncea with γ-TMT gene for increased α-tocopherol content. Transgenic Res
     16:109–113

Zhang M, Fan JL, Taylor DC, Ohlrogge JB (2009) DGAT1 and PDAT1 acyltransferases have
     overlapping functions in Arabidopsis triacylglycerol biosynthesis and are essential for normal
     pollen and flower development. Plant Cell 21:3885–3901

Zhang L, Tan Q, Lee R, Trethewy A, Lee Y-H, Tegeder M (2010) Altered xylem-phloem transfer
     of amino acids affects metabolism and leads to increased seed yield and oil content in
     Arabidopsis. Plant Cell 22:3603–3620

Zhang X, Spiegelman D, Baglietto L, Bernstein L, Boggs DA, van den Brandt PA et al (2012)
     Carotenoid intakes and risk of breast cancer defined by estrogen receptor and progesterone
     receptor status: a pooled analysis of 18 prospective cohort studies. Am J Clin Nutr
     95:713–725

Zheng Z, Xia Q, Cauk M, Shen W, Selvaraj G, Zou J (2003) Arabidopsis *AtGPAT1*, a member of
     the membrane-bound glycerol-3-phosphate acyltransferase gene family, is essential for tape-
     tum differentiation and male fertility. Plant Cell 15:1872–1887

Zheng Q, Li JQ, Kazachkov M, Liu K, Zou J (2012) Identification of *Brassica napus* lysophospha-
     tidylcholine acyltransferase genes through yeast functional screening. Phytochemistry
     75:21–31

Zhu C, Naqvi S, Capell T, Christou P (2009) Metabolic engineering of ketocarotenoid biosynthesis in higher plants. Biochem Biophys 483:182–190

Zou J, Katavic V, Giblin EM, Barton DL, MacKenzie SL, Keller WA et al (1997) Modification of seed oil content and acyl composition in the Brassicaceae by expression of a yeast *sn*-2 acyl-transferase gene. Plant Cell 9:909–923

Zou J, Qi Q, Katavic V, Marillia E-F, Taylor DC (1999) Effects of antisense repression of an *Arabidopsis thaliana* pyruvate dehydrogenase kinase cDNA on plant development. Plant Mol Biol 41:837–849

Chapter 8
Metabolism and Detoxification of Phytoalexins from Crucifers and Application to the Control of Fungal Plant Pathogens

M. Soledade C. Pedras

Abstract This chapter reviews the chemical structures of phytoalexins of crucifers, elicitors and plant sources, biological activities and metabolic pathways, with special focus on biosynthesis from primary building blocks and biotransformation by fungal pathogens. A new strategy that uses paldoxins to control phytopathogenic fungi is discussed.

Keywords Brassicaceae • Cruciferae • Brassinin • Camalexin • Cyclobrassinin • Detoxifying enzyme • Erucalexin • Paldoxin • Phytoalexin • Phytoanticipin

Abbreviations

DAD	Diode array detector
ESI	Electrospray ionization
GLCB	Glucobrassicin
HPLC	High performance/pressure liquid chromatograph/y
IAN	Indolyl-3-acetonitrile
IAO	Indolyl-3-acetaldoxime
IMIT	Indole-3-methylisothiocyanate
MS	Mass spectrometry
NMR	Nuclear magnetic resonance
UPLC	Ultra high pressure liquid chromatograph/y
UV	Ultraviolet

M.S.C. Pedras, Lic, Ph.D., D.Sc. (✉)
University of Saskatchewan, S7N 5C9 Saskatoon, Saskatchewan, Canada
e-mail: s.pedras@usask.ca

S.K. Gupta (ed.), *Biotechnology of Crucifers*, DOI 10.1007/978-1-4614-7795-2_8,
© Springer Science+Business Media, LLC 2013

8.1 Introduction

Crucifers (Brassicaceae) have all sorts of applications, from edible oils and vegetables to condiments, fodder, ornamentals, industrial oils and fuels (Warwick 2011). In addition, some wild crucifers such as *Arabidopsis thaliana* and *Thellungiella salsuginea* are used as model systems in scientific research. For these reasons, there are numerous research publications dealing with cruciferous species. In this article, as the title suggests, I will summarize work covering mainly the metabolism of phytoalexins of crucifers related with their biosynthesis and biotransformation by plant pathogens. A comprehensive review of cruciferous phytoalexins has appeared recently (Pedras et al. 2011).

Plants and many living organisms including microbes and animals produce an incredible array of secondary metabolites, also known as natural products (Hartmann 2007). Such products are characteristic of individual groups of organisms and are important to the overall fitness of these organisms. Secondary metabolites within each group of organisms (e.g., genera, species) are normally classified according to their ecological roles, as for example phytoalexin and phytoanticipin. Unfortunately, these terms are not used consistently across the various research areas dealing with natural products, causing some confusion and misunderstandings. For this reason, this review will point out situations within the Brassicaceae where the terms phytoalexin and phytoanticipin have been misused, with the intent of preventing future misconceptions and contributing to a better understanding of the function and classification of phytoalexins and phytoanticipins.

8.2 Plant Defense Metabolites

Plants produce mainly two groups of secondary metabolites involved in defense, phytoalexins (Bailey and Mansfield 1982) and phytoanticipins. The current definition of phytoalexins states that these plant metabolites are low molecular weight antimicrobial compounds produced *de novo* in response to pathogen attack. By contrast, phytoanticipins are constitutive low molecular weight antimicrobial compounds, although their concentration may increase when plants are under stress (VanEtten et al. 1994). Phytoalexins and phytoanticipins may also be produced under other types of stress including exposure to UV radiation and heavy salts. Although only phytoalexins are the subject of this review, it is pertinent to mention that the cruciferous metabolites known as glucosinolates are considered phytoanticipins because they are constitutive metabolites and precursors of various metabolites with antimicrobial activity such as their corresponding isothiocyanates and nitriles. However, contrary to certain misconceptions (Nongbri et al. 2012), it is important to emphasize that glucosinolates are not phytoalexins. Phytoalexins are not present in healthy plant tissues, their primary precursor(s) is recruited into the phytoalexin biosynthetic pathway after elicitation by biotic or abiotic elicitors.

The clear distinction between the definitions of phytoalexin and phytoanticipin does not prevent a plant metabolite from being a phytoalexin (elicited) in a certain species and a phytoanticipin (constitutive) in another species, as in the case of indolyl-3-acetonitriles. This is not a failure of the definitions, but an outcome reflecting the evolution of secondary metabolism in different groups or organisms. Although the ecological significance and defensive roles of phytoalexins have been widely demonstrated in various plant families, challenges continue to exist making this research area a great opportunity for novel discoveries with enormous potential applications (Essenberg 2001).

8.3   Cruciferous Phytoalexins

The detection of phytoalexins can be challenging due the very small quantities produced relative to the amounts of constitutive metabolites. However, with highly sensitive analytical instruments such as high performance liquid chromatographs (HPLC) or ultra high pressure LC (UPLC) with diode array detector (DAD, very good when compounds have chromophores) and mass detector, phytoalexin detection is less problematic, but is highly dependent on the particular plant species and tissue. Mass spectrometry (MS) ionization techniques, as for example electrospray ionization (ESI), are known to yield reasonable molecular or quasi-molecular ions of many phytoalexins. However, when mass detectors are used, it is important to establish the detector responses using authentic phytoalexin standards, which usually are not commercially available (Pedras et al. 2006). Perhaps for this reason, typical metabolomics analyses do not appear to be concerned with this class of metabolites. Regardless of the analytical instrument type used, screening for phytoalexins requires the analysis of both elicited and control tissues in parallel. As well, it is necessary to use reasonably polar solvents to extract quantitatively these metabolites. Comparison of chromatograms of control tissues with those of elicited tissues can show additional peaks due to elicitation, and in some cases a chemical structure can be proposed based on the molecular mass of component and UV data. Nonetheless, a hypothetical chemical structure is insufficient to reveal a new phytoalexin, that is, the antimicrobial activity of any putative phytoalexin must be proven through the use of antimicrobial bioassays. Hence, a milligram or more of the compound is initially required to carry out bioassays before any phytoalexin claim can be published.

   Due to the availability of highly sensitive MS instruments, it is possible to detect ions in elicited tissues that are not present in control tissues. For this reason, there are a few examples in the current literature in which the presence of elicited compounds detected by MS has led to erroneous claims that new phytoalexins were discovered (Böttcher et al. 2009), an unfortunate confusion. As mentioned above, it is likely that a substantial amount of time may be necessary between the detection of a putative phytoalexin and its isolation in sufficient amounts to obtain spectroscopic data (nuclear magnetic resonance, NMR, is in most of the cases crucial) for structure characterization and bioassays. Nonetheless, NMR spectroscopic data may suggest more than one structure, i.e. ambiguity is possible, hence derivatization or total synthesis of the compound may be required to determine its structure unambiguously. An additional analytical method that can provide detection and structural information on phytoalexins, without carrying out compound isolation, involves coupling of HPLC to a MS detector and NMR instrument. This is a very powerful but rather expensive method. In my group, new phytoalexins are published only after each structure has been confirmed by total synthesis (usually, the binomial combination "total synthesis" is technically applicable to larger and more complex carbon skeletons). Synthesis is necessary to confirm the correctness of the structure and to obtain sufficient amounts of each compound to carry out antimicrobial bioassays.

The analyses and screening of phytoalexins in extracts of cruciferous tissues has also been carried out by TLC, with biodetection utilizing spores of *Cladosporium* or *Bipolaris* species (Pedras et al. 2011). This bio-TLC method is sensitive, but time consuming and may not work well in all situations since other antifungal compounds present in control extracts can mask the bioactivity of phytoalexins, as was the case of *Wasabia japonica* (Pedras and Sorensen 1998). Similarly, comparison of HPLC chromatograms of extracts of elicited tissues with those of control samples can also be misleading because complete separation of all metabolites may be difficult. Indeed, some extracts obtained from crucifers are very complex and an overlap of peaks of phytoalexins with peaks of constitutive metabolites can occur. It is necessary to use at least two eluting systems to ensure that a maximum number of components are detected. To date, 45 phytoalexins have been reported from various cruciferous species (Fig. 8.1, Table 8.1). The newest member is the phytoalexin isocyalexin A (Pedras and Yaya 2012), which was discovered after the last comprehensive review of the cruciferous phytoalexins was published (Pedras et al. 2011). Simple inspection of the chemical structures of cruciferous phytoalexins shows an indole or indole-related nucleus with substituents at C-3 (e.g. brassinins and brassenins) or C-2 and C-3 (e.g. cyclobrassinins, brassicanals). In addition, 3-indolylacetonitriles have been reported as phytoalexins in some cruciferous species and phytoanticipins in other species (Pedras et al. 2011; Pedras and Yaya 2010). Chemical syntheses of the 45 cruciferous phytoalexins have been published in reasonable yields, which has facilitated many biological studies (Kutschy and Mezencev 2008; Pedras et al. 2011).

The biological role of phytoalexins is to protect plants from stress. In some cases phytoalexins act as plasma membrane disruptors of microbial pathogens and inhibitors of respiration (Smith 1996). Yet, the great variety of chemical structures suggests that cruciferous phytoalexins have multiple modes of action and that inside a microbial cell may react with several targets (Pedras et al. 2011). The current range of bioassays used to test the bioactivity of phytoalexins allows comparison of the bioactivity of various structures (Table 8.2). In general, phytoalexins are tested at concentrations of 0.1–0.5 mM in agar or liquid medium using mycelial cultures of fungi or bacterial cultures. These assays have shown that the biological activity of each phytoalexin is different, particularly in the case of antifungal activity, but this

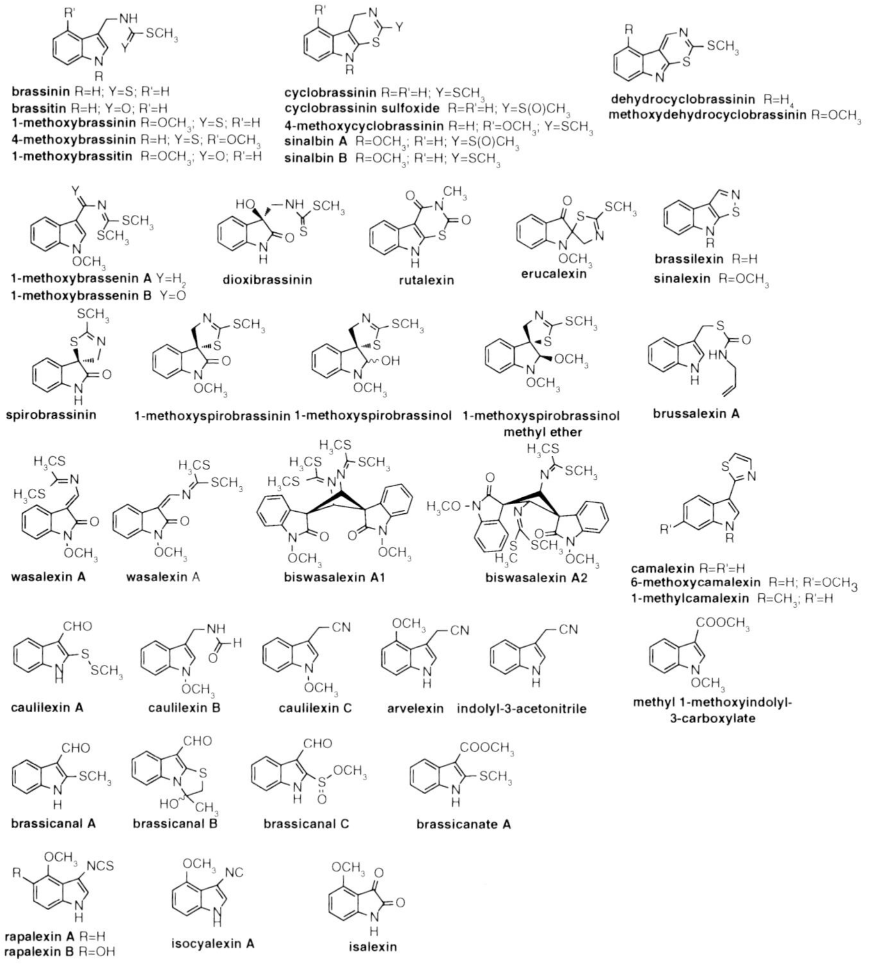

Fig. 8.1 Chemical structures of cruciferous phytoalexins

is not surprising considering their chemical structures. For example, rapalexin A inhibited completely the mycelial growth of *Leptosphaeria maculans* at 0.50 mM (Pedras and Sarma-Mamillapalle 2012), whereas erucalexin and brussalexin A caused only ca. 40 % inhibition at similar concentration. Camalexin also completely inhibited mycelial growth of different plant pathogens at 0.50 mM, including *Alternaria brassicicola*, *Botrytis cinerea* and *L. maculans*. As shown in Table 8.2, some phytoalexins have anticarcinogenic activity against human cell lines. Nonetheless, since phytoalexins are produced in plant tissues in relatively low quantities, it is unlikely that their effects on human or mammalian diets containing naturally produced phytoalexins will have much impact on health, but they might find application as medicinal drugs (Mezencev et al. 2009). Other biological

Table 8.1 Cruciferous species investigated for production of phytoalexins, elicitors and references

Plant species (common name, elicitor)	Phytoalexin (Pedras et al. 2011)
Arabidopsis thaliana (mouse-ear cress, AgNO₃, *Pseudomonas syringae*)	Camalexin Rapalexin A
Arabis lyrata (lyrata rock cress, *Cochliobolus carbonum, P. syringae*)	Camalexin
Brassica adpressa (CuCl₂, *Leptosphaeria maculans*)	1-Methoxybrassinin
B. atlantica (CuCl₂, *L. maculans*)	1-Methoxybrassinin
B. carinata (Abyssinian cabbage, CuCl₂, *L. maculans*)	1-Methoxybrassinin; brassilexin; cyclobrassinin; spirobrassinin
B. juncea (brown mustard, CuCl₂, AgNO₃, *Alternaria brassicae, L. maculans*)	Brassilexin; cyclobrassinin; cyclobrassinin sulfoxide; cyclobrassinin sulfoxide; spirobrassinin; indolyl-3-acetonitrile
B. montana (CuCl₂, *L. maculans*)	1-Methoxybrassinin
B. napus ssp. *oleifera* (canola, *Plasmodiophora brassicae*)	Dehydrocyclobrassinin
B. napus (rapeseed, CuCl₂, *L. maculans*)	1-Methoxybrassinin; brassilexin; cyclobrassinin; cyclobrassinin sulfoxide; spirobrassinin
B. napus ssp. *rapifera* (rutabaga, UV, *Rhizoctonia solani*)	1-Methoxybrassinin; brassicanal A; brassicanate A; brassinin; isalexin; isocyalexin A; spirobrassinin; rutalexin
B. nigra (black mustard, CuCl₂, *L. maculans*)	Brassilexin; cyclobrassinin; cyclobrassinin sulfoxide
B. oleracea (CuCl₂, *L. maculans*)	Brassilexin
B. oleracea var. *capitata* (white cabbage, *P. cichori*, CuCl₂)	1-Methoxybrassenin A; 1-methoxybrassenin B; 1-methoxybrassinin; brassicanal A; brassinin; cyclobrassinin; dioxibrassinin
B. oleraceae var. *botrytis* (cauliflower, UV)	1-Methoxybrassinin; 1-methoxybrassitin; brassicanal C; caulilexin A; caulilexin B; caulilexin C; isalexin; spirobrassinin
B. oleracea var. *capitata* (white cabbage, *P. cichori*)	1-Methoxybrassitin; 4-methoxybrassinin; brassicanal C; spirobrassinin
B. oleracea var. *gemmifera* (Brussels sprouts, UV)	Brussalexin A
B. oleracea var. *gongylodes* (kohlrabi, UV)	1-Methoxybrassinin; 1-methoxybrassitin; 1-methoxyspirobrassinin; spirobrassinin
B. rapa (canola, *Albugo candida*)	Rapalexin A; rapalexin B
B. rapa syn. *B. campestris* (Chinese cabbage, UV, *Erwinia carotovora, P. cichorii*)	1-Methoxybrassinin; 1-methoxybrassitin; brassicanal A; brassicanal B; brassinin; cyclobrassinin; spirobrassinin
B. rapa (rapeseed, CuCl₂)	1-Methoxybrassinin; cyclobrassinin sulfoxide
B. rapa (turnip, UV, *R. solani, L. maculans*)	1-Methoxybrassinin; 4-methoxy dehydrocyclobrassinin; brassilexin; cyclobrassinin; spirobrassinin
Camelina sativa (false flax, *A. brassicae*)	6-Methoxycamalexin; camalexin
Capsella bursa-pastoris (shepherd's purse, *A. brassicae*)	1-Methylcamalexin; 6-methoxycamalexin; camalexin
Crambe abyssinica (Abyssinian mustard, CuCl₂)	Rapalexin B
Diplotaxis tenuifolia (sand rocket, CuCl₂)	Arvelexin

(continued)

Table 8.1 (continued)

Plant species (common name, elicitor)	Phytoalexin (Pedras et al. 2011)
D. muralis (wallrocket, CuCl$_2$)	Rapalexin A; 1,4-dmethoxyindolyl-3-acetonitrile
Erucastrum gallicum (dog mustard, *Sclerotinia sclerotiorum*)	1-Methoxyspirobrassinin; erucalexin
Raphanus sativus var. *hortensis* (Japanese radish, *P. cichorii*)	1-Methoxybrassinin; 1-methoxybrassitin; 1-methoxyspirobrassinol; 1-methoxyspiro-brassinol methyl ether; brassicanal A; brassinin; brassitin; spirobrassinin
Sinapis alba (white mustard, destruxin B, CuCl$_2$, A. *brassicae*, *L. maculans*)	Sinalbin A; sinalbin B; sinalexin
Sinapis arvensis (white mustard, *L. maculans*)	Brassilexin; cyclobrassinin sulfoxide
Sisymbrium officinale (hedge mustard, CuCl$_2$)	Methyl 1-methoxyindole-3-carboxylate
Thlaspi arvense (stinkweed, pennycress, CuCl$_2$, *L. maculans*)	Arvelexin; wasalexin A; wasalexin B
Thellungiella salsuginea (salt cress, UV, CuCl$_2$, NaCl, *L. maculans*)	Biswasalexin A1; biswasalexin A2; wasalexin A; wasalexin B; rapalexin A
Wasabia japonica, syn. Eutrema wasabi (wasabi, CuCl$_2$, *L. maculans*, *P. wasabiae*)	Methyl 1-methoxyindole-3-carboxylate; wasalexin A; wasalexin B

Table 8.2 Cruciferous phytoalexins: biological activity and microbial transformation

Phytoalexin	Biological activity and species	Microbial transformation: organism, product (reference)
1-Methoxybrassenin A	Antimicrobial against *Bipolaris leersiae*	Not reported
1-Methoxybrassenin B	Antimicrobial against *B. leersiae*	Not reported
1-Methoxybrassinin	Antimicrobial against *Pyricularia oryzae, B. leersiae, Alternaria brassicae, Botrytis cinerea, Fusarium nivale, L. maculans, S. sclerotiorum, Pythium ultimum, R. solani, Cladosporium cucumerinum*; antiproliferative; antitrypanosomal	*Sclerotinia sclerotiorum*, 7-O-glucosyl-1-methoxybrassinin (Pedras et al. 2011)
1-Methoxybrassitin	Antimicrobial against *B. leersiae, C. cucumerinum*	Not reported

(continued)

Table 8.2 (continued)

Phytoalexin	Biological activity and species	Microbial transformation: organism, product (reference)
1-Methoxyspirobrassinin	Antimicrobial against *C. cucumerinum, R. solani, S. sclerotiorum*; antiproliferative activity	*S. sclerotiorum*, 1-methoxyspiro[3H-indole-3,5′-thiazolidin]-2(1H),2′-dione; *L. maculans*, 1-methoxy-2′-thioxospiro[3H-indole-3,5′-thiazolidin]-2(1H)-one (Pedras et al. 2011)
1-Methoxyspirobrassinol	Antimicrobial against *B. leersiae*; antiproliferative activity	Not reported
1-Methoxyspirobrassinol methyl ether	Antimicrobial against *B. leersiae*; antiproliferative activity	Not reported
1-Methylcamalexin	Antimicrobial against *C. cucumerinum*	*Rhizoctonia solani*; 2-(1-methyl-3-indolyl)-oxazoline; 1-methylindole-3-carboxamide; 1-methylindole-3-carbonitrile (Pedras et al. 2011)
4-Methoxydehydrocyclobrassinin	Antimicrobial against *B. leersiae*	Not reported
4-Methoxybrassinin	Antimicrobial against *B. leersiae*; oviposition stimulant	Not reported
4-Methoxycyclobrassinin	Antimicrobial against *R. solani, S. sclerotiorum*	Not reported
6-Methoxycamalexin	Antimicrobial against *A. brassicae, C. cucumerinum, P. syringae, R. solani, Bacillus subtilis, Escherichia coli, F. oxysporum, Listeria monocytogenes, Saccharomyces cerevisiae, Xanthomonas campestris, S. sclerotiorum*	*S. sclerotiorum*; 6-O-glucosylcamalexin; 1-glucosyl-6-methoxycamalexin (Pedras et al. 2011)
Arvelexin	Antimicrobial against *A. brassicicola, L. maculans, R. solani, S. sclerotiorum*; mutagen precursor	*A. brassicicola, R. solani, S. sclerotiorum*, 4-methoxyindole-3-carboxylic acid (Pedras et al. 2011)
Biswasalexin A1	Antimicrobial against *L. maculans*	Not reported

Table 8.2 (continued)

Phytoalexin	Biological activity and species	Microbial transformation: organism, product (reference)
Biswasalexin A2	Antimicrobial against *S. sclerotiorum, L. maculans*	Not reported
Brassicanal A	Antimicrobial against *B. leersiae, L. maculans, R. solani, S. sclerotiorum*	*S. sclerotiorum*, 3-hydroxymethylindole-2-methylsulfoxide; *L. maculans*, 3-methylindole-2-methylsulfoxide (Pedras et al. 2011)
Brassicanal B	Antimicrobial against *B. leersiae*	Not reported
Brassicanal C	Antimicrobial against *B. leersiae, L. maculans, R. solani, S. sclerotiorum*	Not reported
Brassicanate A	Antimicrobial against *L. maculans, R. solani, S. sclerotiorum, C. cucumerinum*	Not reported
Brassilexin	Antimicrobial against *L. maculans, A. brassicae, R. solani, S. sclerotiorum;* cytotoxic	*L. biglobosa*, 3-aminomethyleneindole-2-thione; *S. sclerotiorum*, 1-glucosylbrassilexin (Pedras et al. 2011)
Brassinin	Antimicrobial against *P. oryzae, L. maculans, S. sclerotiorum;* oviposition stimulant; antiproliferative; cytotoxic; indoleamine 2,3-dioxygenase inhibitor; antitrypanosomal	*L. maculans*, indole-3-carboxylic acid; *L. biglobosa, A. brassicicola and B. cinerea*, indole-3-methanamine; *S. sclerotiorum*, 1-glucosylbrassinin (Pedras et al. 2011)
Brassitin	Antimicrobial against *B. leersiae;* oviposition stimulant	Not reported
Brussalexin A	Antimicrobial against *S. sclerotiorum*	*L. maculans*, decomposition (Pedras and Sarma-Mamillapalle 2012)

(continued)

Table 8.2 (continued)

Phytoalexin	Biological activity and species	Microbial transformation: organism, product (reference)
Camalexin	Antimicrobial against *A. brassicae, C. cucumerinum, P. syringae, B. subtilis, E. coli, F. oxysporum, L. monocytogenes, S. cerevisiae, X. campestris, Erwinia carotovora, P. cichorii, L. maculans, R. solani, S. sclerotiorum*; antiproliferative; antitrypanosomal	*B. cinerea*, indole-3-carboxylic acid; *R. solani*, 2-formamidophenyl-5-hydroxy-2′-thiazolylketone; 5-hydroxy-indole-3-carbonitrile; *S. sclerotiorum*, 6-O-glucosylcamalexin (Pedras et al. 2011)
Caulilexin A	Antimicrobial against *L. maculans, R. solani, S. sclerotiorum*	Not reported
Caulilexin B	Antimicrobial against *L. maculans, R. solani, S. sclerotiorum*	Not reported
Caulilexin C	Antimicrobial against *L. maculans, R. solani, S. sclerotiorum*	*A. brassicicola, R. solani, S. sclerotiorum*, 1-methoxyindole-3-carboxylic acid (Pedras and Hossain 2011)
Cyclobrassinin	Antimicrobial against *P. oryzae, A. brassicae, B. cinerea, C. cucumerinum, F. nivale, L. maculans, P. ultimum, R. solani, S. sclerotiorum*; cytotoxic; antitrypanosomal	*L. maculans*, dioxibrassinin; *S. sclerotiorum*, 1-glucosyl cyclobrassinin; *R. solani*, 5-hydroxybrassicanal A (Pedras et al. 2011)
Cyclobrassinin sulfoxide	Antimicrobial against *C. cucumerinum*	Not reported
Dehydrocyclobrassinin	Antimicrobial against *R. solani, S. sclerotiorum*	Not reported
Dioxibrassinin	Antimicrobial against *B. leersiae, L. maculans*	*L. maculans*, 2′-thioxospiro[3H-indole-3,5′-oxazolidin]-2(1H)-one (Pedras et al. 2011)
Erucalexin	Antimicrobial against *L. maculans, R. solani, S. sclerotiorum*	*L. maculans*, dihydroerucalexin (Pedras and Sarma-Mamillapalle 2012)

(continued)

Table 8.2 (continued)

Phytoalexin	Biological activity and species	Microbial transformation: organism, product (reference)
Indolyl-3-acetonitrile	Antimicrobial against *A. brassicae, A. brassicicola, L. maculans, R. solani, S. sclerotiorum*	*Beauveria bassiana, L. maculans, R. solani, S. sclerotiorum,* indole-3-acetic acid (Pedras et al. 2011), *A. brassicicola,* indole-3-carboxylic acid (Pedras and Hossain 2011)
Isalexin	Antimicrobial against *L. maculans, R. solani, S. sclerotiorum, C. cucumerinum*	Not reported
Isocyalexin A	Antimicrobial against *A. brassicicola, L. maculans, R. solani, S. sclerotiorum*	Not reported
Methyl 1-methoxyindole-3-carboxylate	Antimicrobial against *C. cucumerinum, L. maculans, P. wasabiae*	Not reported
Rapalexin A	Antimicrobial against *A. candida, L. maculans*	*L. maculans,* not transformed (Pedras and Sarma-Mamillapalle 2012)
Rapalexin B	Antimicrobial against *A. candida*	Not reported
Rutalexin	Antimicrobial against *L. maculans, R. solani, S. sclerotiorum, C. cucumerinum*	Not reported
Sinalbin A	Antimicrobial against *L. maculans*	Not reported
Sinalbin B	Antimicrobial against *L. maculans*	Not reported
Sinalexin	Antimicrobial against *C. cucumerinum; A. brassicae, R. solani, S. sclerotiorum, L. maculans*	*L. maculans,* 1-methoxy-3-amino-methyline-2-indolinethione; *S. sclerotiorum,* 7-O-glucosylsinalexin (Pedras et al. 2011)
Spirobrassinin	Antimicrobial against *P. oryzae; C. cucumerinum, L. maculans, S. sclerotiorum;* cytotoxic	*S. sclerotiorum,* spiro[3H-indole-3,5′-thiazolidin]-2(1H),2′-dione; *L. maculans;* 2′-thioxospiro[3H-indole-3,5′-thiazolidin]-2(1H)-one (Pedras et al. 2011)
Wasalexin A	Antimicrobial against *L. maculans, P. wasabiae*	*L. maculans,* 3-methyl-1-methoxy-2-oxindole (Pedras et al. 2011)
Wasalexin B	Antimicrobial against *L. maculans, P. wasabiae*	*L. maculans,* S-methyl 1-methoxy-3-aminomethyl-2-oxindole dithiocarbamate (Pedras and Suchy 2006)

effects such as antitrypanosomal (Mezencev et al. 2009) and inhibitors of fungal detoxifying enzymes such as brassinin oxidase (Pedras et al. 2008) and brassinin hydrolase (Pedras et al. 2009) have been reported. The later examples will be discussed in Sect. 10.2.2 as they represent recently discovered effects of phytoalexins. On the other hand, camalexin was found to be an inducer of brassinin oxidase and brassinin hydrolases in fungal cultures, an interesting effect that has facilitated the isolation of inducible fungal enzymes (Pedras et al. 2005, 2008, 2009). Phytoalexin toxicity to plant tissues is also expected, but has not been systematically determined; potential synergistic activities of phytoalexin blends are likely to exist, but not simple to quantify.

8.4   Metabolism of Phytoalexins

An understanding of secondary metabolic pathways of plants must include intermediates, final products, enzymes and genes involved in all steps, as well as signalling and regulatory elements. Biosynthetic pathways that involve numerous products, as in the case of cruciferous phytoalexins can be particularly complex. Different approaches are used to dissect biosynthetic pathways, but complementary biosynthetic investigations can start from either the metabolite end, using isotopically labeled intermediates, or from the gene cloning end, using molecular genetics methodologies. As discussed below, to date genes of the camalexin pathway have been cloned, but not from any other phytoalexins. This is not surprising since camalexin is the major phytoalexin produced in *A. thaliana*, and mutants that can facilitate the task are available, but not for other crucifers. Since *T. salsuginea* produces a much larger variety of phytoalexins than *A. thaliana*, it appears to be the ideal model to uncover phytoalexin biosynthetic genes, as several metabolic intermediates and products are known (Pedras et al. 2010). It is expected that development of mutants of *T. salsuginea* will greatly facilitate future gene cloning. For this reason, this area of crucifer research is expected to advance greatly in the next decade.

8.4.1   Biosynthesis

Cruciferous phytoalexins are alkaloids having as primary building block the amino acid L-tryptophan (Trp). Elicitation by an external stimulus, called "elicitor", is necessary for initiation of phytoalexin biosynthesis. In crucifers, a wide variety of elicitors have been used for phytoalexin elicitation, as summarized in Table 8.1. The elicitors of phytoalexins on their own have generated great interest, due to potential applications in crop protection (Walters et al. 2005). Oxidative decarboxylation of L-Trp to indolyl-3-acetaldoxime (IAO) is a step common to the biosynthesis of camalexin and most cruciferous phytoalexins. The pathway from IAO to camalexins involves indolyl-3-acetonitrile (IAN) as intermediate, while brassinins and

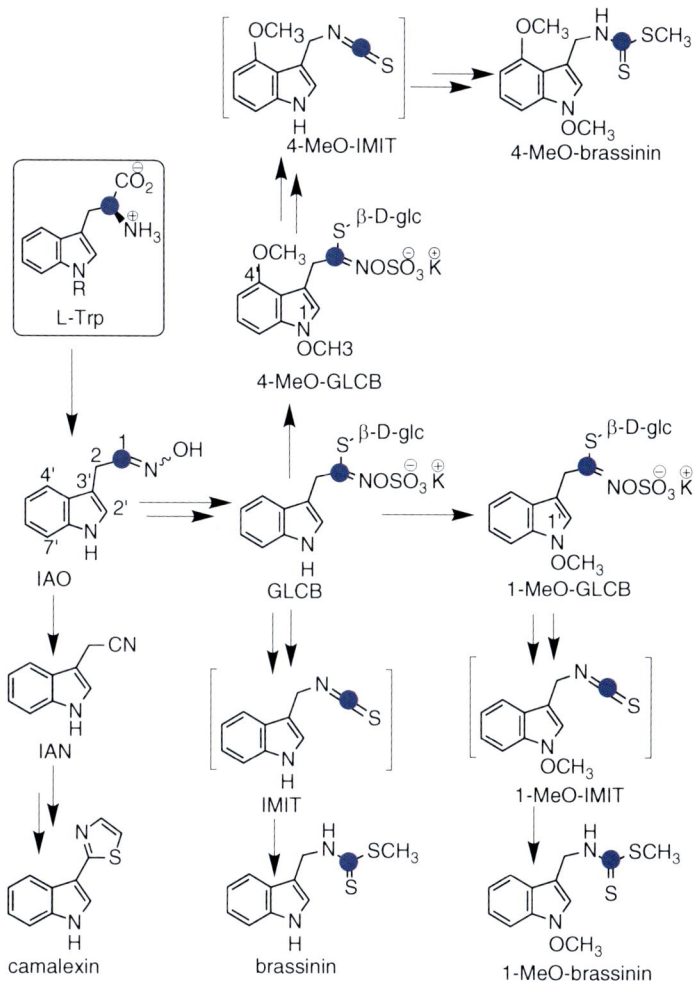

Fig. 8.2 Biosynthetic correlations among L-Trp, IAO, IAN, glucobrassicins (GLCBs), IMITs, brassinin, IAN and camalexin (two *arrows* indicate multiple steps, *square brackets* indicate unstable intermediates)

related are synthesized via glucobrassicins. The biosynthetic relationships among the various crucifer phytoalexins have been established using isotopically labeled precursors containing deuterium (^{2}H), carbon-13 (^{13}C), carbon-14 (^{14}C), sulfur-34 (^{34}S) and sulfur-35 (^{35}S) (Pedras et al. 2011). Based on data obtained over many years in different research groups (Pedras et al. 2011), a biosynthetic scheme is shown in Fig. 8.2; the transformation of L-Trp to glucobrassicins (GLCBs) is not included in this article. Indole-3-methylisothiocyanate (IMIT) is a likely intermediate between glucobrassicin and brassinin, however due to its high reactivity it has

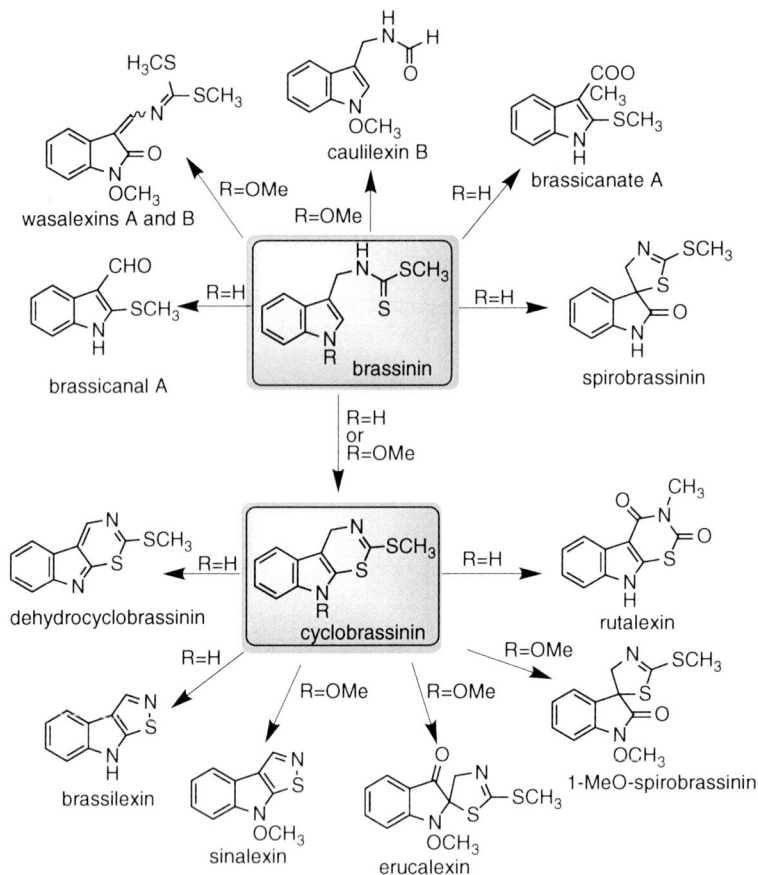

Fig. 8.3 Biosynthetic correlations among cruciferous phytoalexins brassinin, cyclobrassinin and derivatives

not been isolated (Pedras et al. 2011). Similar pathways for 1-methoxybrassinin and 4-methoxybrassinin via the corresponding 1-methoxy or 4-methoxy IMITs are also shown (Pedras et al. 2011). Brassinins and cyclobrassinins are the biosynthetic precursors of a good number of phytoalexins, hence important on their own right to afford diversity in the phytoalexin content of crucifers. While rutalexin is derived from cyclobrassinin, brassicanate A appears to derive from brassinin. Interestingly, both 1-methoxyspirobrassinin and erucalexin, a unique structure among all metabolites since it has the carbon substituent at C-2 rather than at C-3, are both derived from 1-methoxybrassinin (Fig. 8.3). The biosynthetic precursors of brussalexin A, rapalexins A and B, isocyalexin A and isalexin are not known yet, although rapalexin A was shown to incorporate perdeuterated Trp, hence is likely to be derived from an yet unknown 4-methoxy precursor downstream from L-Trp.

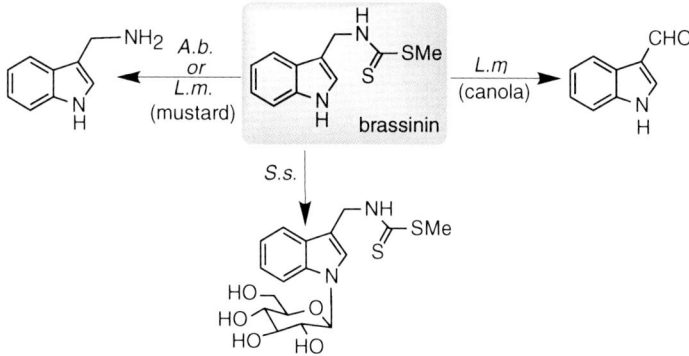

Fig. 8.4 Detoxification of the cruciferous phytoalexin brassinin by four plant fungal pathogens: *A. brassicicola (A.b.)*, *L. maculans (L.m.)* virulent on canola, *L. maculans (L.m.)* virulent on mustard and *S. sclerotiorum (S.s.)*

8.4.2 Detoxification by Cruciferous Fungal Pathogens

Crucifers are susceptible to microbial diseases, namely fungal diseases cause the largest yield losses and their control requires both application of fungicides and crop rotations on a regular basis (Russell 2005). In some cases, the susceptibility of cruciferous species to fungal diseases appears to be related with detoxification of their phytoalexins (VanEtten et al. 1989). It has been shown that plant pathogenic fungi are able to detoxify plant defenses and that these reactions correlated with the virulence of the particular species (Pedras and Ahiahonu 2005). Some very economically significant cruciferous pathogens such as *Alternaria brassicicola* (Schwein.) Wiltshire, *Botrytis cinerea* Pers. Fr. (teleomorph *Botryotinia fuckeliana* (de Bary) Whetzel), *Leptosphaeria maculans* (Desm.) Ces. et Not. [asexual stage *Phoma lingam* (Tode ex Fr.) Desm.], *L. biglobosa*, *Rhizoctonia solani* Kuhn and *Sclerotinia sclerotiorum* (Lib.) de Bary were shown to detoxify cruciferous phytoalexins using different enzymes. This work suggests that these fungal transformations depend on the fungal species, which appear to produce inducible enzymes with relatively high substrate specificity. In addition to the detoxification processes summarized below, transformation of other phytoalexins shown in Table 8.2 have been analyzed and others are under investigation.

Brassinin was detoxified by four plant pathogens to different products, except in the case of *A. brassicicola* and *L. maculans* virulent on mustard, which afforded the same product, as shown in Fig. 8.4 (Table 8.2). It is particularly curious to find that the four fungal enzymes catalyzing these transformation (BHAb from *A. brassicicola*, BOLm from *L. maculans* virulent on canola, BHLm from *L. maculans* virulent on mustard, and SsBGT1 from *S. sclerotiorum*) were all inducible (Pedras et al. 2011). That is no detoxification activity was detected in cell-free extracts of mycelia not induced with phytoalexins or related structures. In addition, although BHAb and

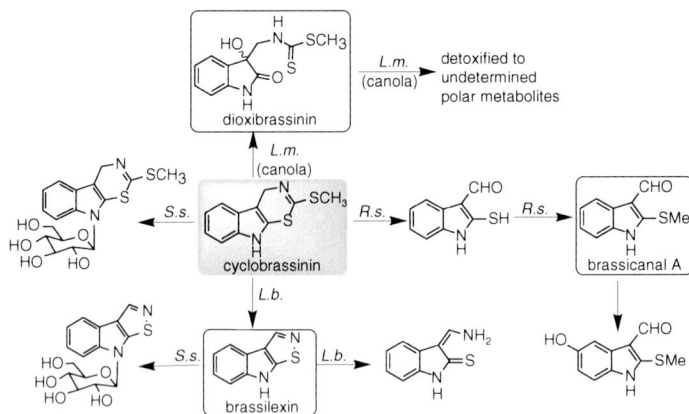

Fig. 8.5 Detoxification of the cruciferous phytoalexin cyclobrassinin by four plant fungal pathogens: *L. biglobosa (L.b.)*, *L. maculans (L.m.)* virulent on canola, *R. solani (R.s.)* and *S. sclerotiorum (S.s.)*

BHLm catalyze the same reaction (hydrolysis of brassinin to the corresponding amine) they have different properties and substrate specificities, suggesting that these enzymes have a specific purpose.

Similar to brassinin, the metabolism of cyclobrassinin by different fungal species was found to yield different products (Fig. 8.5, Table 8.2). Surprisingly, some of these products were shown to be also phytoalexins (brassilexin, dioxibrassinin and brassicanal A), all of which were further metabolized to nontoxic products. Because brassilexin was more inhibitory to *L. biglobosa* than cyclobrassinin, it was suggested that *L. biglobosa* might have acquired a more effective mechanism for detoxifying cyclobrassinin using biosynthetic enzymes from *planta* (Pedras et al. 2011). Although exciting, this hypothesis remains to be demonstrated since none of these enzymes from either the plant or the pathogen have been isolated. *S. sclerotiorum* detoxified cyclobrassinin and brassilexin to the glucosylated derivatives, as in the case of brassinin. For this reason, it was concluded that detoxification of the most inhibitory cruciferous phytoalexins by *S. sclerotiorum* was carried out by glucosyl transferases (Pedras and Hossain 2006).

The phytoalexin camalexin, which is produced only by wild cruciferous species, was metabolized by *B. cinerea, R. solani* and *S. sclerotiorum*, but not by *A. brassicae* or *L. maculans*. As in the case of brassinin and cyclobrassinin, the products of transformation of camalexin by each fungal species were different (Fig. 8.6, Table 8.2) and found to be significantly less toxic to these fungal species than camalexin itself. It is noteworthy that, contrary to direct glucosylation of the indole nitrogen observed during detoxification of the phytoalexins brassinin, cyclobrassinin and brassilexin, *S. sclerotiorum* oxidized C-6 of camalexin before glucosylation (Pedras et al. 2011).

In related work, the metabolism of glucobrassicins and their derivatives was recently investigated to establish pathways by which fungal pathogens interact with

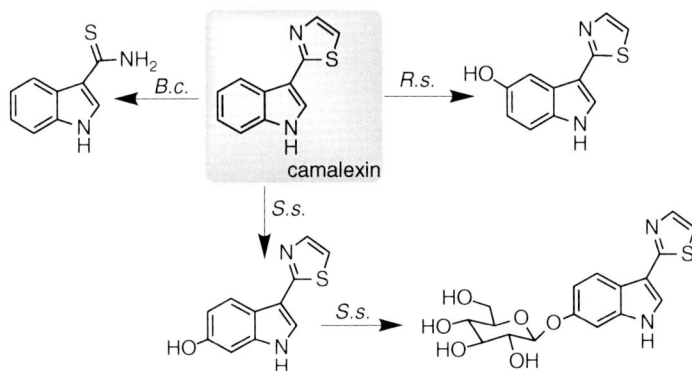

Fig. 8.6 Detoxification of the cruciferous phytoalexin camalexin by three plant fungal pathogens: *B. cinerea (B.c.), R. solani (R.s.)* and *S. sclerotiorum (S.s.)*

these phytoanticipins (Pedras and Hossain 2011). Despite much work and substantial speculation in this area, it was the first time that such transformations were directly analyzed. The metabolism of indolyl glucosinolates, their corresponding desulfo-derivatives and derived metabolites, by three fungal species pathogenic on crucifers showed that glucobrassicin, 1-methoxyglucobrassicin, 4-methoxyglucobrassicin were not metabolized by *A. brassicicola*, *R. solani* and *S. sclerotiorum*. However, the corresponding desulfo-derivatives were metabolized to indolyl-3-acetonitrile, cauli-lexin C (1-methoxyindolyl-3-acetonitrile) and arvelexin (4-methoxyindolyl-3-aceto-nitrile) by *R. solani* and *S. sclerotiorum*, but not by *A. brassicicola*. As summarize in Table 8.2, indolyl-3-acetonitrile, caulilexin C and arvelexin were metabolized to the corresponding indole-3-carboxylic acids. Indolyl-3-acetonitriles display higher inhibitory activity than glucobrassicins and desulfoglucobrassicins, hence the trans-formations of the latter are not detoxification reactions.

8.5 Paldoxins: Designer Molecules to Control of Plant Pathogens

As discussed above, crucifers like other plants synthesize phytoalexins to protect themselves against pathogen attack, while to facilitate invasion, fungi produce enzymes that metabolize and detoxify these metabolites. On going elucidation of these detoxification mechanisms, followed by isolation and characterization of the enzymes responsible for these processes have provided a better understanding of the interactions between crucifers and their pathogens (Pedras et al. 2011). Importantly, these detoxifications appear to be catalyzed by enzymes that are both substrate and pathogen specific. Hence, based on work carried out over the past two decodes, a

Fig. 8.7 Paldoxins, inhibitors designed to disrupt the cycle crucifers-phytoalexins–pathogenic fungi-detoxifying enzymes

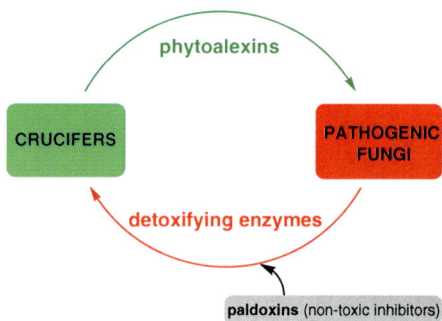

new strategy to control fungal pathogens has emerged. Since specific fungal pathogens produce specific enzymes to detoxify phytoalexins, it follows that selective inhibitors of these enzymes could be designed to prevent phytoalexin degradation *in planta*. These **phyto**alexin **det**oxyfication **in**hibitors were coined **paldoxins** (Fig. 8.7) (Pedras et al. 2003). Furthermore, it is possible to optimize mixtures of paldoxins to act synergistically with the natural disease resistance factors of plants, including their phytoalexins and/or phytoanticipins, to be more effective crop protection agents with minimal environmental impact (Pedras 2004). It is envisioned that the ideal paldoxins will be selective inhibitors to allow accumulation of a plant's own defenses and lead to an environmentally safer control of pathogens like *L. maculans* or *A. brassicicola*. Selective inhibitors are less likely to affect non-targeted organisms and thus are anticipated to have lower impact on the cultivated ecosystem. As such, paldoxins cannot be toxic to living organisms, whether animal, plant or microbe.

The design, synthesis, biological activity, screening and evaluation of potential brassinin and brassilexin detoxification inhibitors have been reported (Pedras et al. 2011). Potential inhibitors of brassinin detoxification by *L. maculans* were designed by replacement of the dithiocarbamate group with various functional groups (carbamate, dithiocarbonate, urea, thiourea, sulfamide, sulfonamide and dithiocarbazate) and the indolyl moiety was replaced with naphthalenyl and phenyl (Pedras and Jha 2006; Pedras et al. 2007). Screening of this potential paldoxin library of brassinin analogues with cultures of *L. maculans* and purified enzyme BOLm showed that none of these compounds inhibited BOLm, although many were not degraded either (Pedras et al. 2008). Ironically, several of these synthetic compounds displayed stronger antifungal activity than brassinin (Pedras and Jha 2006). Most rewardingly, during that work it was discovered that the cruciferous phytoalexins cyclobrassinin and camalexin were competitive inhibitors of BOLm (Pedras et al. 2008). Hence, an additional role for phytoalexins was discovered that suggests partly why plants under stress produce complex phytoalexin blends instead of one or two compounds. Since that work was published, cyclobrassinin was found to inhibit also BHLm and BHAb (Pedras et al. 2009); a few other cruciferous phytoalexins (brassilexin and wasalexins) were found to inhibit the BOLm but not BHLm or BHAb (Pedras et al. 2011). A library of synthetic compounds recently screened for the inhibition of BHAb led to a new generation of very promising inhibitors that do not contain the indole nucleus (Pedras et al. 2012). Those results indicated that the catalytic and potential inhibitory sites of BHAb are highly selective, a desirable feature from the designer's perspective, albeit more difficult to find new potent inhibitors.

Overall, substantial work on fungal detoxifying enzymes indicated that they are produced in extremely small quantities, hence crystallization and X-ray crystallography studies are virtually impossible. Unfortunately, although several groups are working on this problem, to date no heterologous expression systems seem to be available to obtain the corresponding recombinant enzymes. The development of effective expression systems for phytoalexin detoxifying enzymes will be of great assistance to understand the interaction of each enzyme with their substrates and inhibitors and to produce paldoxins that protect crucifers against blackleg, black spot and other major fungal diseases of economically important crops.

8.6   Conclusion and Prospects

Although the first cruciferous phytoalexins were reported in 1986 (Takasugi et al. 1986), there is no question that work on these metabolites is still in its infancy. For example, within the Brassicaceae, only ca. 35 species have been investigated (Table 8.1), but this family contains ca. 3,700 species (Warwick 2011). Further biosynthetic studies to isolate enzymes and clone the genes involved in transformation of glucobrassicins to brassinins are essential, hence the mobilization of plant molecular biologists to this area is extremely important. Genetic manipulation of these complex and unique crucifer defense pathways requires a complete metabolic understanding rather far from our current knowledge.

The commercial availability of paldoxins appears to be a desirable strategy to boost plant defenses to fight fungal diseases, but this strategy is not expected to be applicable to all fungal diseases. It is likely that such a strategy will be applicable to those fungal pathogens that are specific to certain crops. This group of pathogens may include fungi responsible for major diseases of *Brassica* species such as blackleg (*L. maculans* and *L. biglobosa*) and blackspot (*A. brassicicola* and *A. brassicae*). However, it is also important to analyze detoxifying enzymes from non-specific pathogens like *B. cinerea*, as this is a major pathogen that could lead to another use of paldoxins. Cloning of the fungal genes encoding phytoalexin detoxifying enzymes would clarify their role fungal diseases and also facilitate the design of effective paldoxins. In addition, bioinformatic and phylogenetic analyses using sequences of detoxifying enzymes is expected to lead to an understanding of co-evolutionary metabolic pathways of crucifers and their pathogens. No doubt much work remains to be done to advance the current knowledge to a level where immediate commercial applications are feasible.

References

Bailey JA, Mansfield JW (1982) Phytoalexins. Blackie and Son, Glasgow, p 334

Böttcher C, Westphal L, Schmotz C, Prade E, Scheel D, Glawischnig E (2009) The multifunctional enzyme CYP71B15 (phytoalexin deficient3) converts cysteine-indole-3-acetonitrile to camalexin in the indole-3-acetonitrile metabolic network of *Arabidopsis thaliana*. Plant Cell 21:1830–1845

Essenberg M (2001) Prospects for strengthening plant defenses through phytoalexin engineering. Physiol Mol Plant Pathol 59:71–81

Hartmann T (2007) From waste products to ecochemicals: fifty years research of plant secondary metabolism. Phytochemistry 68:2831–2846

Kutschy P, Mezencev R (2008) Indole phytoalexins from Brassicaceae: synthesis and anticancer activity. Targ Heterocycl Syst 12:120–148

Mezencev R, Galizzi M, Kutschy P, Docampo R (2009) *Trypanosoma cruzi*: antiproliferative effect of indole phytoalexins on intracellular amastigotes in vitro. Exp Parasitol 122:66–69

Nongbri PL, Johnson JM, Sherameti I, Glawischnig E, Halkier BA, Oelmüller R (2012) Indole-3-acetaldoxime-derived compounds restrict root colonization in the beneficial interaction between *Arabidopsis* roots and the endophyte *Piriformospora indica*. Mol Plant Microbe Interact 25:1186–1197

Pedras MSC (2004) Prospects for controlling plant fungal diseases: alternatives based on chemical ecology and biotechnology. Can J Chem 82:1329–1335

Pedras MSC, Adio AM, Suchy M (2006) Detection, characterization and identification of crucifer phytoalexins using high-performance liquid chromatography with diode array detection and electrospray ionization mass spectrometry. J Chromatogr A 1133:172–183

Pedras MSC, Ahiahonu PWK (2005) Metabolism and detoxification of phytoalexins and analogs by phytopathogenic fungi. Phytochemistry 66:391–411

Pedras MSC, Hossain M (2006) Metabolism of crucifer phytoalexins in *Sclerotinia sclerotiorum*: detoxification of strongly antifungal compounds involves glucosylation. Org Biomol Chem 4:2581–2590

Pedras MSC, Hossain S (2011) Interaction of phytoanticipins with plant fungal pathogens: indole glucosinolates are not metabolized but the corresponding desulfo-derivatives and nitriles are. Phytochemistry 72:2308–2316

Pedras MSC, Jha M (2006) Toward the control of *Leptosphaeria maculans*: design, syntheses, biological activity and metabolism of potential detoxification inhibitors of the crucifer phytoalexin brassinin. Bioorg Med Chem 2006(14):4958–4979

Pedras MSC, Jha M, Ahiahonu PWK (2003) The synthesis and biosynthesis of phytoalexins produced by cruciferous plants. Curr Org Chem 7:1635–1647

Pedras MSC, Jha M, Okeola OG (2005) Camalexin induces detoxification of the phytoalexin brassinin in the plant pathogen *Leptosphaeria maculans*. Phytochemistry 66:2609–2616

Pedras MSC, Jha M, Minic Z, Okeola OG (2007) Isosteric probes provide structural requirements essential for detoxification of the phytoalexin brassinin by the fungal pathogen *Leptosphaeria maculans*. Bioorg Med Chem 15:6054–6061

Pedras MSC, Minic Z, Jha M (2008) Brassinin oxidase, a fungal detoxifying enzyme to overcome a plant defense: purification, characterization and inhibition. FEBS J 275:3691–3705

Pedras MSC, Minic Z, Sarma-Mamillapalle VK (2009) Substrate specificity and inhibition of brassinin hydrolases, detoxifying enzymes from the plant pathogens *Leptosphaeria maculans* and *Alternaria brassicicola*. FEBS J 276:7412–7428

Pedras, MSC, Minic Z, Hossain, S (2012) Discovery of inhibitors and substrates of brassinin hydrolase: Probing selectivity with dithiocarbamate bioisosteres. Bioorg Med Chem 20:225–233

Pedras MSC, Sarma-Mamillapalle VK (2012) The phytoalexins rapalexin A, brussalexin A and erucalexin: chemistry and metabolism in *Leptosphaeria maculans*. Bioorg Med Chem 20:3991–3996

Pedras MSC, Sorensen JL (1998) Phytoalexin accumulation and production of antifungal compounds by the crucifer wasabi. Phytochemistry 49:1959–1965

Pedras MSC, Suchy M (2006) Design, synthesis, evaluation and antifungal activity of inhibitors of brassilexin detoxification in the plant pathogenic fungus *Leptosphaeria maculans*. Bioorg Med Chem 14:714–723

Pedras MSC, Yaya EE (2010) Phytoalexins from Brassicaceae: news from the front. Phytochemistry 71:1191–1197

Pedras MSC, Yaya EE (2012) The first isocyanide of plant origin expands functional group diversity in cruciferous phytoalexins: synthesis, structure and bioactivity of isocyalexin A. Org Biomol Chem 10:3613–3616

Pedras MSC, Yaya EE, Hossain S (2010) Unveiling the phytoalexin biosynthetic puzzle in salt cress: unprecedented incorporation of glucobrassicin into wasalexins A and B. Org Biomol Chem 8:5150–5158

Pedras MSC, Yaya EE, Glawischnig E (2011) The phytoalexins from cultivated and wild crucifers: chemistry and biology. Nat Prod Rep 28:1381–1405

Russell PE (2005) A century of fungicide evolution. J Agric Sci 143:11–25

Smith CJ (1996) Accumulation of phytoalexins: defence mechanism and stimulus response system. New Phytol 132:1–45

Takasugi M, Katsui N, Shirata A (1986) Isolation of three novel sulphur-containing phytoalexins from the Chinese cabbage *Brassica campestris* L. ssp. *pekinensis* (Cruciferae). Chem Commun 1077–1078

Walters D, Walsh D, Newton A, Lyon G (2005) Induced resistance for plant disease control: maximizing the efficacy of resistance elicitors. Phytopathology 95:1368–1373

VanEtten HD, Mansfield JW, Bailey JA, Farmer EE (1994) Two classes of plant antibiotics: phytoalexins versus "phytoanticipins". Plant Cell 6:1191–1192

VanEtten HD, Matthews DE, Matthews PS (1989) Phytoalexin detoxification: importance for pathogenicity and practical implications. Annu Rev Phytopathol 27:143–164

Warwick SI (2011) Brassicaceae in agriculture. In: Schmidt R, Bancroft I (eds) Genetics and genomics of the Brassicaceae, vol 9, Plant genetics and genomics: crops and models. Springer Science, New York, p 34

Chapter 9
Molecular Basis of Cytoplasmic Male Sterility

Jinghua Yang and Mingfang Zhang

Abstract Cytoplasmic male sterility (CMS), a maternally inherited trait failing to produce functional pollens, is an alternative approach for utilization of hybrid vigor and hybrid seed production in Crucifer crops. CMS and its fertility restoration lines are also good system to explore the interaction between nuclear and mitochondrial genomes. Several cytoplasmic male sterility-associated genes, namely mitochondrial novel open reading frames (*ORFs*), had been identified from different sources of CMS, some of which were confirmed to be associated with CMS phenotypes by transgenic experiments. There were many molecular events occurring at transcriptional and/or editing levels pertaining to energy-related mitochondrial genes. Here, we reviewed the current advances in the research of genetic and molecular basis of CMS in crucifer crops. Thus, the understandings of CMS could help us to recognize the vital role of mitochondria and the manipulation of organellar genetics in practical breeding programs.

Keywords Crucifer crop • Cytoplasmic male sterility • Fertility restoration • Mitochondrial rearrangement • Nuclear-cytoplasmic communication

9.1 Background

Cytoplasmic male sterility (CMS), as a maternally inherited trait that prevents the production of functional pollens, was first termed by Rhoades (Rhoades 1933), and is widely applied in hybrid breeding and currently, is observed in >150 plant species (Laser and Lersten 1972). It was firstly documented that mitochondrial genes

J. Yang, Ph.D. • M. Zhang, Ph.D. (✉)
Institute of Vegetable Science, Yuhangtang Road, Hangzhou, Zhejiang Province 310058, China
e-mail: mfzhang@zju.edu.cn

S.K. Gupta (ed.), *Biotechnology of Crucifers*, DOI 10.1007/978-1-4614-7795-2_9,
© Springer Science+Business Media, LLC 2013

contributed to CMS in terms of their molecular aspect in maize (Dewey et al. 1986). Generally, mutations in mitochondrial genes might induce severe defects in respiration and be lethal. Another kind of mitochondrial gene mutants, chimeric plants, containing wild-type mitochondria (Karpova et al. 2002 and refs therein), usually lead to CMS. In most cases, the CMS systems were developed by intra-specific, inter-specific or inter-generic crosses with alien cytoplasm from other species or genera or by natural mutation (Schnable and Wise 1998; Budar and Pelletier 2001). Most researches considered that the alloplasmic CMS usually affected reproductive development but not other developmental events, however, an increasing number of evidence proofed that some vegetative developments were also impacted in CMS (Leino et al. 2003; Liu et al. 2012; Yang et al. 2012).

To date, CMS has been associated with expressions of mitochondrial novel open reading frames (*ORFs*) that arise from rearrangements of mitochondrial genomes. Since the first case of mitochondrial CMS-associated *ORF, urf13*, was identified from T-maize (Dewey et al. 1987), many CMS-associated *ORFs* have been identified from CMS crops. Such *ORFs* are often located adjacent to genes encoding components of the ATPase complex, and co-transcribed with these genes (Hanson and Bentolila 2004 and refs therein). Fertility of CMS can be recovered by restorer genes encoded by nuclear genome (Schnable and Wise 1998). Coupled with the identification of restorer genes, the mechanism of fertility recovery and incongruity between nucleus and mitochondria has been revealed in CMS systems (Hanson and Bentolila 2004). Here, we reviewed some major progresses in the genetic and molecular basis of cytoplasmic male sterility in crucifer crops, mainly in *Brassica* and *Raphanns* of crucifer crops.

9.2 Identification of Mitochondrial Novel *ORFs* from CMS Crucifer Crops

In crucifer crops, CMS has been paid more attention in important economic crops, like *Brassica* and *Raphanns*. Several types of CMS lines were defined by different sources of sterile cytoplasm with distinct genetic features. These types of CMS include *pol* (Singh and Brown 1991) and *nap* (Brown and Mona 1998) CMS in *Brassica napus, orf220* (Zhang et al. 2003), *hau* (Wan et al. 2008), *berthautii* (Bhat et al. 2008), *moricandia* (Prakash et al. 1998), *erucoides* (Bhat et al. 2006), *siifolia* (Rao et al. 1994), *erucastrum* (Prakash et al. 2001), *oxyrrhina* (Prakash and Chopra 1990) and *tournefortii* (Pradhan et al. 1991) CMS in *Brassica juncea* and *Ogura* CMS in *Raphanns sativus* (Ogura 1968).

9.2.1 Cloning of Novel Mitochondrial CMS-Associated ORFs

Since the first case of mitochondrial CMS-associated *orf, urf13*, was identified from T-maize (Dewey et al. 1986, 1987), many reports were focused on the isolation of

Table 9.1 Examples of novel *ORFs* and mitochondrial genes involved in CMS crucifer crops

Plant species	*ORFs*	Mitochondrial genes	References
Raphanns sativus	orf138	orf158	Bonhomme et al. (1992)
Brassica nap	orf222	atp8, nad5c, orf139	L'Homme et al. (1997)
Brassica pol	orf224	atp6, rps3	Singh and Brown (1991), Handa and Nakajima (1992)
Brassica juncea	orf263	atp6	Landgren et al. (1996)
Brassica juncea	orf220	atpA, nad2, atp9	Zhang et al. (2003), Yang et al. (2009a, b, 2010)
Brassica juncea	orf193	atp9	Dieterich et al. (2003)
Brassica juncea	orf288	atp6	Wan et al. (2008)
Brassica juncea	orf108	atpA	Ashutosh et al. (2008), Kumar et al. (2012)

CMS-associated mitochondrial genes through comparisons of CMS and its maintainer fertile (MF) lines. In crucifer crops, several *ORFs* were isolated from CMS. These include *orf138* from CMS *Raphanns sativus* (Bonhomme et al. 1992), *orf222* from *nap* CMS of *Brassica napus* (L'Homme et al. 1997), *orf224* from *pol* CMS of *Brassica napus* (Singh and Brown 1991), *orf263* from *tournefortii Brassica juncea*, *orf220* from *orf220*-type *Brassica juncea* (Zhang et al. 2003), *orf193* from *tournefortii-Stiewe Brassica juncea*, *orf288* from *hau Brassica juncea* (Wan et al. 2008) and *orf108* from *moricandia Brassica juncea* (Ashutosh et al. 2008). Examples of such novel mitochondrial *ORFs* associated with CMS crucifer crops are listed in Table 9.1. Usually, such *ORFs* are located at the flanking ends of mitochondrial encoding genes and are co-transcribed with these genes (Hanson and Bentolila 2004). Furthermore, comprehensive *ORFs* and genomic structure of mitochondria are identified in CMS based on mitochondrial genome sequencing, which could be quite helpful for cloning more candidate CMS-associated *ORFs* (Chen et al. 2011).

9.2.2   Functional Analysis of Mitochondrial CMS-Associated ORFs

Common strategies to discover CMS-associated mitochondrial factors through comparative research have not been absolutely powerful. Sometimes, the differences between two cytoplasms might arise from evolutionary divergence of different mitochondrial genomes in alloplasmic cytoplasm. Likewise, it should be emphasized that some chimeric mitochondrial *ORFs* clearly do not exhibit CMS phenotypes in *Arabidopsis* (Marienfeld et al. 1997). Usually, there are two routine ways to prove the relationship between specific mitochondrial genes and CMS. One approach is to study the expression patterns of those genes under the control of the restorer genes to compare their characterizations in CMS, maintainer fertile and restored lines (Desloire et al. 2003; Koizuka et al. 2003). Another approach is to directly validate their functions by transgenic engineering, although plant mitochondria could not be easily genetically manipulated. Only some CMS-associated *ORFs* were confirmed to functionally cause male sterility in crucifer crops, for instances, *ORF220* (Yang et al. 2010), *ORF108* (Kumar et al. 2012) and *ORF288* (Jing et al. 2012) in *Brassica juncea*. In those successful cases, mitochondrial targeted pre-sequence is often needed to be fused into forepart of *ORFs*, of which CMS-associated *ORFs* could be guided into mitochondria and cause CMS phenotype when expressed in nuclear genome (Yang et al. 2010; Kumar et al. 2012; Jing et al. 2012). Nevertheless, mitochondrial-targeted expression of *ORFs* failed to induce male sterility in some cases (Chaumont et al. 1995; Duroc et al. 2006; Wintz et al. 1995). The cause is probably due to the failure of sub-mitochondrial location (Duroc et al. 2006), the expression period of *ORFs* (Wintz et al. 1995) or the substoichiometric levels of *ORFs* (Chaumont et al. 1995) in transgenic plants. Thus, precise locations of alien CMS-associated *ORFs* and their adequate dose of gene expressions seem to be quite vital to pinpoint their functions.

9.2.3   Occurrence and Origination of Mitochondrial CMS-Associated ORFs

In many cases, CMS-associated genes were caused by rearrangements of mitochondrial genomes resulting in the birth of new *ORFs* composed of fragments of other mitochondrial respiratory-related genes or non-coding sequences (Schnable and Wise 1998; Hanson and Bentolila 2004). However, the exact mechanism of the origination of CMS-associated *ORFs* is still largely unknown in CMS. In plants, an unusual nature of mitochondrial genome undergoes genomic recombination (Mackenzie and McIntosh 1999; Mackenzie 2005). This type of mitochondrial recombination appears to play a key role in plant mitochondrial genome evolution, generating novel mitotypes (Small et al. 1989), and also serves as a possible mechanism for fertility reversion (Fauron et al. 1995). Dramatic changes of mitochondrial DNA molecule stoichiomtries, a phenomenon termed substoichiometric shifting (SSS), often accompanies recombination in mitochondrial genome (Janska et al. 1998). Experimental evidence indicates that SSS of mitochondrial genome leads to male sterility and spontaneous reversion to fertility (Janska et al. 1998; Feng et al. 2009). One of the nuclear genes, *MSH1* (*MutS* Homolog 1), controls the mitochondrial genome recombination (Shedge et al. 2007). When *MSH1* gene is suppressed, it dramatically alters mitochondrial and plastid properties and plant response to environment, meanwhile, triggers developmental reprogramming (Xu et al. 2011, 2012).

9.3 Transcriptional Regulation of Mitochondrial Genes in CMS

Usually, chimeric *ORFs* are located at the flanking end of genes encoding subunits of mitochondrial complexes. Consequently, the expressions of those genes are altered in the CMS systems. Moreover, RNA editings of mitochondrial genes were observed to be changed in CMS compared with its MF line (Hanson and Bentolila 2004).

9.3.1 Mitochondrial Genes are Co-Transcribed with ORFs

So far, numerous mitochondrial rearrangement regions associated with the CMS phenotype have been identified indicating the striking manner and frequency of recombination events in mitochondrial genome. Most rearrangements on mitochondrial loci were focused on subunit genes of ATP synthesis such as ATP synthesis subunit 4, 6, 8 and 9 genes in CMS-associated loci (Schnable and Wise 1998; Hanson and Bentolila 2004). Some other subunit genes of the mitochondrial respiratory complexes were also displayed to be associated with CMS including NDAH complex in some CMS species. In crucifer crops, several co-transcribed CMS-associated *ORFs* and mitochondrial genes were observed from CMS. These include *orf138* co-transcription with *orf158* in CMS *Raphanns sativus* (Bonhomme et al. 1992), *orf222* co-transcription with *nad5c* in *nap* CMS *Brassica* napus (L'Homme et al. 1997), *orf224* co-transcription with *atp6* in *pol* CMS *Brassica napus* (Singh and Brown 1991), *orf263* co-transcription with *atp6* in *tournefortii* CMS *Brassica juncea* (Landgren et al. 1996), *orfB* co-transcription with *nad2* in *orf220*-type *Brassica juncea* (Yang et al. 2009a), *orf193* co-transcription with *atp9* in *tournefortii-Stiewe Brassica juncea* (Dieterich et al. 2003), *orf288* co-transcription with *atp6* in *hau* CMS *Brassica juncea* (Wan et al. 2008) and *orf108* co-transcription with *atpA* in CMS *Brassica juncea* (Ashutosh et al. 2008). Examples of co-transcribed CMS-associated *ORFs* and mitochondrial genes in crucifer crops are listed in Table 9.1. Actually, more evidences are needed to clarify whether the event of *ORFs* co-transcription with mitochondrial genes is causal or phenomenal factors in CMS.

9.3.2 RNA Editings of Mitochondrial Genes Are Altered in CMS

RNA editing, as a crucial post-transcriptional step for RNA processing in higher plant organelle, regulates most mitochondrial and chloroplast gene expression in plant (Maier et al. 1996). Through comparative study, alterations on RNA editing sites were observed in CMS crucifer crops. Three silent RNA editing sites were reported for mitochondrial *nad3* transcripts in the CMS line of carrot (Rurek 2001).

Temporal and spatial characteristics of RNA editing for *atp9* gene were found in *orf220*-type CMS *Brassica juncea* (Yang et al. 2007). Different RNA editing patterns of mitochondrial *nad3/rps12* gene were identified in CMS *Brassica oleracea* (Wang et al. 2007). When RNA editing is specifically altered in some types of CMS, several attempts were made to correlate RNA editings with the occurrence of CMS. Expression of an unedited mitochondrial *atp9* gene in a fertile line caused male sterile in *Arabidopsis*, of which unedited mitochondrial *atp9* gene led to mitochondrial dysfunction (Gomez-Casati et al. 2002). Obviously, CMS trait could be induced at the post-transcriptional level through RNA editings and this result points to an alternative approach for generating engineered male sterile plants.

9.4 Molecular Control of Nuclear-Cytoplasmic Communication

In plant cells, mitochondria and chloroplast are semi-autonomous organelles that encode partial genetic information, with the majority being derived and imported from the nucleus (Unseld et al. 1997). Wide inter-organellar communications among the three organelles, in which signals from nucleus to mitochondria and chloroplast is termed anterograde regulation and signals from mitochondria and chloroplast to nucleus is termed retrograde regulation correspondingly (Woodson and Chory 2008). Because of the nature of nuclear-cytoplasmic interaction, it is a good model of CMS to study the anterograde and retrograde regulation among the organelles. To date, two types of nuclear-cytoplasmic regulation pathway have been found in CMS including restorer gene in mediation of CMS-associated *ORFs* and mitochondrial retrograde regulation of nuclear genes involved in reproductive development.

9.4.1 Identification of Restorer Gene in Mediation of Nuclear Regulation of Mitochondria

Recovery of fertility mediated by nuclear restorer gene has been well described in terms of nuclear restorer genes which suppress the function of CMS-associated *ORFs* in the fertile restored line. The majority of nuclear restorer genes operate at post-transcriptional level, such as RNA editing, processing, and polyadenylation, acting by controlling copy numbers at the DNA level, post-translational modification of CMS-associated proteins and compensation of mitochondrial dysfunction at metabolic level (Hanson and Bentolila 2004). In crucifer crops, several restorer genes or loci had already been identified or mapped, including *Rfo* gene from *Ogura* CMS *Raphanus sativus* (Koizuka et al. 2003; Yasumoto et al. 2009), *Rfd1* loci from *DCGMS* CMS *Raphanus sativus* (Cho et al. 2012), *Rfp* loci from *pol* CMS *Brassica napus* (Formanová et al. 2010; Liu et al. 2012), *Rfk1* loci from *Ogu*-INRA *Brassica rapa* (Niemelä et al. 2012), *Rf* loci from *Moricandia* CMS *Brassica juncea*

Table 9.2 Examples of mapping or identification of restorer genes in CMS crucifer crops

Restorer gene	Encoded protein/ function	CMS type	References
Rfp	mapping	*pol Brassica napus*	Formanová et al. (2010), Liu et al. (2012)
Rfk1	mapping	*Ogu*-INRA *Brassica rapa*	Niemelä et al. (2012)
Rf	mapping	*Moricandia Brassica juncea*	Ashutosh et al. (2007)
Rfd1	mapping	*DCGMS Raphanus sativus*	Cho et al. (2012)
Rfo	PPR protein/RNA processing	*Ogura Raphanus sativus*	Koizuka et al. (2003), Yasumoto et al. (2009)

(Ashutosh et al. 2007). Examples of mapping or identification of restorer genes in CMS crucifer crops are listed in Table 9.2. The *Rfo* gene from *Ogura* CMS *Raphanus sativus* encodes a pentatricopeptide repeat (PPR) protein like most restorer genes from other CMS crops (Koizuka et al. 2003; Yasumoto et al. 2009; Hanson and Bentolila 2004).

PPR proteins constitute a large family, more than 400 members in plants, of which about 60 % were predicted to be targeted in mitochondria and involved in post-transcriptional processes (Lurin et al. 2004). PPR proteins had been suggested to function as sequence-specific adaptors for a variety of other RNA-associated proteins (Lurin et al. 2004), which were primarily and definitively supported by some experimental evidence (Wang et al. 2006b). Hence, PPR proteins were considered as probable candidates for molecules of nuclear-mitochondrial interactions with essential effectors in CMS systems.

9.4.2 Identification of Candidate Nuclear Targeted Gene in Mitochondrial Retrograde Regulation

Until now, there has been no evidence supporting the idea that mitochondrial genes are directly involved in floral organ development, microsporogenesis, or other reproductive development. We thought that all types of abnormal phenotypes in CMS should originate from alterations in the expression of nuclear gene signals regulated by mitochondria, which would lead indirectly, but specifically to male sterility. Indeed, two groups of nuclear genes had been reported to be potential target genes: genes involved in programmed cell death (PCD) and transcriptional factor genes needed for development of floral organs as well as pollen.

In tapetal cell degeneration inducing the male sterile type of CMS, PCD of tapetal cells were subsequently extended to other tissues of anthers and were shown to be activated by the partial release of cytochrome c from the mitochondria into the above cells (Balk and Leaver 2001). Studies on the homeotic-type of CMS clearly demonstrated the probable pathway for the effect of mitochondria on the expression of specific nuclear homeotic genes for floral organ development. In higher plants, floral organ development has been intensively researched in dicotyledonary plants, especially in *Arabidopsis* and *Antirrhinum*, which are mainly controlled by the homeotic genes (Theissen 2001). One classical genetic model, the ABC model, in developmental biology, can explain and predict flower organ families based on three classes of nuclear homeotic genes, termed A, B, and C. Any alterations of transcription or mutation in these genes could lead to global variations in four whorl structures of a flower, of which a certain type of flower organ would be replaced by the another (Coen and Meyerowitz 1991). Interestingly, such dramatic variations in flower organ were observed in alloplasmic cytoplasmic male sterile (CMS) tobacco (Kofer et al. 1991), carrot (Linke et al. 1999), wheat (Murai and Tsunewaki 1993), *Brassica juncea* (Yang et al. 2005) and *Brassica napus* (Teixeira et al. 2005), which, in most cases, exhibited a complete conversion of stamens into other floral organs. In recent research, nuclear MADS-box transcriptional factor (TF) genes, *AGAMOUS* (*AG*), *APETALA3* (*AP3*), *PISTILLATA* (*PI*), *GLOBOSA-* and *DEFICIENS*-like genes, were found to be transcriptionally down-regulated in CMS carpelloid tobacco flowers (Zubko et al. 2001), CMS pistillody carrot (Linke et al. 2003), CMS *Brassica napus* (Teixeira et al. 2005), CMS wheat (Murai et al. 2002; Hama et al. 2004) and CMS *Brassica juncea* (Yang et al. 2008a). Insightful studies of microsporogensis in *Arabidopsis* reveals *SPOROCYTELESS* (*SPL*, also known as *NOZZLE*, *NZZ*) gene, encoding a novel nuclear protein related to MADS-box transcription factor, was required to promote the differentiation of the primary sporogenous cells and cells of the anther wall (Schiefthaler et al. 1999; Yang et al. 1999). The research about putative target genes of *AG* showed that the homeotic protein *AG* controlled microsporogenesis by regulation of the *SPL* gene in *Arabidopsis* (Ito et al. 2004). In cytoplasmic male sterile plants, pollen development was halted at a very early or late developmental stage depending on the CMS system (Hanson and Bentolila 2004). In CMS *Brassica juncea* with failure of microsporogenesis, the absent expression of *SPL* gene was considered as the failure of pollen development (Yang et al. 2008b).

In addition, other candidate nuclear genes manipulated by mitochondrial retrograde regulation were identified, including retrograde regulating of *CTR1* (a negative regulator of ethylene signaling pathway) gene in ethylene response and retrograde regulating of *RCE1* (Related to ubiquitin1-conjugating enzyme) gene in auxin response (Liu et al. 2012; Yang et al. 2012). All the above down-regulated nuclear TF or other genes in CMS abnormal reproductive and vegetative development allow us to hypothesize the pathway of the molecular mechanism of mitochondrial retrograde regulation in CMS.

9.5   Discussion and Perspectives

CMS provides a path to explore the role of mitochondria in vegetative and reproductive development and interactions between the mitochondria and nucleus, apart from its agronomic importance in hybrid production. Indeed, so many mitochondrial

CMS-associated causal factors have been identified to date. Likewise, some restorer genes in mediating of CMS-associated *ORFs* expression and potential nuclear targeted genes regulated by mitochondria have also been studied in some CMS types. Hence, mitochondrial genes could not directly operate on nuclear targeted genes, and thus there must be a signal pathway from mitochondria to nucleus inducing male sterility and affecting other traits. However, our understanding of CMS remains limited, in part because many of the genes involved is still not known including the genes of controlling mitochondrial recombination, and how their functions are controlled by nucleus/organelle and how to place anterograde or retrograde signaling. Moreover, from some breeders' personal communications, we were puzzled that they were unable to observe heterosis in some CMS crops, especially in vegetative growth. Thus, in certain CMS sources applied to crops with vegetative organs as economic trait, CMS could probably only contribute to the seed hybrid production, but not heterosis vigor in hybrids.

Indeed, when *MSH1* gene that controls organellar genomic recombination is suppressed, some research groups observed extremely similar phenotypes to CMS, including male sterility, alterations on phytohormone metabolism and others (Sandhu et al. 2007; Xu et al. 2011, 2012). However, most of the previous researches were mainly focused on genes dysfunction from mitochondria and their reversions by restorer genes in nucleus. The striking findings of the CMS-inducing function of *MSH1* gene provide us with a new window and shed light on further clarifying how nuclear genes cause mitochondrial recombination in anterograde regulation, and vice versa, how mitochondrial genes affect responsive nuclear genes expressions in retrograde regulation, ultimately leading to CMS occurrence.

Acknowledgements This work was supported by serial grants from the National Natural Science Foundation of China (30971994).

References

Ashutosh, Sharma PC, Prakash S et al (2007) Identification of AFLP markers linked to the male fertility restorer gene of CMS *(Moricandia arvensis) Brassica juncea* and conversion to SCAR marker. Theor Appl Genet 114:385–392

Ashutosh, Kumar P, Kumar DK et al (2008) A novel *orf108* co-transcribed with the *atpA* gene is associated with cytoplasmic male sterility in *Brassica juncea* carrying *Moricandia arvensis* cytoplasm. Plant Cell Physiol 49(2):284–289

Balk J, Leaver CJ (2001) The PET1-CMS mitochondrial mutation in sunflower is associated with premature programmed cell death and cytochrome c release. Plant Cell 13:1803–1818

Bhat SR, Vijayan P, Ashutosh et al (2006) Diplotaxis erucoides induced cytoplasmic male sterility in *Brassica juncea* is rescued by the *Moricandia arvensis* restorer: genetic and molecular analyses. Plant Breed 125:150–155

Bhat SR, Kumar P, Prakash S (2008) An improved cytoplasmic male sterile (*Diplotaxis berthautii*) *Brassica juncea*: identification of restorer and molecular characterization. Euphytica 159:145–152

Bonhomme S, Budar F, Lancelin D, Small I, Defrance MF, Pelletier G (1992) Sequence and analysis of *Nco*2.5 Ogura-specific fragment correlated with male sterility in *Brassica* cybrids. Mol Genomics Genet 235:240–248

Brown GG, Mona D (1998) Molecular analysis of *Brassica* CMS and its application to hybrid seed production. Acta Hortisci 459:265–274

Budar F, Pelletier G (2001) Male sterility in plant: occurrence, determinism, significance and use. Life Sci 324:543–550

Chaumont F, Bernier B, Buxant R et al (1995) Targeting the maize T-*urf13* product into tobacco mitochondria confers methomyl sensitivity to mitochondrial respiration. Proceedings of the National Academy of Sciences USA 92:1167–1171

Chen J, Guan R, Chang S et al (2011) Substoichiometrically different mitotypes coexist in mitochondrial genomes of *Brassica napus* L. PLoS One 6:e17662

Cho Y, Lee Y, Park B et al (2012) Construction of a high-resolution linkage map of Rfd1, a restorer-of-fertility locus for cytoplasmic male sterility conferred by DCGMS cytoplasm in radish (*Raphanus sativus* L.) using synteny between radish and *Arabidopsis* genomes. Theor Appl Genet 125:467–477

Coen ES, Meyerowitz EM (1991) The war of the whorls: genetic interactions controlling flower development. Nature 353:31–37

Desloire SH, Gherbil W, Laloui S et al (2003) Identification of the fertility restoration locus, *Rfo*, in radish, as a member of the pentatricopeptide-repeat protein family. EMBO Rep 4:588–594

Dewey RE, Levings CS III, Timothy DH (1986) Novel recombinations in the maize mitochondrial genome produce a unique transcriptional unit in the Texas male-sterile cytoplasm. Cell 44:439–449

Dewey RE, Timothy DH, Levings CS III (1987) A mitochondrial protein associated with cytoplasmic male sterility in the T-cytoplasm of maize. Proc Natl Acad Sci U S A 84:5374–5378

Dieterich JH, Braun HP, Schmitz UK (2003) Alloplasmic male sterility in *Brassica napus* (CMS 'Tournefortii-Stiewe') is associated with a special gene arrangement around a novel *atp9* gene. Mol Genet Genomics 269:723–731

Duroc Y, Gaillard C, Hiard S et al (2006) Nuclear expression of a cytoplasmic male sterility gene modifiers mitochondrial morphology in yeast and plant cell. Plant Science 170:755–767

Fauron C, Casper M, Gao Y et al (1995) The maize mitochondrial genome: dynamic, yet functional. Trends Genet 11:228–235

Feng X, Kaur AP, Mackenzie SA et al (2009) Substoichiometric shifting in the fertility reversion of cytoplasmic male sterility pearl millet. Theor Appl Genet 118:1361–1370

Formanová N, Stollar R, Geddy R et al (2010) High-resolution mapping of the *Brassica napus Rfp* restorer locus using *Arabidopsis*-derived molecular markers. Theor Appl Genet 120:843–851

Gomez-Casati DF, Busi MV, Gonzalez-Schain N et al (2002) A mitochondrial dysfunction induces the expression of nuclear-encoded complexIgenes in engineered male sterile *Arabidopsis thaliana*. Fed Eur Biochem Soc Lett 532:70–74

Hama E, Takumi S, Ogihara Y et al (2004) Pistillody is caused by alteration to the class-B MADS-box gene expression pattern in alloplasmic wheats. Planta 218:712–720

Handa H, Nakajima K (1992) Different organization and altered transcription of the mitochondrial *atp6* gene in the male-sterile cytoplasm of rapeseed (*Brassica napus* L.). Curr Genet 21:153–159

Hanson MR, Bentolila S (2004) Interactions of mitochondrial and nuclear genes that affect male gametophyte development. Plant Cell 16S:154–169

Ito T, Wellmer F, Yu H et al (2004) The homeotic protein *AGAMOUS* controls microsporogenesis by regulation of *SPOROCYTELESS*. Nature 430:356–360

Janska H, Sarria R, Woloszynska M et al (1998) Stoichiometric shifts in the common bean mitochondrial genome leading to male sterility and spontaneous reversion to fertility. Plant Cell 10:1163–1180

Jing B, Heng S, Tong D et al (2012) A male sterility-associated cytotoxic protein ORF288 in *Brassica juncea* cause aborted pollen development. J Exp Bot 63:1285–1295

Karpova OV, Kuzmin EV, Elthon TE, Newton KJ (2002) Differential expression of alternative oxidase genes in maize mitochondrial mutants. Plant Cell 14:3271–3284

Kofer W, Glimelius K, Bonnett HT (1991) Modification of mitochondrial DNA cause changes in floral development in homeotic-like mutants of tobacco. Plant Cell 3:759–769

Koizuka NR, Imai R, Fujimoto H et al (2003) Genetic characterization of a pentatricopeptide repeat protein gene, *orf687*, that restores fertility in the cytoplasmic male-sterile *Kosena* radish. Plant J 34:407–415

Kumar P, Vasupalli N, Srinivasan R et al (2012) An evolutionarily conserved mitochondrial *orf108* is associated with cytoplasmic male sterility in different alloplasmic lines of *Brassica juncea* and induces male sterility in transgenic *Arabidopsis thaliana*. J Exp Bot 63:2921–2932

L'Homme Y, Stahl RJ, Li XQ, Hammeed A, Brown GG (1997) *Brasicca nap* cytoplasmic male sterility is associated with expression of a mtDNA region containing a chimeric gene similar to the *pol* CMS-associated *orf224* gene. Curr Genet 31:325–335

Landgren M, Zetterstrand M, Sundberg E et al (1996) Alloplasmic male-sterile *Brassica* lines containing *B. tournefortii* mitochondria express an ORF 3' of the *atp6* gene and a 32 kDa protein. Plant Mol Biol 32:879–890

Laser KD, Lersten NR (1972) Anatomy and cytology of microsporogenesis in cytoplasmic male-sterile angiosperms. Bot Rev 38:425–454

Leino M, Teixeira R, Landgren M, Glimelius K (2003) *Brassica napus* lines with rearranged Arabidopsis mitochondria display CMS and a range of development aberrations. Theor Appl Genet 106:1156–1163

Linke B, Nothnagel T, Börner T (1999) Morphological characterization of modified flower morphology of three novel alloplasmic male sterile carrot sources. Plant Breeding 118:543–548

Linke B, Nothnagel T, Borner T (2003) Flower development in carrot cms plant: mitochondria affect the expression of MADS box genes homologous to *GLOBOSA* and *DEFICIENS*. Plant J 34:27–37

Liu Z, Liu P, Long F et al (2012) Fine mapping and candidate gene analysis of the nuclear restorer gene *Rfp* for *pol* CMS in rapeseed (*Brassica napus* L.). Theor Appl Genet 125:773–779

Lurin C, Andres C, Aubourg S et al (2004) Genome-wide analysis of Arabidopsis pentatricopeptide repeat proteins reveals their essential role in organelle biogenesis. Plant Cell 16:2089–2103

Mackenzie SA (2005) The mitochondrial genome of higher plants: A target for natural adaptation. Diversity and Evolution of Plants, R. J. Henry, ed. CABI Publishers, Oxon, UK. pp. 69–80

Mackenzie S, McIntosh L (1999) Higher plant mitochondria. Plant Cell 11:571–585

Maier RM, Zeltz P, Kossel H et al (1996) RNA editing in plant mitochondria and chloroplasts. Plant Mol Biol 32:343–365

Marienfeld JR, Unseld M, Brandt P et al (1997) Mosaic open reading frames in the *Arabidopsis thaliana* mitochondrial genome. Biological Chemistry 378:859–862

Murai K, Tsunewaki K (1993) Photoperiod-sensitive cytoplasmic male sterility in wheat with *Aegilops crassa* cytoplasm. Euphytica 67:41–48

Murai K, Takumi S, Koga H et al (2002) Pistillody, homeotic transformation of stamens into pistil-like structures, caused by nuclear-cytoplasm interaction in wheat. Plant J 29:169–181

Niemelä T, Seppanen M, Badakshi F et al (2012) Size and location of radish chromosome regions carrying the fertility restorer *Rfk1* gene in spring turnip rape. Chromosome Res 20:35–361

Ogura H (1968) Studies on the new male sterility in Japanese radish with special references to the utilization of this sterility towards the practical raising of hybrid seed. Mem Fac Agr Kogoshima Univ 6:39–78

Pradhan AK, Mukhopadhyay A, Pental D (1991) Identification of the putative cytoplasmic donor of a CMS system in *Brassica juncea*. Plant Breed 106:204–208

Prakash S, Chopra VL (1990) Male sterility caused by cytoplasm of *Brassica oxyrrhina* in *B. campestris* and *B. juncea*. Theor Appl Genet 79:285–287

Prakash S, Kirti PB, Bhat SR et al (1998) A *Moricandia arvensis*-based cytoplasmic male sterility and fertility restoration system in *Brassica juncea*. Theor Appl Genet 97:488–492

Prakash S, Ahuja I, Upreti HC et al (2001) Expression of male sterility in alloplasmic *Brassica juncea* with *Erucastrum canariense* cytoplasm and the development of a fertility restoration system. Plant Breed 120:479–482

Rao GV, Batra-Sarup V, Prakash S, Shivanna KR (1994) Development of a new cytoplasmic male-sterile system in *Brassica juncea* through wide hybridization. Plant Breed 112:171–174

Rhoades MM (1933) The cytoplasmic inheritance of male sterility in *Zea mays*. J Genet 27:71–93

Rurek M, Szklarczyk M, Adamczyk N et al (2001) Differences in editing of mitochondrial *nad3* transcripts from cms and fertile carrots. Acta Biochimica Polonica 48:711–717

Sandhu APS, Abdelnoor RV, Mackenzie SA (2007) Transgenic induction of mitochondrial rearrangements for cytoplasmic male sterility in crop plants. Proc Natl Acad Sci U S A 104:1766–1770

Schiefthaler U, Balasubramanian S, Sieber P et al (1999) Molecular analysis of *NOZZLE*, a gene involved in pattern formation and early sporogenesis during sex organ development in *Arabidopsis thaliana*. Proc Natl Acad Sci U S A 96:11664–11669

Schnable PS, Wise RP (1998) The molecular basis of cytoplasmic male sterility and fertility restoration. Trends Plant Sci 3:175–180

Shedge V, Arrieta-Montiel M, Christensen AC et al (2007) Plant mitochondrial recombination surveillance requires unusual *RecA* and *MutS* homologs. Plant Cell 19:1251–1264

Singh M, Brown GG (1991) Suppression of cytoplasmic male sterility by nuclear genes alters expression of a novel mitochondria gene region. Plant Cell 3:1349–1362

Small I, Suffolk R, Leaver CJ (1989) Evolution of plant mitochondrial genomes via substoichiometric intermediates. Cell 58:69–76

Teixeira RT, Farbos I, Glimelius K (2005) Expression levels of meristem identity and homeotic genes are modified by nuclear-mitochondrial interactions in alloplasmic male-sterile lines of *Brassica napus*. Plant J 42:731–742

Theissen G (2001) Development of floral organ identity: stories from the MADS house. Curr Opin Plant Biol 4:75–85

Unseld M, Marienfeld JR, Brandt P et al (1997) The mitochondrial genome of *Arabidopsis thaliana* contains 57 genes in 366,924 nucleotides. Nat Genet 15:57–61

Wan Z, Jing B, Tu J et al (2008) Genetic characterization of a new cytoplasmic male sterility system (hau) in *Brassica juncea* and its transfer to B-napus. Theor Appl Genet 116:355–362

Wang C, Chen X, Lan T et al (2006a) Cloning and transcript analysis of the chimeric gene associated with cytoplasmic male sterility in cauliflower (*Brasscia oleracea* var. Botrytis). Euphytica 151:111–121

Wang ZH, Zou YJ, Li XY et al (2006b) Cytoplasmic male sterility of rice with *boro* II cytoplasm is caused by a cytotoxic peptide and is restored by two related PPR motif genes via distinct modes of mRNA silencing. Plant Cell 18:676–687

Wang C, Chen X, Li H et al (2007) RNA editing analysis of mitochondrial *nad3/rps12* genes in cytoplasmic male sterility and male-fertile cauliflower (*Brassica oleracea* var. *botrytis*) by cDNA-SSCP. Bot Stud 48:13–23

Wintz H, Chen HC, Sutton CA et al (1995) Expression of the CMS-associated *urfS* sequence in transgenic petunia and tobacco. Plant Molecular Biology 28:83–92

Woodson JD, Chory J (2008) Coordination of gene expression between organellar and nuclear genomes. Nat Rev Genet 9:383–395

Xu Y, Arrieta-Montiel M, Virdi KS et al (2011) Muts homolog1 is a nucleoid protein that alters mitochondrial and plastid properties and plant response to high light. Plant Cell 23:3428–3441

Xu Y, Santamaria R, Virdi KS et al (2012) The chloroplast triggers developmental reprogramming when MUTS homolog1 is suppressed in plants. Plant Physiol 159:710–720

Yang WC, Ye D, Xu J et al (1999) The *SPOROCYTELESS* of *Arabidopsis* is required for initiation of sporogenesis and encodes a novel nuclear protein. Genes Dev 13:2108–2117

Yang JH, Zhang MF, Yu JQ et al (2005) Identification of alloplasmic cytoplasmic male-sterility line of leaf mustard synthesized by intra-specific hybridization. Plant Sci 168:865–871

Yang JH, Zhang MF, Yu JQ (2007) Alterations of RNA editing for mitochondrial *atp9* gene in new *orf220*-type cytoplasmic male-sterile line of stem mustard (*Brassica juncea* Coss. var. *tumida* Tsen et Lee). J Integr Plant Biol 49:672–677

Yang JH, Zhang MF, Yu JQ (2008a) MADS-box genes are associated with cytoplasmic homeosis in cytoplasmic male-sterile stem mustard as partially mimicked by specifically inhibiting mtETC. Plant Growth Regul 56:191–201

Yang JH, Zhang MF, Yu JQ (2008b) Relationship between cytoplasmic male sterility and *SPL-like* gene expression in stem mustard. Physiol Plant 133:426–434

Yang JH, Yan H, Zhang MF (2009a) Mitochondrial *atpA* gene is altered in a new orf220-type cytoplasmic male-sterile line of stem mustard (*Brassica juncea*). Mol Biol Rep 36:273–280

Yang JH, Zhang MF, Yu JQ (2009b) Mitochondrial *nad2* gene is co-transcripted with CMS-associated *orfB* gene in cytoplasmic male-sterile stem mustard (*Brassica juncea*). Mol Biol Rep 36:345–351

Yang JH, Liu XY, Yang XD, Zhang MF (2010) Mitochondrially-targeted expression of a cytoplasmic male sterility-associated *orf220* gene causes male sterility in *Brassica juncea*. BMC Plant Biol 10:231

Yang XD, Liu XY, Lv WH et al (2012) Reduced expression of *BjRCE1* gene modulated by nuclear-cytoplasmic incompatibility alters auxin response in cytoplasmic male-sterile *Brassica juncea*. PLoS One 7:e38821

Yasumot K, Terachi T, Yamagishi H (2009) A novel *Rf* gene controlling fertility restoration of Ogura male sterility by RNA processing of orf138 found in Japanese wild radish and its STS markers. Genome 52:495–504

Zhang MF, Chen LP, Wang BL et al (2003) Characterization of *atpA* and *orf220* genes distinctively present in a cytoplasmic male-sterile line of tuber mustard. J Hortic Sci Biotech 78:837–841

Zubko MK, Zubko E, Ryban AV et al (2001) Extensive development and metabolic alteration in cybrids *Nicotiana tabacum* (+*Hyoscyamus niger*) are caused by complex nuclear-cytoplasmic incompatatibility. Plant J 25:627–639

Chapter 10
Self-Incompatibility

Hiroyasu Kitasiba and Takeshi Nishio

Abstract Most of wild species and some crops in Brassicaceae have self-incompatibility, which is a mechanism to prevent self-fertilization after self-pollen recognition by stigmas to avoid inbreeding depression and to maintain genetic variations in populations. Genetics and molecular biology of self-incompatibility have advanced by the studies mainly using *Brassica* crops, such as *Brassica rapa* and *Brassica oleracea*, and this biological trait is used for practical breeding of *Brassica* crops. In this chapter, the study of self-incompatibility in Brassicaceae species and its application to practical breeding program of *Brassica* crops will be reviewed and discussed.

Key words Self-incompatibility • Brassicaceae • *S* haplotype

10.1 Introduction

Self-incompatibility is a system which inhibits germination and pollen tube growth of self-pollen. Many flowering plants have this system (de Nettancourt 2001) and Darwin (1876) investigated this trait in detail. Self-incompatibility has been considered to have biological roles for avoiding inbreeding depression and for maintaining genetic variation in populations. Many species of Brassicaceae plants such as *Brassica rapa* and *Brassica oleracea*, members of which are major vegetables, and genus *Arabidopsis*, which includes the biological model plant *Arabidopsis thaliana*, have the self-incompatibility system. Genetic studies have revealed that this trait is controlled by a single locus, the *S* locus, with multiple alleles (Bateman 1955). The self-incompatibility phenotype of the pollen side is determined sporophytically by

H. Kitasiba, Ph.D. • T. Nishio, Ph.D. (✉)
Graduate School of Agricultural Science, Tohoku University, Sendai, Miyagi 981-8555, Japan
e-mail: nishio@bios.tohoku.ac.jp

S.K. Gupta (ed.), *Biotechnology of Crucifers*, DOI 10.1007/978-1-4614-7795-2_10,   187
© Springer Science+Business Media, LLC 2013

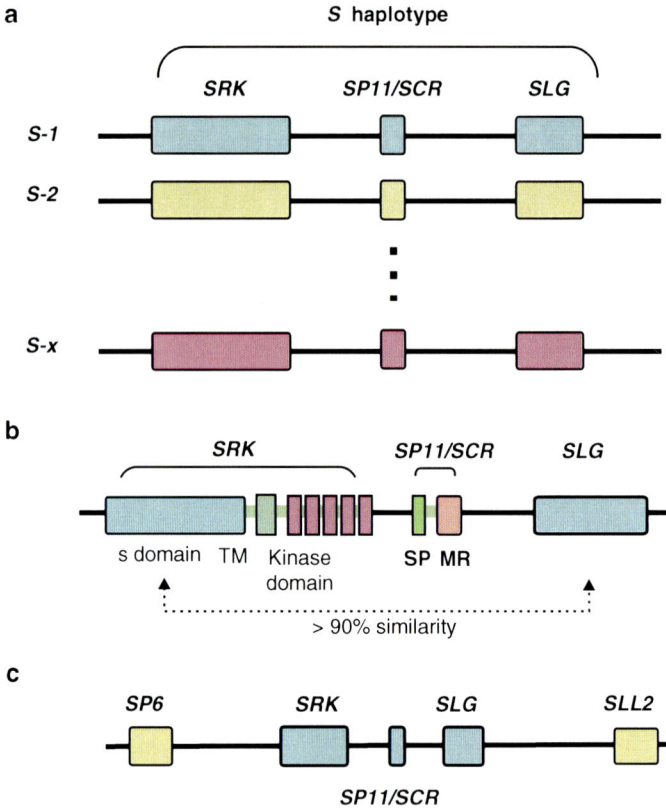

Fig. 10.1 Linkage of *SRK, SP11/SCR*, and *SLG* at the *S* locus. (**a**) Three genes, i.e., *SRK, SP11/ SCR*, and *SLG*, which are located on the *S* locus, and a set of alleles of these genes are called *S* haplotype. (**b**) Schematic structure of *SRK, SP11/SCR*, and *SLG*. *TM* transmembrane domain, *SP* signal peptide, *MR* mature protein. (**c**) Schematic drawing of the *S* locus and the flanking regions

a diploid genotype of a parent plant. Dominance relationships between *S* alleles are observed on both pollen and stigma sides and determine the *S* phenotype (Thompson and Taylor 1966). When the *S* phenotype of pollen is identical to that of a stigma, pollen grains are rejected by the stigma through inhibition of pollen germination and pollen tube penetration into stigma papilla cells. Molecular genetic studies have revealed three important genes located at the *S* locus, i.e., *S*-receptor kinase (*SRK*) as a female determinant, *S*-locus protein 11/ *S*-locus cystein rich protein (*SP11/ SCR*) as a male determinant, and *S*-locus glycoprotein (*SLG*) highly similar to the extracellular domain (*S*-domain) of SRK (Nasrallah et al. 1988; Stein et al. 1991; Schopfer et al. 1999; Suzuki et al. 1999). *SRK, SP11/SCR*, and *SLG* are closely linked to each other at the *S* locus, and alleles of these three genes are inherited by progeny as one set (Fig. 10.1a). Therefore, a set of alleles of these genes is termed "*S* haplotype" (Nasrallah and Nasrallah 1993). Studies of self-incompatibility in Brassicaceae plants were developed mainly in *Brassica* crops and have been applied to breeding programs. Since the early 2000s, self-incompatibility in the genus *Arabidopsis* has also been studied with a focus on the molecular basis of self-incompatibility and the evolution of *S* haplotypes. In this chapter, we review molecular genetics of self-incompatibility in Brassicaceae species and its application to practical breeding program of *Brassica* crops.

S alleles and *S* haplotypes have been commonly represented by numerical subscripts, e.g., S_1, S_2, and S_3. However, at present, alleles are represented with + or – under the standard nomenclature, e.g., *S+1* or *S-1*. Nonmutant dominant alleles are shown with + and mutant recessive alleles are shown with -. However, dominance relationships between *S* haplotypes are complicated. Therefore, in this review, we use – to represent *S* haplotypes and alleles of *S* locus genes. *S* haplotypes in *B. rapa* and *B. oleracea* are shown as *BrS-1* and *BoS-1*, respectively.

10.2 Genes Controlling Self-Incompatibility: *SLG*, *SRK* and *SP11/SCR*

10.2.1 SLG

Biochemical analyses of self-incompatibility in *Brassica* crops have revealed *S*-specific glycoproteins in the stigma (Nasrallah and Wallence 1967; Nishio and Hinata 1977), later designated *S*-locus glycoprotein, SLG. SLGs show allele-specific pI values and cosegregation with *S* alleles (Hinata and Nishio 1978; Nou et al. 1993). SLG is abundant in the cell wall of mature stigma papilla cells, where self-incompatibility reaction occurs (Nasrallah et al. 1988). Clones of *SLG* cDNA have been isolated and sequenced (Nasrallah et al. 1985, 1987), and the amino acid sequence of SLG has been determined (Takayama et al. 1987). This sequence information has enabled sequencing of many alleles of *SLG* in *Brassica* crops such as *B. rapa, B. oleracea*, and *Raphanus sativus* (Trick and Flavell 1989; Chen and Nasrallah 1990; Kusaba et al. 1997; Sakamoto et al. 1998). SLG has a hydrophobic signal peptide at the N-terminus for secretion to the outside of cells, several *N*-glycosylation sites, three hypervariable regions, and 12 conserved cysteine residues.

10.2.2 SRK

SRK has been found as an *S*-linked gene encoding *S* receptor kinase (SRK) (Stein et al. 1991). SRK protein has an extracellular domain (*S*-domain) highly similar to SLG, a transmembrane domain, and a serine/threonine kinase domain toward the C-terminus (Fig. 10.1b). Additionally, SLG and the *S*-domain of SRK of the same *S* haplotype exhibit high amino acid sequence similarity to each other, more than

90 % (Stein et al. 1991; Watanabe et al. 1994; Hatakeyama et al. 1998b; Sato et al. 2002). The *SRK* gene is also expressed in stigma papilla cells before flower anthesis, coincident with the timing of the self-incompatibility reaction (Watanabe et al. 1994; Delorme et al. 1995).

10.3   Which Is Responsible for Self-Incompatibility, *SLG* or *SRK*?

Which is responsible for self-incompatibility reaction as a female determinant, SLG or SRK, or both? To answer this question, gain-of-function studies have been conducted. In *B. rapa*, a transgenic plant transformed with an *SRK-9* allele rejected pollen grains of an *S-9* homozygote, while that with *SLG-9* allele did not (Takasaki et al. 2000). This experiment demonstrated that *SRK* is the sole female determinant for the *S*-haplotype specificity in the self-incompatibility response. Supporting this idea, despite a defect in the SLG-coding sequence and a lack of an *SLG* gene found in several *Brassica S* haplotypes, e.g., *BrS-32, BrS-36, BoS-18*, and *BoS-60*, plants having these *S* haplotypes have been found to exhibit a strong self-incompatibility reaction (Suzuki et al. 2000; Sato et al. 2002).

There are two views as to whether SLG is unnecessary for the self-incompatibility reaction. A transgenic *B. rapa* plant harboring the *SRK-9* transgene along with the *SLG-9* transgene has been reported to produce fewer seeds than a transgenic plant having the *SRK-9* transgene alone, when pollen grains of the *BrS-9* homozygote were pollinated. This result suggests that SLG enhances the self-incompatibility response (Takasaki et al. 2000). In contrast, another gain-of-function study using *Brassica napus* revealed no evidence for enhancement of self-incompatibility reaction by *SLG* (Silva et al. 2001). Gene conversion of the *S* domain of *SRK* with the *SLG* sequence in *S-54* of *B. rapa* has been found to result in loss of self-incompatibility (Fujimoto et al. 2006a), indicating that SLG of *S-54* does not recognize the *S* determinant of *S-54* pollen. Taken together these findings indicate that SRK is necessary in the self-incompatibility reaction and SLG is not. The role of SLG in enhancement of self-incompatibility is probably limited to some *S* haplotypes such as *BrS-9*, in which SLG exhibits high similarity (98 % amino acids) to the *S*-domain of SRK.

10.3.1   SP11/SCR

In 1999, two groups, Suzuki et al. (1999) and Schopfer et al. (1999) succeeded in identification of the male determinant gene named *SP11* (*S*-locus protein 11) or *SCR* (*S*-locus cysteine rich) of *B. rapa* by cloning and sequencing of the *S*-locus region. *SP11/SCR* is located near *SLG* and *SRK* on the *S* locus. Protein encoded by *SP11/SCR* was predicted to be a secreted protein with a signal peptide at the

N-terminus. Some characteristic sequences are observed in mature proteins after cleavage of the signal peptide: (1) a basic cysteine-rich protein, (2) eight conserved cysteine residues (designated C1 to C8), and (3) glycine residue and an aromatic amino acid residue present between C1 and C2 and between C3 and C4, respectively (Schopfer et al. 1999; Takayama et al. 2000; Watanabe et al. 2000). The sequence similarity of mature protein without the signal peptide is very low (below 50 %) between alleles within species (Watanabe et al. 2000; Sato et al. 2002; Okamoto et al. 2004). Transcripts are specifically detected in the anthers, especially in tapetum cells and microspores (Suzuki et al. 1999; Schopfer et al. 1999; Takayama et al. 2000; Shiba et al. 2002). Furthermore, pollen grains from transgenic plants expressing *SP11/SCR* were found to be rejected on the stigma of the *S* haplotype having the same *SP11/SCR* allele (Schopfer et al. 1999; Shiba et al. 2001). Thus, *SP11/SCR* is considered to be the male determinant of self-incompatibility.

10.4   Interaction Between *SP11/SCR* and *SRK*

Takayama et al. (2001) and Kachroo et al. (2001) have reported that an interaction of SP11/SCR with SRK from the same *S* haplotype induces autophosphorylation of SRK in an *S*-haplotype-specific manner. SRK tends to form a dimer (or oligomer) *in vivo* in the absence of SP11/SCR from the same *S* haplotype (Giranton et al. 2000; Shimosato et al. 2007; Naithani et al. 2007). An SRK molecule interacts more strongly with an SRK molecule derived from the same *SRK* allele than that derived from another *SRK* allele (Naithani et al. 2007). These results suggest that homodimerization of SRK is required for binding of an SP11/SCR protein from the same *S* haplotype. Three hypervariable regions have been identified in SLG and the *S*-domain of SRK (Kusaba et al. 1997; Sato et al. 2002). Domain swapping using several functional SRKs of self-incompatible *Arabidopsis lyrata* has confirmed that hypervariable regions are responsible for recognition of SP11/SCR (Boggs et al. 2009). Several amino acids have been found to be important for self-recognition in some *S* haplotypes, but a general rule on the key amino acids for self-recognition has not been formulated.

10.5   Genomic Structure of the *S* Locus

Sequencing of the *S* locus region containing *SLG, SRK,* and *SP11/SCR* has been carried out in *B. rapa* (Suzuki et al. 1999). In the flanking regions of the *S* core region containing *SRK, SP11/SCR,* and *SLG*, several genes of unknown function, such as *SP6* (*S*-locus protein 6) at one side and *SLL2* (*S*-locus-linked gene 2) at the opposite side, were found (Fig. 10.1c). Structural and transcriptional comparative analyses of *S* haplotypes in *B. rapa* and *B. napus* (Cui et al. 1999; Shiba et al. 2003) have suggested that the regions outside of the *SP6* and *SLL2* genes have high similarity between different S haplotypes, whereas the region delimited by these two genes is highly polymorphic and rich with S-haplotype-specific intergenic sequences.

On the basis of nucleotide sequences of SLG alleles or SRK alleles, S haplotypes have been classified into two groups in *Brassica*, class I and class II (Nasrallah et al. 1991). In class-II S haplotypes, the order of SRK and $SP11/SCR$ was the reverse of that in the class-I S haplotypes (Fukai et al. 2003). Comparative analysis has revealed that the direction of transcription of SRK, $SP11/SCR$, and SLG of four class-II S haplotypes in *B. rapa* is completely conserved, whereas the region between SRK and $SP11/SCR$ is highly diverse (Kakizaki et al. 2006). These diversities observed in class-I and class-II S core regions might contribute to the suppression of the recombination at the S locus. On the other hand, detailed phylogenetic analysis using S-locus sequences of similar S haplotypes between different species has demonstrated that although recombination is suppressed in $SP11/SCR$ and the S-domain of SRK, the kinase domain of SRK as well as other genes not responsible for self-recognition in the S locus have experienced recombination between S haplotypes (Takuno et al. 2007).

10.6 Dominance Relationships Between S Haplotypes

The nucleotide sequence similarities between SLG and SRK alleles are about 80–90 % within class I or class II; on the other hand, the similarities between classes are low, about 65 %. On the pollen side, S haplotypes of class I are generally dominant to those of class II (Thompson and Taylor 1966; Hatakeyama et al. 1998a). Analyses of transcripts of $SP11$ gene in developing anther have suggested that a dominant class-I S haplotype suppresses transcription of a class-II $SP11/SCR$ allele at the RNA level (Shiba et al. 2002). In addition, transcription of recessive $SP11/SCR$ alleles is suppressed epigenetically by *de novo* methylation of 5' promoter sequences in tapetum cells before the initiation of transcription of the recessive $SP11/SCR$ alleles (Shiba et al. 2006). Interestingly, the suppression of recessive $SP11/SCR$ alleles requires no expression of dominant $SP11/SCR$ alleles (Fujimoto et al. 2006b). What factor causes such a suppression of class-II $SP11/SCR$? Tarutani et al. (2010) have found an inverted genomic sequence on the S locus that was similar to sequences in class-II $SP11$ promoters and gave rise to a trans-acting small non-coding RNA expressed specifically in anthers. A small RNA induces methylation of the promoter of recessive class-II $SP11/SCR$ alleles and suppresses their transcription. However, the precise epigenetic mechanism of interaction between dominant and recessive $SP11/SCR$ alleles is still unclear.

In class II, four and three S haplotypes are known in *B. rapa* and *B. oleracea* to date, respectively, and linear dominance relationships have been observed (Thompson and Taylor 1966; Kakizaki et al. 2003). Similar *de novo* methylation and transcription suppression of recessive $SP11/SCR$ alleles have been observed between class-II S haplotypes (Kakizaki et al. 2003; Shiba et al. 2006). However, the trans-acting non-coding RNA sequence has not been found in the S locus of

BrS-44, which is the most dominant among class-II *S* haplotypes in *B. rapa*, and there was little transcription of the other *S* haplotypes (Tarutani et al. 2010). This suggested the possibility that another consensus sequence is responsible for epigenetic control or that a different molecular genetic system is involved. In the class-I *S* haplotype, non-linear dominance relationships among *S* haplotypes have been observed. However, the correlation between the suppression and the *de novo* methylation of recessive alleles is unknown. Together with class-II *S* haplotypes, further analyses on the pollen side are required.

Thompson and Taylor (1966) and Hatakeyama et al. (1998a) have studied the dominance relationship between *S* haplotypes on the stigma side, but a general relationship like that on the pollen side has not been uncovered. The expression level of a transgene of a dominant *SRK* allele has been found to be lower than a recessive endogenous allele, but the dominance relationship between these *SRK* alleles was not changed (Hatakeyama et al. 2001), suggesting that the dominance relationships between *S* haplotypes on the stigma side are determined by SRK protein itself, not by the difference of their relative expression levels. The kinase domain of *BrSRK-54* is highly similar to those of *BrSRK-8* and *BrSKR-46*, but *BrS-54* is codominant and recessive to *BrS-8* and *BrS-46*, respectively (Takuno et al. 2007, and our unpublished data), implying the importance of the *S*-domain for the dominance relationships of *SRK*. Naithani et al. (2007) have shown a preferential homodimerization by an yeast two-hybrid system using eSRKs, i.e., SRK without the kinase domain, suggesting the possibility that its strong affinity might cause codominance relationships, which are exhibited by a majority of *SRK* alleles (Naithani et al. 2007). However, these inferences are not sufficient to explain the dominance relationships of *SRK* alleles. Further genetic and biochemical studies are necessary.

10.7 Self-Compatibility of Amphidiploid Species in *Brassica*

Most mono-genomic *Brassica* species are self-incompatible. For example, *B. rapa*, *Brassica nigra*, and *B. oleracea* having A, B, and C genomes, respectively (U 1935), have self-incompatibility. However, di-genomic species in this genus, i.e., amphidiploid species with two different genomes, exhibit self-compatibility (SC) (Takahata and Hinata 1980). *B. napus*, *Brassica juncea*, and *Brassica carinata* are amphidiploid species having AC, AB, and BC genomes, respectively (U 1935), and the natural species exhibit self-compatibility. However, artificially synthesized amphidiploids are self-incompatible (Hinata and Nishio 1980). This discrepancy between natural amphidiploid species and artificial amphidiploids has been unresolved for a long time.

Analysis of *S* haplotypes in 45 lines of *B. napus* identified six *S* genotypes, most of them being combinations of a class-I dominant *S* haplotype and a class-II recessive one (Okamoto et al. 2007). In self-compatible *B. napus* 'Westar', suppression of transcription of a dominant *SP11/SCR* allele due to an insertion mutation in its promoter region as well as suppression of a recessive *SP11/SCR* allele caused by the dominant *S* haplotype have been observed (Okamoto et al. 2007). Such suppression of a recessive *SP11/SCR* allele by a dominant *S* haplotype having a nonfunctional *SP11/SCR* allele has also been observed in a monogenomic *Brassica* species (Fujimoto et al. 2006b). This self-compatible trait has been found to be complemented by a transgene of a functional homoeologous dominant *SP11/SCR* allele from *B. rapa* (Tochigi et al. 2011). In two other genotypes, *SRK* alleles of dominant *S* haplotypes have been revealed to have knockout mutations. These results suggest that a single mutation event which occurs in a dominant *S* haplotype can cause self-compatibility in amphidiploid *Brassica* plants. Causative factors responsible for self-compatibility in the other three genotypes have not yet been revealed.

10.8   *S* Haplotypes in *Arabidopsis* Species

Arabidopsis lyrata L. and *Arabidopsis halleri* have a self-incompatibility trait determined by recognition between *SRK* and *SCR*. In the past decade, many full or partial sequences of *SRK* and *SCR* alleles have been cloned in both species (Schierup et al. 2001; Prigoda et al. 2005; Bechsgaard et al. 2006; Tsuchimatsu et al. 2010), and, furthermore, genomic structures of some *S* haplotypes have been revealed (Kusaba et al. 2001; Goubet et al. 2012). The sequence conservation in flanking regions on both sides of the *S* locus is observed among different *S* haplotypes. *ARK3* and U-box genes are located at each boundary, respectively, with high sequence-similarity among *S* haplotypes, while within the region delimited by these genes, the order and orientation of *SRK* and *SCR* as well as their physical sizes are different among *S* haplotypes (Kusaba et al. 2001; Goubet et al. 2012) and micro-synteny has been found to be very low because of insertions of variable numbers and types of transposable elements in an *S*-haplotype-dependent manner, as reported previously in *Brassica* (Fujimoto et al. 2006c). Comparison of the flanking genes of the *S* locus by phylogenetic analysis has shown much deeper genealogies for *SCR* and *SRK*, and four phylogenetic classes have been defined based on *SRK* sequences (Prigoda et al. 2005), which correlate with classes of *SCR* sequences. This suggests coevolution of *SRK* alleles and *SCR* alleles. The dominance relationships among *S* haplotypes on both pollen- and stigma sides in *A. lyrata* have been observed to be related to four phylogenetic classes (Prigoda et al. 2005), like the relationship observed between two classes in *Brassica*, i.e., class I and class II.

A. *thaliana*, which is a relative of *A. lyrata* and *A. harreli*, is a self-compatible species. *A. thaliana* has non-functional alleles of *SRK* (psudo-*SRK*, *ΨSRK*) and *SCR* (*ΨSCR*) (Kusaba et al. 2001), and their sequences from respective *A. thaliana* accessions show polymorphisms (Nasrallah et al. 2004; Boggs et al. 2009). Based on the analysis of *SRK* and *SCR* sequence divergence, three distinct *ΨS* haplotypes, designated *ΨSA*, *ΨSB*, and *ΨSC*, have been identified (Kusaba et al. 2001; Tang et al. 2007; Shimizu et al. 2008). Transformation with just two transgenes, namely, *SRK-b* and *SCR-b* isolated from the *S-b* haplotype of *A. lyrata*, conferred a self-incompatibility phenotype ("self-incompatible *A. thaliana*" hereafter) (Nasrallah et al. 2002; Boggs et al. 2009). This result implies that the signal transduction pathway downstream from SRK is active even in *A. thaliana*. However, self-incompatibility phenotypes of some ecotypes in *A. thaliana*, such as Col-0, RLD, and Ws-0, transformed with *SRK-b* and *SCR-b* have been found to be limited to a short period from mature buds to young flowers in floral development (flower stage 13 and early stage 14 as referred to in Smyth et al. 1990), while in other ecotypes such as C24, Cvi-0, and Sha, transgenic plants with these genes were revealed to sustain a stable self-incompatibility phenotype almost identical to that of naturally self-incompatible *Arabidopsis* species (Nasrallah et al. 2004; Boggs et al. 2009). A recent study on the difference of self-incompatibility reaction among transformants of these ecotypes has identified a candidate factor, PUB8, responsible for modification of the degree of the self-incompatibility reaction (Liu et al. 2007). This *PUB8* gene is closely linked to the *S* locus and encodes an uncharacterized ARM repeat and a U box-containing protein that regulates *SRK* transcript levels (Liu et al. 2007).

10.9 Signaling Pathway for Self-Incompatibility Reaction

As described above, self-recognition on the stigma begins with the interaction of SP11/SCR, a pollen ligand, and SRK, a stigma receptor kinase. The signaling pathway following activation of SRK and an immediate cause of self-pollen rejection are poorly understood. However, several candidate genes have been proposed to be downstream effectors in the SRK signaling pathway in *Brassica* species. Three of them, i.e., *MLPK, ARC1*, and *Exo70A1*, have been studied in detail.

10.9.1 MLPK

Hinata and Okazaki (1986) have reported a self-compatible *B. rapa* 'Yellow sarson', which is cultivated in India, and two independent loci, i.e., *S* and *M*, controlling this trait have been analyzed by classical genetic study. A molecular genetic study on the *S* locus of 'Yellow sarson' has identified an insertion of a retrotransposon-like sequence in the first intron of SRK and deletion in the promoter of *SP11*, resulting in suppression of transcription of both mutated genes (Fujimoto et al. 2006b). On the other hand, the *M* locus has been surveyed by positional cloning (Murase et al. 2004). A gene encoding a membrane-anchored cytoplasmic protein kinase, designated *MLPK*, was found to be located on this locus. The kinase domain of the mutated *MLPK* has been found to have a single base nonsynonymous substitution causing no autophosphorylation activity and a loss of localization at the cell membrane (Murase et al. 2004). Two isoforms from the same allele by alternative splicing can interact with SRK (Kakita et al. 2007). Although there has been no stable complementation experiment reported, transient expression of a wild-type *MLPK* allele by particle bombardment in the stigma papilla cells of plants with *mlpk/mlpk* genotype has been reported to complement the self-incompatibility phenotype of these cells (Kakita et al. 2007). Thus, MLPK is thought to function in SRK-mediated signaling.

10.9.2   ARC1

ARC1 (ARM-repeat containing 1) has been isolated in a yeast two-hybrid screen as the protein which interacts with and is phosphorylated by the SRK-kinase domain (Gu et al. 1998). *ARC1* is expressed specifically in stigma tissues. Suppression of *ARC1* transcripts by antisense gene transfer in a self-incompatible *B. napus* line has been found to cause a partial breakdown of self-incompatibility (Stone et al. 1999). Since ARC1 is an E3 ubiquitin ligase, compatibility factors are considered to be ubiqitinated by ARC1 and degraded through a proteasomal system (Stone et al. 2003), causing pollen rejection in *Brassica*.

10.9.3   Exo70A1

Exo70A1 has been isolated as a factor interacting with ARC1 in *Brassica napus* (Samuel et al. 2009). Exo70A1 is a putative component of the exocyst complex, which generally functions in polarized secretion in yeast and animals (Munson and Novick 2006; Synek et al. 2006). Overexpression of Exo70A1 in epidermal cells of self-incompatible *B. napus* has been found to confer partial breakdown of the trait, whereas suppression of *Exo70A1* in self-compatible *B. napus* and *A. thaliana* by an RNA-i method and T-DNA insertion, respectively, inhibited pollen adhesion, hydration, and germination (Samuel et al. 2009). These results suggest the possibility that an exocyst complex is involved in the secretion of compatibility factors.

10.10   Function of Orthologs of *MLPK*, *ARC1*, and *Exo70A1* in *Arabidopsis*

An *Arabidopsis MLPK* ortholog, *AtAPK1b*, is located on the genomic region of *A. thaliana* with high synteny to that of *B. rapa*. Two isoforms from the same *AtAPK1b* allele have also been predicted by alternative splicing and the expression is preferentially detected in the stigma (Kakita et al. 2007). However, transforma- tion of *A. thaliana* plants having an inactivated *AtAPK1b* allele due to T-DNA inser- tion with *SRK-b* and *SCR-b* alleles from *A. lyrata* has shown self-incompatibility (Kitashiba et al. 2011), suggesting that functional *AtAPK1b* is not required for the self-incompatibility function of the stigmas.

A part of *A. thaliana* chromosome 2 has been found to share extensive synteny with the ARC1-containing region of A04 linkage group in *B. rapa* (Kitashiba et al. 2011). In the *A. thalina* syntenic region, fragmented *ARC1* sequences have been found to be interspersed with other neighboring genes by comparative genomic sequence analysis (Kitashiba et al. 2011). This fragmentation has also been observed even in *A. lyrata,* a self-incompatible species. Additionally, a null allele of its closest paralog, *PUB17*, has been reported to be unable to disrupt self-incompatibility in self-incompatible *A. thaliana* (Rea et al. 2010). Thus, there is no evidence for the involvement of *ARC1* and an *ARC1*-like gene in the *Arabidopsis* species. Similarly, overexpression of *AtExo70A1* in stigma tissue under self-incompatible *A. thaliana* was shown not to weaken the self-incompatibility response (Kitashiba et al. 2011), contrary to the report of *B. napus* (Samuel et al. 2009). Thus, *MLPK/ARC1/Exo70A1*-based model of signaling pathway proposed for *Brassica* is not applied to self-incompatible *A. thaliana.* Why are these discrepancies between these two genera observed? Is it because of different signaling pathways in the two genera or because of different cumulative outcomes of multiple SRK-mediated signaling pathways? *A. thaliana* is a model plant utilized to understand many biological processes in plants. Since both *Brassica* species and *Arabidopsis* species belong to the same Brassicaceae family, it is difficult to support the above-described ideas. To resolve the discrepancies, reexamination of the roles of the three genes in *Brassica* species will be necessary.

10.11 Collection of *S* Haplotypes

S tester lines, which are *S* homozygous lines with well-characterized *S* haplotypes used for *S* haplotype identification by pollination tests, were originally developed in *B. oleracea* by Thompson and Taylor (1966). They increased the number of *S* haplotypes to 36 from *S-1* to *S-47*, 11 missing *S* haplotypes of which are due to synonyms, and the collection of *S* haplotypes was further expanded by Ockendon (2000), who added 14 *S* haplotypes, giving *S* haplotype names to *S-67*, the total number of *S* haplotypes becoming 50.

Thirteen *S* tester lines from *S-1* to *S-13* in *B. rapa* were developed by Nishio and Hinata (1978) and were used for analysis of *S* glycoproteins in stigmas. Three of them, i.e., *S-8, S-9,* and *S-12,* were later used as materials for molecular genetic study of self-incompatibility (Schopfer et al. 1999; Suzuki et al. 1999; Takayama et al. 2001). *S* haplotype collection was further expanded by Nou et al. (1993), who collected 22 *S* haplotypes, giving them new names from *S-21* to *S-49*, because seeds of the original *B. rapa* *S* tester lines except *S-8, S-9,* and *S-12* had lost their germination ability. *S-43, S-28,* and *S-24* of Nou et al. (1993) are synonyms of *S-8, S-9,* and *S-12,* respectively. Eleven new *S* haplotypes from *S-53* to *S-72* were added by Takuno et al. (2010). The total number of *S* haplotypes in *B. rapa* identified so far is 39. We determined *SLG* sequences of old dead seeds of the original *S* tester lines and revealed correspondence between *S* haplotypes of Nishio and Hinata (1978) and those with deposited nucleotide sequences (Table 10.1). Since there are

S haplotypes	Corresponding published *SLG* sequences
S-1	*S-60*
S-2	*S-40*
S-3	*S-26*
S-4	*S-56*
S-5	*S-44*
S-6	*S-44*
S-7	*No corresponding sequence*
S-10	*S-27*
S-11	*S-12*
S-13	*S-8*

Table 10.1 Correspondence of *SLG* sequences of the original *S* tester lines in *Brassica rapa* to the published sequences

many missing numerals of *S* haplotypes and even *S-99* without identification of *S* haplotypes from *S-73* to *S-98*, renaming might be required.

S tester lines from *S-1* to *S-21* in *R. sativus* have been reported by Sakamoto et al. (1998). Niikura and Matsuura (1999) have collected 37 *S* haplotypes and named them *S-201* to *S-237*. There may be some synonyms between them. Furthermore, Lim et al. (2002) have independently collected ten *S* haplotypes in *R. sativus* and named them from *S-1* to *S-10*, causing confusion of *S* haplotype names. *S* haplotypes from *S-1* to *S-10* of Lim et al. (2002) are different from those of Sakamoto et al. (1998). Different nucleotide sequences of *SLG* alleles have been deposited in DNA databases under the same *S* haplotype names, producing homonyms. To decrease such confusion, exchange of materials is required.

Based on his longtime efforts to collect new *S* haplotypes in *B. oleracea*, Ockendon (2000) reported that the total number of *S* haplotypes in *B. oleracea* is unlikely to go much beyond 50. Comparing *S* haplotypes in natural populations of *B. rapa* in Japan and Turkey, Nou et al. (1993) have estimated the total number of *S* haplotypes in *B. rapa* to be more than 100. Since the methods for estimation of the total number of *S* haplotypes in a species are different between these studies, it may not be appropriate to simply compare the total numbers between them. If it is allowed, *B. rapa* can be considered to have a much larger number of *S* haplotypes than *B. oleracea*. In fact, *B. oleracea* has only three class-II *S* haplotypes, i.e., *S-2*, *S-5*, and *S-15*, which are pollen recessive *S* haplotypes more frequently found in populations than other class-I *S* haplotypes, while *B. rapa* has four class-II *S* haplotypes, i.e., *S-29*, *S-40*, *S-44*, and *S-60*. *S-40*, *S-44*, and *S-60* in *B. rapa* have nucleotide sequences similar to those of *S-5*, *S-2*, and *S-15* in *B. oleracea*, respectively (Sato et al. 2006). Furthermore, we have recently identified two new class-II *S* haplotypes in *B. rapa* (Kawanabe et al. unpublished data). Although identified class-I *S* haplotypes in *B. rapa* are fewer than those in *B. oleracea*, new *S* haplotypes can still be identified with relative ease, suggesting that the number of *S* haplotypes in *B. rapa* is more than those in *B. oleracea*.

Although there is some confusion regarding *S* haplotypes, i.e., synonyms and homonyms, the *S* tester lines whose nucleotide sequences have been determined and deposited earlier in the DNA database will become the international standards of *S* haplotypes. Determination and deposition of *SP11, SRK*, and *SLG* sequences of many *S* haplotypes are important. However, in some *S* haplotypes, only partial sequences of *SLG* alleles or *S* domain of *SRK* alleles have been deposited. It will be difficult to make such *S* haplotypes the international standards without nucleotide sequence data of the coding regions of *SP11, SRK*, and *SLG* alleles.

International standards of *S* tester lines available to the public are required. However, the production of *S* homozygous lines results in inbreeding depression. Since these lines are maintained by hand pollination, contamination of different *S* haplotypes sometimes occurs. Determination of the nucleotide sequences of *S*-locus genes enables development of a rapid reliable method for *S* haplotype identification by which contamination of different *S* haplotypes can be efficiently eliminated from the *S* tester lines.

10.12   *S* Haplotype Sequences

At the beginning of the study of nucleotide sequencing of alleles of the *S*-locus genes, cloned DNA isolated from cDNA libraries or genomic DNA libraries were used for sequencing (Nasrallah et al. 1985, 1988; Trick and Flavell 1989; Stein et al. 1991). However, such a method is much too laborious and does not reveal variation of nucleotide sequences of the *S*-locus genes because of the presence of many *S* haplotypes in a species. Although polymerase chain reaction (PCR) is an efficient method for isolation of specific sequences, high homology between *SLG* and *SRK* makes it difficult to amplify one of them. Use of the 3'-UTR sequence of *SLG* as a primer has enabled specific amplification of *SLG*, and nucleotide sequences of many *SLG* alleles have been determined in *B. rapa, B. oleracea* (Kusaba et al. 1997), and *R. sativus* (Sakamoto et al. 1998). Because of the presence of a highly variable long intron in *SRK* and *SP11*, specific amplification of their genomic sequences have been unsuccessful except in class-II *SP11* (Sato et al. 2006). Reverse transcription-PCR (RT-PCR) has been used for amplification of *SRK* (Sato et al. 2002) and *SP11* (Watanabe et al. 2000; Sato et al. 2002), and many of their alleles have been sequenced. Nucleotide sequences and deduced amino acid sequences of the *S* domain of *SRK* are highly similar to those of *SLG*, especially within the same *S* haplotypes, suggesting frequent gene conversion between *SRK* and *SLG* at the same *S* locus (Nishio and Kusaba 2000; Sato et al. 2002). Variations of nucleotide sequences and deduced amino acid sequences of *SP11* are higher than those of *SRK* and *SLG* (Watanabe et al. 2000; Sato et al. 2002).

An interesting finding in a sequencing study of a large number of *SLG, SRK*, and *SP11* alleles is the presence of interspecific pairs of *S* haplotypes between *B. rapa* and *B. oleracea* (Kusaba et al. 1997; Kusaba and Nishio 1999; Kimura et al. 2002; Sato et al. 2003). Comparison of a large number of *SLG* sequences in *B. rapa* and *B. oleracea* has revealed high sequence variation of *SLG* within a species and highly similar sequences between these two species, e.g., *SLG-25* in *B. rapa* and *SLG-14* in *B. oleracea* (Kusaba et al. 1997). Combinations of *S* haplotypes having similar *SLG* sequences between *B. rapa* and *B. oleracea* have also been revealed to have similar sequences in *SRK* and *SP11* (Kusaba and Nishio 1999; Kimura et al. 2002), and these combinations of *S* haplotypes are called interspecific pairs. Because of the presence of interspecific incompatibility between *B. rapa* and *B. oleracea*, it is not easy to compare recognition specificities of *S* haplotypes between the two species. Use of interspecific hybrids between *B. rapa* and *B. oleracea*, *B. rapa* transformants with *B. oleracea SP11* alleles, and bioassay of SP11 proteins synthesized in bacteria have revealed the interspecific pairs of *S* haplotypes, i.e., *BrS-46/BoS-7*, *BrS-8/BoS-32*, *BrS-36/BoS-24*, *BrS-47/BoS-12*, and *BrS-41/BoS-64*, to have the same recognition specificities (Kimura et al. 2002; Sato et al. 2003). In class-II *S* haplotypes, *BrS-60*, *BrS-40*, and *BrS-44* have sequences of *SLG*, *SRK*, and *SP11* highly similar to those of *BoS-15*, *BoS-5*, and *BoS-2b*, respectively, suggesting these pairs of *S* haplotypes to be interspecific pairs, but recognition specificity might have changed after speciation of these species. Although *BrS-60* has been found to have the same recognition specificity as *BoS-15*, the recognition specificity of *BrS-40* has been revealed to be slightly different from that of *BoS-5*, and that of *BrS-44* has been shown to be completely different from that of *BoS-2b* (Sato et al. 2006). A species in a different genus, *R. sativus*, has also been found to have an *S* haplotype similar to that in *B. rapa* (Okamoto et al. 2004). Recognition specificity of SP11 of *S-21* in *R. sativus* has been revealed to be the same as that of SP11 of *S-9* in *B. rapa* (Sato et al. 2004).

10.13 Evolutionary Diversification of *S* Haplotypes in *Brassica* Species

Distribution of the same *S* haplotypes in different species in *Brassica* or in different genera suggests that the diversification of *S* haplotypes occurred before divergence of species and the same ancestral *S* haplotypes are shared by different species or genera in the evolution of *Brassica* species as shown in Fig. 10.2. Except for a few *S* haplotypes, e.g., *BrS-44* and *BoS-2b*, recognition specificities of *S* haplotypes may not have changed for a long time.

The presence of interspecific pairs of *S* haplotypes having the same recognition specificities is probably not due to recent introgression of *S* haplotypes. Determining *S* locus sequences of interspecific pairs between *B. rapa* and *B. oleracea*, Fujimoto et al. (2006c) have revealed that *B. oleracea S* haplotypes have longer sequences than those of *B. rapa S* haplotypes in three interspecifc pairs. Although genome sequences covering *SLG*, *SRK*, and *SP11* have been determined in *B. rapa*, determination of the sequence of the region covering *SLG*, *SRK*, and *SP11* in *B. oleracea* has not been possible because of the presence of many retrotransposon-like sequences. It has been hypothesized that many transposable elements have been inserted into the *S* locus of *B. oleracea* but not so much into that of *B. rapa* after speciation.

Common Ancestral Species

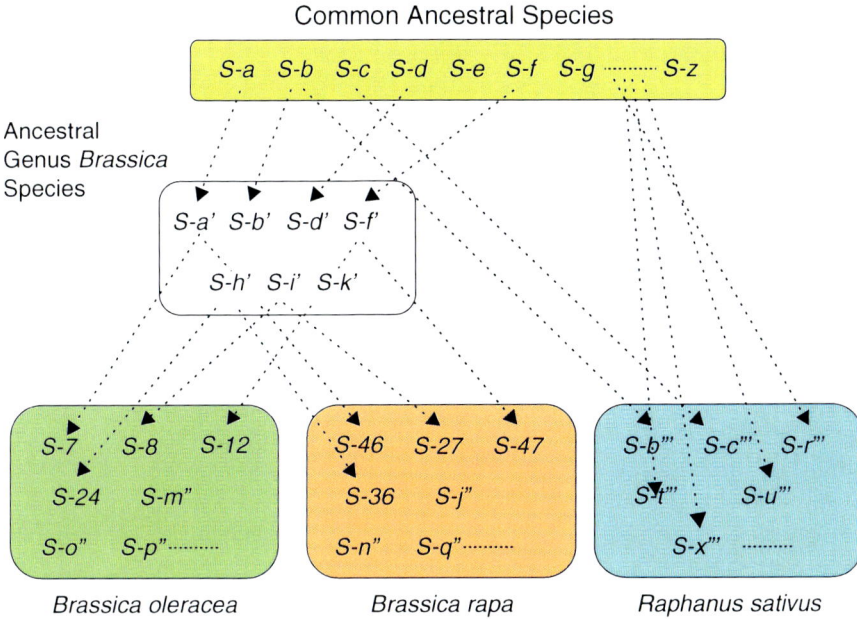

Fig. 10.2 A model of *S* haplotype evolution in *Brassica* and *Raphanus*

How new *S* haplotypes were generated has long been an enigma of self-incompatibility study. Comparison of nucleotide sequences and deduced amino acid sequences of *SRK* and *SP11* between different *S* haplotypes in a species suggest that new *S* haplotypes were generated not by a single mutation event but by accumulation of amino acid changes in both *SRK* and *SP11* alleles (Chookajorn et al. 2004; Sato et al. 2004). Mutations that greatly change the recognition specificity of only *SRK* or *SP11* in an *S* haplotype cause unsuccessful recognition between SRK and SP11 and result in the loss of self-incompatibility.

10.14 Development of Methods for *S* Haplotype Identification and Their Application to *Brassica* Genetic Study and Breeding

S haplotypes have been identified by test crossing with the *S* tester lines, and incompatibility between a test plant and an *S* tester line investigated by observation of seed set or pollen tube behavior indicates these plants to have the same *S* haplotype. This is the standard method for *S* haplotype identification, but handling of many *S* tester lines is extremely laborious. Furthermore, environmental conditions have an influence on the incompatibility reaction. Isoelectric focusing analysis of *S* glycoprotein

(=SLG) has been used for *S* haplotype identification (Nishio and Hinata 1980; Nou et al. 1993), but this method is costly. Genomic Southern blot analysis of *SLG* can also be used for *S* haplotype identification (Nasrallah et al. 1988; Okazaki et al. 1999). However, this analysis is time consuming and requires skill for DNA analysis. A simple rapid method for *S* haplotype identification is required.

PCR-RFLP analysis of *SLG* has been reported to be usable for *S* haplotype identification (Brace et al. 1993; Nishio et al. 1994). This method is easy and reliable, and therefore commonly used for basic studies of self-incompatibility, F₁ hybrid breeding of Brassicaceae vegetables, and seed purity tests. However, although discrimination of *S* genotypes is easy with this method, identification of *S* genotypes of heterozygotes is difficult. Furthermore, there are some *S* haplotypes lacking *SLG* (Sato et al. 2002). The same method can be applied for polymorphism analysis of *S-RNase* in *S* haplotype identification of Rosaceae fruit trees (Ishimizu et al. 1999; Takasaki et al. 2004; Tao et al. 1999).

Since alleles of *SP11* are highly variable in a species, they can be identified by allele-specific hybridization to dot-blotted DNA (Fujimoto and Nishio 2003). In this method, plant genomic DNA is dot-blotted onto a nylon membrane, and therefore intensities of detected signals are sometimes not high enough, and analysis requires skill as in the case of genomic Southern blot analysis. Dot-blotting of DNA amplified by PCR using multiple primer pairs for many *SP11* alleles has been found to enable allele-specific detection of *SP11* with high signal intensities (Takuno et al. 2010; Oikawa et al. 2011). Dot-blot hybridization of *SP11* alleles enables identification of *S* haplotypes, and analysis of *S* genotypes of heterozygotes and a large number of plants at one time. This method may replace PCR-RFLP of *SLG*.

10.15   Use of Self-Incompatibility in F₁ Hybrid Breeding

In self-incompatible Brassicaceae vegetables, self-incompatibility has been used for seed production of F₁ hybrids. Although male sterility is generally used for this purpose in many self-compatible crops, use of self-incompatibility is simpler and more efficient than the use of male sterility. In the use of self-incompatibility, two inbred lines having different *S* haplotypes are cultivated side by side, and all obtained seeds are F₁ hybrids. In the use of male sterility, a male sterile line is used as a maternal parent, and F₁ seeds can be harvested only from the male sterile line. For developing self-incompatible inbred lines, selfings are repeated by overcoming self-incompatibility. Bud pollination is commonly used for selfing. For large scale seed production of inbred lines, CO_2 gas is supplied to a plastic house, where each inbred line is grown, at flowering time to enhance CO_2 concentration to about 4 % (Nakanishi and Hinata 1975). Spraying of salt solution is also effective in overcoming self-incompatibility in *Brassica*, but effect is unstable. In Brassica vegetables, a single cross is commonly used for hybrid seed production, while double cross is common in radish, which sets a small number of seeds per pod. The biggest problem of F₁ hybrid seed production using self-incompatibility is contamination of selfed seeds in F_1 seeds, which results from weak self-incompatibility of inbred lines. Selection of inbred lines having strong self-incompatibility is required for production of F_1 seeds with high purity. Since the strength of self-incompatibility depends on S haplotypes, selection of S haplotypes using the S haplotype identification method is important.

In self-compatible *B. napus*, cytoplasmic male sterility is used conventionally for efficient seed production of F_1 hybrids. However, male-sterile lines are not preferred by insect pollinators because of the absence of pollen grains and abnormality of nectaries or flower morphology caused by male sterility cytoplasm. Tochigi et al. (2011) exhibited a model for F_1 hybrid breeding in *B. napus* using a dominant class-II S haplotype, which shows strong self-incompatibility. An artificially synthesized *B. napus* line with *BrS-44,* which is the most dominant among class-II S haplotypes, and a recessive S haplotype exhibiting strong self-incompatibility was developed (Tochigi et al. 2011). This artificial line as the maternal side was crossed with natural *B. napus* 'Westar' as the pollen side, which harbors a non-functional class-I S haplotype. The developed F_1 hybrid did not express any *SP11/SCR* genes, and then showed a self-compatible phenotype resulting in high seed production. Additionally, the natural *B. napus* plants with two other combinations, in which mutations occurred in class-I *SRK* alleles, are also considered to be usable as a paternal parent (Okamoto et al. 2007; Tochigi et al. 2011). Taken together, if an elite self-incompatible *B. napus* parent having a highly dominant class-II S haplotype were developed, the elite parent as the maternal side would be useful in F_1 hybrid breeding by crossing with natural self-compatible *B. napus*. A self-incompatible *B. napus* line with class-II S haplotypes has been developed by interspecific crossing with *B. rapa,* that might be useful in F_1 hybrid breeding (Zhang et al. 2008). The causes of self-compatibility in other amphidiploids, e.g., *B. juncea* and *B. carinata*, are still unclear. If the phenotype is due to mutations in S haplotypes, the same F_1 hybrid breeding system will be usable in these species.

References

Bateman AJ (1955) Self-incompatibility systems in angiosperms. III. Cruciferae. Heredity 9:52–68
Bechsgaard JS, Castric V, Charlesworth D, Vekemans X, Schierup MH (2006) The transition to self-compatibility in *Arabidopsis thaliana* and evolution within *S*-haplotypes over 10 Myr. Mol Biol Evol 23:1741–1750
Boggs NA, Dwyer KG, Nasrallah ME, Nasrallah JB (2009) *In vivo* detection of residues required for ligand-selective activation of the *S*-locus receptor in *Arabidopsis*. Curr Biol 19:1–6
Brace J, Ockendon DJ, King GJ (1993) Development of a method for the identification of *S* alleles in *Brassica oleracea* based on digestion of PCR-amplified DNA with restriction endonucleases. Sex Plant Reprod 6:133–138
Chen CH, Nasrallah JB (1990) A new class of *S* sequences defined by a pollen recessive self-incompatibility allele of *Brassica oleracea*. Mol Gen Genet 222:241–248
Chookajorn T, Kachroo A, Ripoll DR, Clark AG, Nasrallah JB (2004) Specificity determinants and diversification of the *Brassica* self-incompatibility pollen ligand. Proc Natl Acad Sci USA 101:911–917

Cui Y, Brugière N, Jackman L, Bi Y-M, Rothstein SJ (1999) Structural and transcriptional comparative analysis of the *S* locus regions in two self-incompatible *Brassica napus* lines. Plant Cell 11:2217–2231

Darwin C (1876) The effects of cross and self fertilization in the vegetable kingdom. John Murray, London

de Nettancourt D (2001) Incompatibility and incongruity in wild and cultivated plants, 2nd edn. Springer Verlag, Berlin

Delorme V, Giranton J-L, Hetzfeld Y, Friry A, Heizmann P, Ariza MJ, Dumas C, Gaude T, Cock JM (1995) Characterization of the S locus genes, SLG and SRK, of the *Brassica* S3 haplotype: identification of a membrane-localized protein encoded by the *S* locus receptor kinase gene. Plant J 7:429–440

Fujimoto R, Nishio T (2003) Identification of *S* haplotypes in *Brassica* by dot-blot analysis of *SP11* alleles. Theor Appl Genet 106:1433–1437

Fujimoto R, Sugimura T, Nishio T (2006a) Gene conversion from *SLG* to *SRK* resulting in self-compatibility in *Brassica rapa*. FEBS Lett 580:425–430

Fujimoto R, Sugimura T, Fukai E, Nishio T (2006b) Suppression of gene expression of a recessive *SP11/SCR* allele by an untranscribed *SP11/SCR* allele in *Brassica* self-incompatibility. Plant Mol Biol 61:577–587

Fujimoto R, Okazaki K, Fukai E, Kusaba M, Nishio T (2006c) Comparison of the genome structure of the self-incompatibility (*S*) locus in interspecific pairs of *S* haplotypes. Genetics 173:1157–1167

Fukai E, Fujimoto R, Nishio T (2003) Genomic organization of the *S* core region and the *S* flanking regions of a class-II *S* haplotype in *Brassica rapa*. Mol Gen Genom 269:361–369

Giranton J-L, Dumas D, Cock JM, Gaude T (2000) The integral membrane *S*-locus receptor kinase of *Brassica* has serine/threonine kinase activity in a membranous environment and spontaneously forms oligomers *in planta*. Proc Natl Acad Sci USA 97:3759–3764

Goubet PM, Berge`s H, Bellec A, Prat E, Helmstetter N, Mangenot S, Gallina S, Holl A-C, Fobis-Loisy I, Vekemans X, Castric V (2012) Contrasted patterns of molecular evolution in dominant and recessive self-incompatibility haplotypes in *Arabidopsis*. PLoS Genet 8:e1002495

Gu T, Mazzurco M, Sulaman W, Matias DD, Goring DR (1998) Binding of an arm repeat protein to the kinase domain of the *S*-locus receptor kinase. Proc Natl Acad Sci USA 95:382–387

Hatakeyama K, Watanabe M, Takasaki T, Ojima K, Hinata K (1998a) Dominace relationships between *S*-allele in self-incompatible *Brassica campestris* L. Heredity 80:241–247

Hatakeyama K, Takasaki T, Watanabe M, Hinata K (1998b) High sequence similarity between SLG and the receptor domain of SRK is not necessarily involved in higher dominance relationships in stigma in self-incompatible *Brassica rapa* L. Sex Plant Reprod 11:292–294

Hatakeyama K, Takasaki T, Suzuki G, Nishio T, Watanabe M, Isogai A, Hinata K (2001) The *S* receptor kinase gene determines dominance relationships in stigma expression of self-incompatibility in *Brassica*. Plant J 26:69–76

Hinata K, Nishio T (1978) *S*-allele specificity of stigma proteins in *Brassica oleracea* and *B. campestris*. Heredity 41:93–100

Hinata K, Nishio T (1980) Self-incompatibility in crucifers. In: Tunoda S, Hinata K, Gomez-Campo C (eds) Brassica crops and wild allies. Japan Scientific Societies Press, Tokyo, pp 223–234

Hinata K, Okazaki K (1986) Role of stigma in the expressioin of self-incompatibility in crucifers in view of genetic analysis. In: Mulcahy DL, Mulcahy GB, Ottaviano E (eds) Biotechnology and ecology of pollen. Springer, New York, pp 185–190

Ishimizu T, Inoue K, Shimonaka M, Saito T, Terai O, Norioka S (1999) PCR-based method for identifying the *S*-genotypes of Japanese pear cultivars. Theor Appl Genet 98:961–967

Kachroo A, Schopfer CR, Nasrallah ME, Nasrallah JB (2001) Allele-specific receptor-ligand interactions in *Brassica* self-incompatibility. Science 293:1824–1826

Kakita M, Murase K, Iwano M, Matsumoto T, Watanabe M, Shiba H, Isogai A, Takayama S (2007) Two distinct forms of *M*-locus protein kinase localize to the plasma membrane and interact directly with *S*-locus receptor kinase to transduce self-incompatibility signaling in *Brassica rapa*. Plant Cell 19:3961–3973

Kakizaki T, Takada Y, Ito A, Suzuki G, Shiba H, Takayama S, Isogai A, Watanabe M (2003) Linear dominance relationship among four class-II S haplotypes in pollen is determined by the expression of *SP11* in *Brassica* self-incompatibility. Plant Cell Physiol 44:70–75

Kakizaki T, Takada Y, Fujioka T, Suzuki G, Satta Y, Shiba H, Isogai A, Takayama S, Watanabe M (2006) Comparative analysis of the *S*-intergenic region in class-II *S* haplotypes of self-incompatible *Brassica rapa* (syn. *campestris*). Genes Genet Syst 81:63–67

Kimura R, Sato K, Fujimoto R, Nishio T (2002) Recognition specificity of self-incompatibility maintained after the divergence of *Brassica oleracea* and *Brassica rapa*. Plant J 29:215–223

Kitashiba H, Liu P, Nishio T, Nasrallah JB, Narallah ME (2011) Functional test of *Brassica* self-incompatibility modifiers in *Arabidopsis thaliana*. Proc Natl Acad Sci 108:18173–18178

Kusaba M, Nishio T (1999) Comparative analysis of *S* haplotypes with very similar *SLG* alleles in *Brassica rapa* and *Brassica oleracea*. Plant J 17:83–91

Kusaba M, Nishio T, Satta Y, Hinata K, Ockendon DJ (1997) Striking sequence similarity in inter- and intra-specific comparisons of class I *SLG* alleles from *Brassica oleracea* and *Brassica campestris*: implications for the evolution and recognition mechanism. Proc Natl Acad Sci USA 94:7673–7678

Kusaba M, Dwyer K, Hendershot J, Vrebalov J, Nasrallah JB, Nasrallah ME (2001) Self-incompatibility in the genus *Arabidopsis*: characterization of the *S* locus in the outcrossing *A. lyrata* and its autogamous relative *A. thaliana*. Plant Cell 13:627–643

Lim SH, Cho HJ, Lee SJ, Cho YH, Kim BD (2002) Identification and classification of *S* haplotypes in *Raphanus sativus* by PCR-RFLP of the *S* locus glycoprotein (*SLG*) gene and the *S* locus receptor kinase (*SRK*) gene. Theor Appl Genet 104:1253–1263

Liu P, Sherman-Broyles S, Nasrallah ME, Nasrallah JB (2007) A cryptic modifier causing transient self-incompatibility in *Arabidopsis thaliana*. Current Biol 17:734–740

Munson M, Novick P (2006) The exocyst defrocked, a framework of rods revealed. Nat Struct Mol Biol 13:577–581

Murase K, Shiba H, Iwano M, Che F-S, Watanabe M, Isogai A, Takayama S (2004) A membrane-anchored protein kinase involved in *Brassica* self-incompatibility signaling. Science 303:1516–1519

Nagaharu U (1935) Genome analysis in *Brassica* with special reference to the experimental formation of *B. napus* and peculiar mode of fertilization. Jpn J Bot 7:389–452

Naithani S, Chookajorn T, Ripoll DR, Nasrallah JB (2007) Structural modules for receptor dimerization in the *S*-locus receptor kinase extracellular domain. Proc Natl Acad Sci USA 104:12211–12216

Nakanishi T, Hinata K (1975) Self-seed production by CO_2 gas treatment in self-incompatible cabbage. Euphytica 24:117–120

Nasrallah JB, Nasrallah ME (1993) Pollen-stigma signaling in the sporophytic self-incompatibility response. Plant Cell 5:1325–1335

Nasrallah ME, Wallence DH (1967) Immunogenetics of self-incompatibility in *Brassica oleracea* L. Heredity 22:519–527

Nasrallah JB, Kao TH, Goldberg ML, Nasrallah ME (1985) A cDNA clone encoding an *S-locus-specific* glycoprotein from *Brassica oleracea*. Nature 318:263–267

Nasrallah JB, Kao TH, Chen CH, Goldberg ML, Nasrallah ME (1987) Amino-acid sequence of glycoproteins encoded by three alleles of the *S* locus of *Brassica oleracea*. Nature 326:617–619

Nasrallah JB, Yu S-M, Nasrallah ME (1988) Self-incompatibility genes of *Brassica oleracea*: expression, isolation, and structure. Proc Natl Acad Sci USA 85:5551–5555

Nasrallah JB, Nishio T, Nasrallah ME (1991) The self-incompatibility genes of *Brassica*: expression and use in genetic ablation of floral tissues. Annu Rev Plant Physiol Plant Mol Biol 42:393–422

Nasrallah ME, Liu P, Nasrallah JB (2002) Generation of self-incompatible *Arabidopsis thaliana* by transfer of two *S* locus genes from *A. lyrata*. Science 297:247–249

Nasrallah ME, Nasrallah ME, Liu P, Sherman-Broyles S, Boggs NA, Boggs NA, Nasrallah JB (2004) Natural variation in expression of self-incompatibility in *Arabidopsis thaliana*: implications for the evolution of selfing. Proc Natl Acad Sci USA 101:16070–16074

Niikura S, Matsuura S (1999) Genetic variation of the S-alleles and level of self-incompatibility in the Japanese cultivated radish (*Raphanus sativus* L.). Breed Res 1:211–220

Nishio T, Hinata K (1977) Analysis of S specific proteins in stigma of *Brassica oleracea* L. by isoelectric focusing. Heredity 38:391–396

Nishio T, Hinata K (1978) Stigma proteins in self-incompatible *Brassica campestris* L. and self-compatible relatives, with special reference to S-allele specificity. Jpn J Genet 53:27–33

Nishio T, Hinata K (1980) Rapid detection of S-glycoproteins of self-incompatible crucifers using Con-A reaction. Euphytica 29:217–221

Nishio T, Sakamoto K, Yamaguchi J (1994) PCR-RFLP of S locus for identification of breeding lines in cruciferous vegetables. Plant Cell Rep 13:546–550

Nishio T, Kusaba M (2000) Sequence diversity of *SLG* and *SRK* in *Brassica oleracea* L. Ann Bot 85(suppl A):141–146

Nou IS, Watanabe M, Isuzugawa K, Isogai A, Hinata K (1993) Isolation of S-alleles from a wild population of *Brassica campestris* L. at Balcesme, Turkey and their characterization by S-glycoproteins. Sex Plant Reprod 6:71–78

Ockendon DJ (2000) The S-allele collection of *Brassica oleracea*. Acta Hort 539:25–30

Oikawa E, Takuno S, Izumita A, Sakamoto K, Hanzawa H, Kitashiba H, Nishio T (2011) Simple and efficient methods for S genotyping and S screening in genus *Brassica* by dot-blot analysis. Mol Breed 28:1–12

Okamoto S, Sato Y, Sakamoto K, Nishio T (2004) Distribution of similar self-incompatibility (S) haplotypes in different genera, *Raphanus* and *Brassica*. Sex Plant Reprod 17:33–39

Okamoto S, Odashima M, Fujimoto R, Sato Y, Kitashiba H, Nishio T (2007) Self-compatibility in *Brassica napus* is caused by independent mutations in S-locus genes. Plant J 50:391–400

Okazaki K, Kusaba M, Ockendon DJ, Nishio T (1999) Characterization of S tester lines in *Brassica oleracea*: polymorphism of restriction fragment length of *SLG* homologues and isoelectric points of S-locus glycoproteins. Theor Appl Genet 98:1329–1334

Prigoda NL, Nassuth A, Mable BK (2005) Phenotypic and genotypic expression of self-incompatibility haplotypes in *Arabidopsis lyrata* suggests unique origin of alleles in different dominance classes. Mol Biol Evol 22:1609–1620

Rea AC, Liu P, Nasrallah JB (2010) A transgenic self-incompatible *Arabidopsis thaliana* model for evolutionary and mechanistic studies of crucifer self-incompatibility. J Exp Bot 61:1897–1906

Sakamoto K, Kusaba M, Nishio T (1998) Polymorphism of the S-locus glycoprotein gene (*SLG*) and the S-locus related gene (*SLR1*) in *Raphanus sativus* L. and self-incompatible ornamental plants in the Brassicaceae. Mol Gen Genet 258:397–403

Samuel MA, Chong YT, Haasen KE, Aldea-Brydges MG, Stone SL, Goring DR (2009) Cellular pathways regulating responses to compatible and self-incompatible pollen in *Brassica* and *Arabidopsis* stigmas intersect at Exo70A1, a putative component of the exocyst complex. Plant Cell 21:2655–2671

Sato K, Nishio T, Kimura R, Kusaba M, Suzuki T, Hatakeyama K, Ockendon DJ, Satta Y (2002) Coevolution of the S-locus genes *SRK*, *SLG* and *SP11/SCR* in *Brassica oleracea* and *B. rapa*. Genetics 162:931–940

Sato Y, Fujimoto R, Toriyama K, Nishio T (2003) Commonality of self-recognition specificity of S haplotypes between *Brassica oleracea* and *Brassica rapa*. Plant Mol Biol 52:617–626

Sato Y, Okamoto S, Nishio T (2004) Diversification and alteration of recognition specificity of the pollen ligand SP11/SCR in self-incompatibility of *Brassica* and *Raphanus*. Plant Cell 16:3230–3241

Sato Y, Sato K, Nishio T (2006) Interspecific pairs of class II S haplotypes having different recognition specificities between *Brassica oleracea* and *Brassica rapa*. Plant Cell Physiol 47:340–345

Schierup MH, Mable BK, Awadalla P, Charlesworth D (2001) Identification and characterization of a polymorphic receptor kinase gene linked to the self-incompatibility locus of *Arabidopsis lyrata*. Genetics 158:387–399

Schopfer CR, Nasrallah ME, Nasrallah JB (1999) The male determinant of self-incompatibility in *Brassica*. Science 286:1697–1700

Shiba H, Takayama S, Iwano M, Shimosato H, Funato M, Nakagawa T, Che F-S, Suzuki G, Watanabe M, Hinata K, Isogai A (2001) A pollen coat protein, SP11/SCR, determines the pollen *S* specificity in the self-incompatibility of *Brassica* species. Plant Physiol 125:2095–2103

Shiba H, Kakizaki T, Iwano M, Entani T, Ishimoto K, Shimosato H, Che F-S, Satta Y, Ito A, Takada Y, Watanabe M, Isogai A, Takayama S (2002) The dominance of alleles controlling self-incompatibility in *Brassica* pollen is regulated at the RNA level. Plant Cell 14:491–504

Shiba H, Kenmochi M, Sugihara M, Iwano M, Kawasaki S, Suzuki G, Watanabe M, Isogai A, Takayama S (2003) Genomic organization of the *S*-locus region of *Brassica*. Biosci Biotechnol Biochem 67:622–626

Shiba H, Kakizaki T, Iwano M, Tarutani Y, Watanabe M, Isogai A, Takayama S (2006) Dominace relationships between self-incompatibility alleles controlled by DNA methylation. Nat Genet 38:297–299

Shimizu KK, Shimizu-Inatsugi R, Tsuchimatsu T, Purugganan MD (2008) Independent origins of self-compatibility in *Arabidopsis thaliana*. Mol Ecol 17:704–714

Shimosato H, Yokota N, Shiba H, Iwano M, Entani T, Che F-S, Watanabe M, Isogai A, Takayama S (2007) Characterization of the SP11/SCR high-affinity binding site involved in self/nonself recognition in *Brassica* self-incompatibility. Plant Cell 19:107–117

Silva NF, Stone SL, Christie LN, Sulaman W, Nazarian KAP, Burnett LA, Arnoldo MA, Rothstein SJ, Goring DR (2001) Expression of the *S* receptor kinase in self-incompatible *Brassica napus* cv. Westar leads to the allele-specific rejection of self-incompatible *Brassica napus* pollen. Mol Gent Genom 265:552–559

Smyth DR, Bowman JL, Meyerowitz EM (1990) Early flower development in *Arabidopsis*. Plant Cell 2:755–767

Stein JC, Howlett B, Boyes DC, Nasrallah ME, Nasrallah JB (1991) Molecular cloning of a putative receptor protein kinase gene encoded at the self-incompatibility locus of *Brassica oleracea*. Proc Natl Acad Sci USA 88:8816–8820

Stone SL, Arnoldo MA, Goring DR (1999) A breakdown of *Brassica* self-incompatibility in ARC1 antisense transgenic plants. Science 286:1729–1731

Stone SL, Anderson EM, Mullen RT, Goring DR (2003) ARC1 is an E3 ubiquitin ligase and promotes the ubiquitination of proteins during the rejection of self-incompatible *Brassica* pollen. Plant Cell 15:885–889

Suzuki G, Kai N, Hirose T, Fukui K, Nishio T, Takayama S, Isogai A, Watanabe M, Hinata K (1999) Genomic organization of the *S* locus: identification and characterization of genes in *SLG/SRK* region of *S9* haplotype of *Brassica campestris* (syn. *rapa*). Genetics 153:391–400

Suzuki T, Kusaba M, Matsushita M, Okazaki K, Nishio T (2000) Characterization of *Brassica* *S*-haplotypes lacking *S*-locus glycoprotein. FEBS Lett 482:102–108

Synek L, Schlager N, Elias M, Quentin M, Hauser MT, Zarsky V (2006) AtEXO70A1, a member of a family of putative exocyst subunits specifically expanded in land plants, is important for polar growth and plant development. Plant J 48:54–72

Takahata Y, Hinata K (1980) A variation study of subtribe Brassicinae by principal component analysis. In: Tunoda S, Hinata K, Gomez-Campo C (eds) Brassica crops and wild allies. Japan Scientific Societies Press, Tokyo, pp 33–49

Takasaki T, Hatakeyama K, Suzuki G, Watanabe M, Isogai A, Hinata K (2000) The *S* receptor kinase determines self-incompatibility in *Brassica* stigma. Nature 403:913–916

Takasaki T, Okada K, Castillo C, Moriya Y, Saito Y, Sawamura Y, Norioka N, Norioka S, Nakanishi T (2004) Sequence of the *S9*-RNase cDNA and PCR-RFLP system for discriminating *S1*- to *S9* allele in Japanese pear. Euphytica 135:157–167

Takayama S, Isogai A, Tsukamoto C, Ueda YK, Hinata KO, Suzuki A (1987) Sequences of *S*-glycoproteins, products of the *Brassica campestris* self-incompatibility locus. Nature 326:102–105

Takayama S, Shiba H, Iwano M, Shimosato H, Che F-K, Kai N, Watanabe M, Suzuki G, Hinata K, Isogai A (2000) The pollen determinant of self-incompatibility in *Brassica campestris*. Proc Natl Acad Sci USA 97:1920–1925

Takayama S, Shimosato H, Shiba H, Funato M, Che F-K, Watanabe M, Iwano M, Isogai A (2001) Direct ligand-receptor complex interaction controls *Brassica* self-incompatibility. Nature 413:534–538

Takuno S, Fujimoto R, Sugimura T, Sato K, Okamoto S, Zhang S-L, Nishio T (2007) Effects of recombination on hitchhiking diversity in the *Brassica* self-incompatibility locus complex. Genetics 177:949–958

Takuno S, Oikawa E, Kitashiba H, Nishio T (2010) Assessment of genetic diversity of accessions in Brassicaceae genetic resources by frequency distribution analysis of *S* haplotypes. Theor Appl Genet 120:1129–1138

Tang C, Toomajian C, Sherman-Broyles S, Plagnol V, Guo Y-L, Hu TT, Clark RM, Nasrallah JB, Weigel D, Nordborg M (2007) The evolution of selfing in *Arabidopsis thaliana*. Science 317:1070–1072

Tao R, Yamane H, Sugiura A, Murayama H, Sassa H, Mori H (1999) Molecular typing of *S*-alleles through identification, characterization and cDNA cloning for *S*-RNases in sweet cherry. J Am Soc Hortic Sci 124:224–233

Tarutani Y, Shiba H, Iwano M, Kakizaki T, Suzuki G, Watanabe M, Isogai A, Takayama S (2010) Trans-acting small RNA determines dominance relationships in *Brassica* self-incompatibility. Nature 466:983–987

Thompson KF, Taylor JP (1966) Non-linear dominance relationships between *S* alleles. Heredity 21:345–362

Tochigi T, Udagawa H, Li F, Kitashiba H, Nishio T (2011) The self-compatibility mechanism in *Brassica napus* L. is applicable to F$_1$ hybrid breeding. Theor Appl Genet 123:475–482

Trick M, Flavell RB (1989) A homozygous *S* genotype of *Brassica oleracea* expresses two *S*-like genes. Mol Gen Genet 218:112–117

Tsuchimatsu T, Suwabe K, Shimizu-Inatsugi R, Isokawa S, Pavlidis P, Sta¨dler T, Suzuki G, Takayama S, Watanabe M, Shimizu KK (2010) Evolution of self-compatibility in Arabidopsis by a mutation in the male specificity gene. Nature 464:1342–1347

Watanabe M, Takasaki T, Toriyama K, Yamakawa S, Isogai A, Suzuki A, Hinata K (1994) A high degree of homology exists between the protein encoded by *SLG* and the *S* receptor domain encoded by *SRK* in self-incompatible *Brassica campestris* L. Plant Cell Physiol 35:1221–1229

Watanabe M, Ito A, Takada Y, Ninomiya C, Kakizaki Y, Takahata Y, Hatakeyama K, Hinata K, Suzuki G, Takasaki T, Satta Y, Shiba H, Takayama S, Isogai A (2000) Highly divergent sequences of the pollen self-incompatibility (*S*) gene in class-I *S* haplotypes of *Brassica campestris* (syn. *rapa*). FEBS Lett 473:139–144

Zhang XG, Ma CZ, Tang JY, Tang W, Tu JX, Shen JX, Fu TD (2008) Distribution of *S* haplotypes and its relationship with restorer-maintainers of self-incompatibility in cultivated *Brassica napus*. Theor Appl Genet 117:171–179

Chapter 11
Hybrid Technology in Cruciferous Vegetables

Muhammad Awais Ghani, Langlang Zhang, Junxing Li, Bin Liu, and Liping Chen

Abstract Use of male sterile plants has become an important technique in heterosis breeding,which simplify and reduce the cost of hybrid seed production. Cruciferous vegetables are very important crops in the world, and two types of male sterility have been mainly explored in cruciferous vegetables (1) nuclear male sterility, this kind of male sterility is controlled by the dominant or recessive nuclear genes, and its sterility is easy to be restored, but difficult to maintained; (2) cytoplasmic male sterility, in which male sterility is controlled by a particular cytoplasmic male sterile gene (S). Cytoplasmic male sterility is easy to be maintained but complicated to be recovered. Male sterility can be produced by different ways; natural mutation, wide hybridization, and protoplast fusion. Ogu CMS and Polma CMS were found in radish and *B. napus* respectively by natural mutations. Among the male sterile materials, most of them were obtained by the wide hybridization among varieties, species and genera. CMS cabbage was produced by the fusion of leaf protoplasts from fertile cabbage and CMS Ogura broccoli lines. The Pol CMS had been transferred from CMS *B. napus* to Chinese cabbage. Chinese breeders produced many cabbage varieties after introduced male sterility from other materials. A new cabbage hybrid varieties Zhonggan no.16, 17 and 18 were produced by hybridization of dominant genic male sterility line and inbred line in China. A stable CMS line of tuber mustard was developed by distant crosses and subsequent backcrosses and induced 100 % male sterility. To date, extensive efforts have been made on identification of male sterility systems and the possibilities of development of hybrids in application.

Keywords Cruciferous vegetables • Male sterility • Natural mutation • Protoplast fusion • Wide hybridization • Chinese cabbage • Cabbage • Mustard

M.A. Ghani, M.S., M.D. • L. Zhang, B.S. • J. Li, M.D. • B. Liu, B.S. • L. Chen, Ph.D. (✉)
Department of Horticulture, Zhejiang University, 866 Yuhangtang Road,
310058 Hangzhou, Zhejiang Province, China
e-mail: chenliping@zju.edu.cn

S.K. Gupta (ed.), *Biotechnology of Crucifers*, DOI 10.1007/978-1-4614-7795-2_11,   209
© Springer Science+Business Media, LLC 2013

11.1 Introduction

The concept of heterosis was first proposed by Shull (1908). It's synonymous is hybrid vigor. Heterosis or hybrid vigor refers to the phenomenon that progeny of diverse inbred varieties exhibit greater biomass, speed of development, and fertility than the better of the two parents (James et al. 2010). First suggestions to exploit the heterosis were made by Hayes and Jones (1916) in cucumber crops. And later, this phenomenon has been studied in other crops, such as maize, sorghum, sunflower, oil seed, rice, Chinese cabbage and cabbage etc. The practical utilization of heterosis in crop plants has been greatly facilitated during the past years by the use of controlled pollination system for low cost, large scale emasculation of the seed parents of hybrids. The method is now being used in commercial production of hybrid (Duvick 1959). There are two ways to control pollinations in crops: (1) removal of anthers or male flowers, and (2) the use of male sterility systems. Male sterility mean, the inability of the plant to produce fertile pollen, which provides one of the most efficient and direct controlled pollination for hybrid seed production in crops on large scale (Prakash et al. 2009). Therefore, the plant breeders had begun to pay attention to utilizing male sterility system for hybrid seed production in various crops.

Cruciferous vegetables, belonged to the family Brassicaceae, are very important crops, and are widely cultivated in the world, with many genera, species such as cauliflower, cabbage, Chinese cabbage and broccoli. Cruciferous vegetables are rich in nutrients, including several carotenoids (beta-carotene, lutein, zeaxanthin); vitamins C, E, and K; folate; and minerals. In addition, Cruciferous vegetables contain a group of substances known as glucosinolates, which are responsible for the pungent aroma and bitter flavor in Cruciferous vegetables. Many scientific studies have presented the health benefits of these vegetables such as, broccoli and cauliflower which help to prevent bladder, ovarian, prostate, colon, breast and lung cancers. Similarly more eating cabbage can reduce the menstrual pain for endometriosis sufferer due to the anti-inflammation. Radish have folic acid, anthocyanins and Vitamin C, which is very effective in fighting cancer especially in fighting against oral, colon, intestinal, stomach and kidney cancers (http://ingenira.hubpages.com).

Cruciferous vegetables have shown positive response towards heterosis. There are two ways for hybrid seed production in Cruciferous vegetables i.e. (1) use of self-incompatibility line; (2) use of male sterile line. However, there are some problems existed by using self-incompatibility system. One is the parents showed continuous viability recession (depression). The other is the parent need to be manually stripping bud pollination with high cost, and it is very difficult to achieve 100 % purity in the hybrids. The use of male sterile line to produce a generation hybrids can solve these problems. This has prompted people to pay more attention to the genetic sources of male sterility and use of them in heterosis breeding of Cruciferous crops. The main type of male sterility in Cruciferous vegetable is summarized as: (1) nuclear male sterility (NMS), this kind of male sterility is controlled by the dominant or recessive nuclear genes, and its sterility is easy to be restored, but not easy to be maintained; (2) cytoplasmic male sterility (CMS), this type of male sterility is controlled by a particular cytoplasmic male sterile gene (S). The plants show cytoplasm male sterility no matter what is a nuclear gene. It has the characteristics of the cytoplasmic inheritance, maternal inheritance, and its sterility is easy to be kept, but not easy to be recovered

Nuclear male sterile types generally exists in many plants, the types of male sterility is controlled by nucleus sterile gene, which can be divided into dominant and recessive gene. Dominant nuclear male sterile (DNMS, DGMS) have been reported in several Cruciferous crops, e.g. Chinese cabbage and cauliflower (Van Der Meer 1987; Ruffio-Chable et al. 1993). In the cabbage, the dominant male sterility gene Ms-cd1 (c, cabbage; d, dominant) was identified as a spontaneous mutation (Fang et al. 1997). Li et al. (1990) further comfirmed the male sterility controlled by dominate gene in *B. oleracea* and *B. napus*. Zhang et al. (1990) also developed the male sterility system controlled by dominant gene "88-1A" that was used for hybrid seed production in Chinese cabbage. Recessive genic male sterile (RGMS) was discovered as a spontaneous mutant in a *B. napus* in 1991 (Chen et al. 1993). As for recessive nuclear gene male sterility, it can be controlled by a single gene or double genes. The male sterile line of *B. napus* S445AB, Norins3AB and Chinese cabbage Xiaoqingkou 127 were controlled by single recessive nuclear gene (Niu et al. 1980). Some recessive genic male sterile lines contain two recessive genes (Bnms3 and Bnms4) with one epistatic suppressor gene (BnRf or BnEsp) (Chen et al. 1998). Hou et al. (1990) reported double recessive male sterility nuclear genes i.e. "S45AB" and "117AB" in *B. napus*.

Cytoplasmic male sterility (CMS) was first discovered by Correns (1906). As of 1972, CMS had been reported in 140 species from 47 genera and 20 families (Laser and Lersten 1972; Edwardson 1970; Grun 1976). Several types of cytoplasmic male sterility has been identified in Cruciferous vegetables which include Raphanus/ogu (Ogura 1968), tour (Mathias 1985; Stiewe and Röbbelen 1994), polima or pol (Fu 1981), and nap (Shiga and Baba 1971, 1973; Thompson 1972). Pol CMS was discovered in 1972 (Fu 1981), and became the first CMS system to be extensively utilized for hybrid seed production (Fan et al. 1986; Röbbelen 1991). The pol CMS is sensitive to environment in certain nuclear backgrounds leading to breakdown of sterility, which reduces hybridity levels in F_1 hybrids. So far, nine CMS lines in *B. napus* have been characterized into four types: pol, ogu, nap, tour (Yang et al. 1998). The recent reports of CMS in Brassica include CMS 681A of spontaneous origin in a mutant line of a *B. napus* cv. Xiangyu 13. It has the same restorer and maintainer relationship as pol CMS but possessed different mitochondrial DNA sequences (Liu et al. 2005). Another CMS system named "126-1" was identified in a population of doubled haploids of synthetic *B. napus* ISN 706 and subsequently transferred to *B. juncea* (Sodhi et al. 2006). Several CMS systems of alloplasmic origin have also been reported in *B. juncea* containing cytoplasm of several wild species viz. *B. oxyrrhina* (Prakash and Chopra 1990), *B. tournefortii* (Pradhan et al. 1991; Arumugam et al. 1996), *Diplotaxis siifolia* (Rao et al. 1994), *Moricandia arvensis* (Prakash et al. 1998), *Erucastrum canariense* (Prakash et al. 2001), *Diplotaxis erucoides* (Bhat et al. 2006) and *D. berthautii* (Bhat et al. 2008). This chapter will discuss the production of male sterility, development of hybrid technology in major Cruciferous vegetables of the world and their future prospects.

11.2 The Main Way to Produce Male Sterility in Cruciferous Vegetables

11.2.1 Natural Mutations

Natural mutation also occurred in male sterility gene as same way as in other genes in nature. Mutation rates increase up to 0.2 % by the mutagenic effect of cosmic rays, or other factors. Most of natural genetic sterility is due to recessive nuclear mutation, and cytoplasmic or genetic male sterility, while dominant genetic male sterility was less reported. In 1968, Radish cytoplasm infertility source (Ogu CMS) was first found in Kagoshima, Japan (Ogura 1968). In 1970, Dickson found male sterile materials controlled by recessive gene in *B. oleracea* crops. In China, the earliest Pol CMS is found in sterile plants of Polma population of the *B. napus* by Fu and Yang (1990). Fang et al. (1997) obtained 79-399-3 male sterile plants in wild cabbage named 79-399, which the major gene controlled by a single dominant nuclear gene. Zhang found a dominant male sterile gene, and received 100 % sterile groups from Wanquanqin varieties in Chinese cabbage (Zhang et al. 1990). He and Shi (1987) found male sterility resource in a number of local varieties of Chinese radish, and developed "77-01A" and other male sterile lines, which controlled by a nuclear recessive gene, and sterile cytoplasm.

11.2.2 Development of CMS Lines Through Wide Hybridization

CMS plants have appeared during breeding programs in a series of backcrosses resulted from the exchange of the nuclear genome in one breeding line and other cytoplasm line. CMS plants also produced due to successive backcrosses by exchange of nuclear genome of one variety to the nuclear genome other variety. Such process led the concept that male sterility could arise from nuclear cytoplasmic incompatibility. Later, attempts were made to create CMS lines deliberately by a program of crossing one species as the male parent with another species as female parent. For example, such deliberate backcrosses have given rise to sunflower CMS lines (Le-Clercq 1983) and CMS lines of *Nicotiana* that contain nuclear genomes of one specie and cytoplasmic genomes of another (Gerstel 1980).

Among the male sterile materials, most of them were obtained by the hybridization among varieties, species and genera. For example, Bannerot et al. (1977) reported the male sterility was introduced into Cruciferous vegetable by intergeneric hybridization that was practically used for the F_1 seed production. A hybrid can be produced by the cross between *B. napus* and *B. oleracea*, if the *B. oleracea* is used as recurrent parent in backcross. Male sterility continued, *B. oleracea* progenies exchange the cytoplasm with *B. napus* through backcross. Male fertility was fully restored when the C genome was reintroduced in the cytoplasm of *B. oleracea* (Chiang and Crete 1987). The cross between *B. juncea* (2n = 36, AABB) and *B. rapa* (2n = 20, AA) was happened by repeated backcrossings with fertile tuber mustard using as recurrent parents and as corresponding maintainers of CMS which produced the donor of CMS tuber mustard (*B. juncea* var. tumida Tsen et Lee) with an alloplasmic cytoplasm. It was successful to produce a stable male sterile line of tuber mustard showing 100 % in male sterility (Chen et al. 1995).

The *B. campestris*, *B. napus* and *B. juncea* contained the oxyrhina cytoplasm ('oxy') which lead to the male sterility (Prakash and Chopra 1990). Male sterile line was isolated from *B. napus*, *B. oleracea*, and *B. campestris* which are now widely applied in agricultural production (Huang et al. 2009). Due to interactive inheritance of genetic male sterility, that was controlled by two types of genes. 100 % male sterility plants were confirmed by Wei et al. (1992) and Zhang et al. (1990). Later on, Feng et al. (1996) interpreted the new idea that the inheritance was controlled by multiple alleles and also proposed a multiple allele model for genetic male sterility. Cao and Li (1980) was produced the genic male sterile A and B lines i.e. Bcajh97-01A/B from 'Aijiaohuang' (*B. campestris* ssp. chinensis). This sterility is controlled by a recessive mutation at a single nuclear locus (Huang et al. 2008). The stable sterile *B. juncea* line was established by the ten successive backcrosses with *B. juncea* 00-6-102B. This CMS was transferred to *B. napus* through interspecific hybridization (Wan et al. 2008). Stable CMS lines of kale were bred by interspecific or intra-specific hybridization and subsequent backcrosses (Zhu and Wei 2006). Shiga and Baba (1973) found the Nap CMS from the F_4 generation of hybridization between Qianjia and Beilu23 in *B. napus*. Li (1986) found the male sterile plant among the hybrid progeny and then developed male sterile line Shan 2A of *B. napus*. In addition, the Pol CMS had been transferred from *B. napus* to Chinese cabbage (Ke et al. 1992).

11.2.3 Protoplast Fusion

Protoplast fusion technology can generate genetic variability. It allows the formation of new nuclear cytoplasmic combinations by the transfer of gene groups and families, chromosome parts or chromosomes, and cytoplasmic genetic information (Gleba and Shlumukov 1990; Pelletier 1993; Bajaj 1994; Ilcheva and San 1997). Thus, new sources of male sterility are possible. Protoplast fusion has often been suggested as a means of developing unique hybrid plants which cannot be produced by conventional hybridization. Protoplasts can be produced from many plants, including most crop species (Gamborg et al. 1981; Evans and Bravo 1983; Lal and Lal 1990; Feher and Dudits 1994). The conventional method for recovery of cytoplasmic male sterile Cruciferous vegetables for hybrid seed production is laboring and time consuming. It can be substituted by in vitro techniques (Menczel et al. 1983; Kumashiro et al. 1988; Nikova et al. 1991).

Ogura male sterile plant was produced through the protoplast fusion of cauliflower and cabbage (Waiters et al. 1992). CMS were transferred to *B. napus* from Ogura radish by recurrent backcross through *B. oleracea* used as a bridge plant (Bannerot et al. 1974), and also reported by Pelletier et al. (1983), Menczel et al. (1987) and Jourdan et al. (1989a, b). Protoplast fusion of *Raphanus sativus* (CMS line) and *B. napus* were produced sterile hybrid (Sakai and Imamura 1990). With the help of protoplast fusion transfer of a newly found cytoplasm male sterility were in *Raphanus sativus* to *B. napus* in a single step (Sakai and Imamura 1990). Cytoplasmic male sterile (CMS) cabbage (*Brassica oleracea* var. *capitata*) was produced by the fusion of leaf protoplasts from fertile cabbage and CMS Ogura broccoli lines (Sigareva and Earle 1997). One line (7642A) carried the CMS gene in Ogura (R₁) derived from radish; the other line (7642B) carried a normal Brassica cytoplasm. The cytoplasmic traits in *B. oleracea* were stable after one cycle of in-vitro culture and regeneration (Jourdan et al. 1990).

11.3   Male Sterility in Chinese Cabbage

Chinese cabbage (*Brassica campestris* L.ssp.pekinensis) is the most important leafy vegetable with widely growing cultivation areas in China. Because Chinese cabbage is one of the cross-pollination crops, so it has an obvious heterosis. Therefore, for the high yield, the hybrid seed is used on a large scale. The excellent hybrid lines have about 30 % more yield than other general varieties. While the artificial hybrid and the self-incompatibility line can't produce the F₁ hybrid progeny on a large scale. The pure hybrid seeds were produced by utilization of male sterility which makes it to be the most important way to produce hybrid seeds in Chinese cabbage. Therefore, the breeders from China and foreign countries give importance to the studies of male sterility in Chinese cabbage.

In the start of 1970s, Chinese scientists started their research program on Chinese cabbage and successfully bred genetic male sterility lines (double purposes line). These lines were applied to the production of hybrid, which solves the problems of high cost and inbreeding depression. However, it increased breeding costs due to the removal of the 50 % female fertile plants in hybrid seed production. In addition, the seed quality cannot be guaranteed and the hybrid vigor rate decreased with the incomplete removal of plant. To get 100 % of the infertile groups, Li et al. (1995) were to propagate the infertile plants of Chinese cabbage by using tissue culture method and were succeeded, but the application of seed production in large area is still difficult. Hinata and Konno (1979) developed the cytoplasmic male sterility by series backcrossing *Diplotaxismuralis* as female with *B. campestris*. Van Der Meer (1987) found the male sterile plant in F₁ hybrid progeny, and then developed the dominant genic male sterile material through backcrossing and thought it was controlled by single dominant gene. Zhang et al. (1990) proposed the model of interactive genic male sterility of Chinese cabbage, and then developed the Kuaicaiyihao, Shenyangkuaicai, LvxingJ6 and other male sterile series of Chinese cabbage lines for production. Feng et al. (1995) reported the model of multiple allele male sterility in Chinese cabbage, then developed the excellent combination of Chinese cabbage and cabbage. After then, the domestic scientists committed to the research on breeding change of interactive genie male sterility and multiple allele male sterility in different Chinese cabbage varieties. Chen et al. (2000) obtained the nucleo-cytoplasmic interactive Chinese cabbage male sterile line 98-2 with 100 % rate of male sterility. Wen et al. (2003) got the successful transfer and obtained the Chinese cabbage interactive genic male sterile lines of Qingmaye type to develop the combination groups.

11.4 Male Sterility in Cabbage

Cabbage (*Brassica oleracea*) is one of the most important vegetable crops in the family of Brassicaceae and is grown in many countries of the globe, especially in Europe and America. And the annual cultivated area in China has reached 270 thousand hectare (Wang 2000). Like Chinese cabbage, the cabbage also has an obvious heterosis, so using of the heterosis is the main approach for cabbage breeding. That is to produce the F_1 hybrid progeny by using of self-incompatibility line and male sterile line. For a long time, the using of self-incompatibility line is the main way to produce F_1 hybrid in cabbage, and it has made a great achievements by developing a large number of excellent varieties. However, the cabbage breeders took a long time to study the using of male sterile line in breeding and have made a big breakthrough. Despite the late start of male sterility studying in China, but the breeders take a high importance to it. After years of hardly studies, the cabbage breeders in China have made great achievements. Here, we mainly talk about the research of hybrid seed production using male sterility in cabbage.

The cabbage breeders in China started the further research earliest from the introduction of foreign countries materials. From 1980s, Breeders of Institute of Vegetables and Flowers, Chinese Academy of Agricultural Sciences have made a series studies of cabbage. They have introduced the radish cytoplasmic male sterility and *B. nigra* cytoplasmic male sterility, and the improved the radish cytoplasmic male sterility by crossing and series backcrossing with different cabbage inbred lines. However, it is difficult to apply these materials in cabbage breeding because of the decreased hybrid vigor of the male sterile lines after several backcrossing and the limit of temperatures (Fang et al. 2001). In 1998, they introduced 6 improved Radish cytoplasmic male sterility materials include $CMS_{R3}625$ and $CMS_{R3}629$ which showed good prospect in cabbage breeding (Liu et al. 2002).

Fang et al. (1997) found the dominant genic male sterile plant 79-399-3 in the natural population of original material 79-399 in cabbage, and then he discovered that the male sterility was controlled by a dominant nucleic gene CDM_S399-3. Now, they have developed more than 20 homozygous dominant male sterile lines include

318P5 and 99Z522-6 and selected the 01-20M$_S$, 01-216M$_S$ and other two excellent dominant male sterile lines with normal performance and 100 % sterile plants and infertility. By crossing with admirable inbred lines, they have obtained a number of excellent hybrid combinations. Among these, the Zhonggan no.16, Zhonggan no.17 and the Zhonggan no.18 have been approved by the National Crop Variety Approval Committee. This was the first using of dominant genic male sterility to produce the new cabbage varieties (Yang et al. 2004).

11.5   Male Sterility in Mustard

Cytoplasmic male sterility system has developed in mustard (*Brassica juncea*) following repeated backcrossings of the somatic hybrid *Moricandia arvensis* (2n = 28, MM) × *B. juncea* (2n = 36, AABB), carrying mitochondria and chloroplasts from *M. arvensis*, to *B. juncea*. Cytoplasmic male sterile (CMS) plants are similar to normal *B. juncea*. Female fertility was normal. Genetic information for fertility restoration was introgressed following the development of a *M. arvensis* monosomic addition line on CMS *B. juncea*. The putative restorer plant also exhibited severe chlorosis similar to CMS plants but possessed 89 % and 73 % pollen and seed fertility, respectively, which subsequently increased to 96 % and 87 % in the selfed progeny. The progeny of the cross of CMS line with the restorer line MJR-15, segregated into 1 fertile: 1 sterile (Prakash et al. 1998). A stable CMS line of tuber mustard was developed by distant crosses and subsequent backcrosses and induced 100 % male sterility during crossing with more than 600 tester line including 150 *B. juncea* lines and 450 *B. rapa* lines (Yu et al. 2009). Banga (1986) proposed the cytoplasmic male sterility in F$_1$ hybrid progeny of the cross combination between the mustard species RLM198 and EJ-33. Anand and Rawat (1978) also produced the male sterility in mustard, and developed the restorer lines with restoring rate 90–95 % by the transfer of two restorer genes from *B. nigra* to mustard. Shi et al. (1991) found a male sterile plant in mustard name "European-Xinping A". Then after several years of study by test crossing and inter crossing, a set of "three lines" with low erucic acid was accomplished and the F$_1$ hybrid yielded 19.2–34.8 % more than the control "Kunming-Gaoke".

References

Anand IJ, Rawat DS (1978) Male sterility in India mustard. Indian J Genet Plant Breed 39(3):412–418

Arumugam N, Mukhopadhyay A, Gupta V, Pental D, Pradhan AK (1996) Synthesis of hexaploid (AABBCC) somatic hybrids:a bridging material for transfer of 'tour' cytoplasmic male sterility to diVerent *Brassica* species. Theor Appl Genet 92:762–768

Bajaj YPS (1994) Somatic hybridization a rich source of genetic variability. Biotechnol Agric For 27:3–32

Banga S (1986) Foreign agricultural. Oil Crop 2:14–16

Bannerot H, Boulidard L, Cauderon Y, Tempe J (1974) Transfer of cytoplasmic male sterility from *Raphanus sativus* to *Brassica oleracea*. In: Proceedings of Eucarpia Meeting Cruciferae, Dundee, pp 52–54

Bannerot H, Boulidard L, Chupeau Y (1977) Unexpected difficulties met with the radish cytoplasm in *Brassica oleracea*. Eucarpia Cruciferae Newsl 2:16

Bhat SR, Kumar P, Prakash S (2008) An improved cytoplasmic male sterile (*Diplotaxis berthautii*) *Brassica juncea*: identification of restorer and molecular characterization. Euphytica 159:145–152

Bhat SR, Vijayan P, Ashutosh DKK, Prakash S (2006) *Diplotaxis erucoides*-induced cytoplasmic male sterility in *Brassica juncea* is rescued by the *Moricandia arvensis* restorer: genetic and molecular analyses. Plant Breed 125:150–155

Cao SC, Li SJ (1980) Breeding and utilization of the Chinese cabbage 'Aijiaohuang' genic male-sterile AB line. J Nanjing Agric Univ 1:59–67 (in Chinese)

Chen FX, Hu BC, Li C, Li QS, Chen WS, Zhang ML (1998) Genetic studies on GMS in *Brassica napus* L: I. Inheritance of recessive GMS line 9012A. Acta Agron Sin 24:431–438

Chen FX, Hu BC, Li QS (1993) Discovery and study of genic male sterility (GMS) material 9012A in *Brassica napus* L. Acta Agric Univ Pekin 19:57–61 (in Chinese)

Chen WH, Fang SG, Zeng XL (2000) Breeding of Chinese cabbage male sterile line 98-2 with nucleo-cytoplasmic interactive type. Fujian J Agric Sci 15(4):22–25 (in Chinese)

Chen ZJ, Zhang MF, Wang BL, Dong W, Huang S (1995) A study on fertility and agronomic characters of CMS lines for tuber mustard. Acta Hortic Sin 1:40–46 (in Chinese)

Chiang MS, Crete R (1987) Cytoplasmic male sterility in *Brassrca oleracea* induced by *B. napus* cytoplasm-female fertility and restoration of male fertility. Can J Plant Sci 672:891–897

Correns C (1906) Die Verenbung der Geschlectsformen bei den gynodiocischen Pflanzer. Ber Dtsch Bot Ges 24:459–474

Duvick DN (1959) The Use of Cytoplasmic male-sterility in hybrid seed production. Econ Bot 13(3):167–195

Edwardson JR (1970) Cytoplasmic male sterility. Bot Rev 36:341–420

Evans DA, Bravo JE (1983) Protoplast fusion. In: Evans DA, Sharp WR, Ammirato PV, Yamada Y (eds) Handbook of plant cell culture, vol 1. Macmillan, New York, pp 291–321

Fan Z, Stefansson BR, Sernyk JL (1986) Maintainers and restorers for three male-sterility – inducing cytoplasms in rape (*Brassica napus* L.). Can J Plant Sci 66:229–234

Fang ZY, Sun PT, Liu YM (1997) A male sterile line with dominant gene (Ms) in cabbage (*Brassica oleracea* var. *capitata*) and its utilization for hybrid seed production. Euphytica 97:265–268

Fang ZY, Sun PT, Liu YM (2001) Investigation of different types of male sterility and application of dominant male sterility in cabbage. China Veg 1:6–10 (in Chinese)

Feher A, Dudits D (1994) Plant protoplasts and cell fusion and direct DNA uptake: culture and regeneration systems. In: Vasil KI, Thorpe TA (eds.), PlantCell and tissue culture, Dordrecht, The Netherlands, Kluwer pp 71–118

Feng H, Wei Y, Ji S, Jin G, Jin J, Dong W (1996) Multiple allele model for genic male sterility in Chinese cabbage. Acta Hortic 467:133–142 (in Chinese)

Feng H, Wei YT, Zhang SN (1995) Inheritance of and utilization model for genic male sterility in Chinese cabbage (*Brassica pekinensis* Rupr.). Acta Hortic 402:133–140

Fu TD (1981) Production and research of rapeseed in the People's Republic of China. Eucarpia Cruciferae Newsl 6:6–7

Fu YD, Yang GS (1990) Study on Polima cytoplasmic male sterility of *Brassica napus* L. and its utilization. Crop Res 4(3):9–12 (in Chinese)

Gamborg OL, Shylak JP, Shahin EA (1981) Isolation, fusion, and culture of plant protoplasts. In: Thorpe TA (ed.), Plant tissue culture: methods and applications in agriculture. Academic, New York, pp 115–153

Gerstel DU (1980) Cytoplasmic male sterility in Nicotiana (a review). NC Agric Exp Sm Tech Bull 263:1–31

Gleba YY, Shlumukov LR (1990) Somatic hybridization and cybridization. In: Bhojwani SS (ed) Plant tissue culture: applications and limitations, vol 19, Development of crop science. Elsevier, Amsterdam, pp 316–344

Grun P (1976) Cytoplasmic genetics and evolution. Columbia University Press, New York

Hayes HK, Jones DF (1916) First generation crosses in cucumber. Repub Conf Agric Exp Stat 5:319–322

He QW, Shi HL (1987) Studies on the breeding and inheritance of male sterile line of Chinese radish. Sci Agric Sin 20(2):26–33

Hinata K, Konno N (1979) Studies on a male-sterile strain having the *Brassica campestris* nucleus and the *Diplotaxismuralis cytoplasm* I. On the breeding procedures and some characteristics of the male sterile strain. Jpn J Breed 29:305–311

Hou GZ, Wang H, Zhang RM (1990) Genetic study on genic male sterility (GMS) 117A in *Brassica napus*. Chin J Oil Crop Sci 2:7–10 (in Chinese)

http://ingenira.hubpages.com/hub/Cruciferous-Vegetables-Cancer-Prevention-and-Cancer-Fight-Veggies

Huang L, Cao JS, Ye WZ, Liu TT, Jiang L, Ye YQ (2008) Transcriptional differences between the male-sterile mutant bcms and wild-type *Brassica campestris*ssp. chinensis reveal genes related to pollen development. Plant Biol 10:342–355

Huang L, Ye WZ, Liu TT, Cao JS (2009) Characterization of the male-sterile line Bcajh97-01A/B and identification of candidate genes for genic male sterility in Chinese cabbage-pak-choi. J Amer Soc Hortic Sci 134(6):632–640

Ilcheva V, San LH (1997) Hybridationz somatiques chez le gendreNicottana: revue bibligraphique. Annals du Tabac 2(9):19–37 (in French)

James AB, Hong Y, Sivanandan C, Daniel V, Reiner AV (2010) Heterosis. Plant Cell 22:2105–2112

Jourdan PS, Earle ED, Mutscher MA (1989a) Atrazine-resisrant cauliflower obtained by somatic hybridization between *Brassica oleracea* and ATR-*B. napus*. Theor Appl Genet 78:271–279

Jourdan PS, Earle ED, Mutscher MA (1989b) Synthesis of male-sterile, triazine resistant *Brassieanapus* by somatic hybridization between cytoplasmic male-sterile *B. oleracea* and atrazine-resistant *B. eampestris*. Theor Appl Genet 78:445–455

Jourdan PS, Earle ED, Mutschler MA (1990) Improved protoplast culture and stability of cytoplasmic traits in plants regenerated from leaf protoplasts of cauliflower (*Brassica oleracea* ssp. botrytis). Plant Cell Tiss Org Cult 21:227–236

Ke GL, Zhao ZY, Song YZ (1992) The development of allo-cytoplasmic male sterility in Chinese cabbage and its utilization. Acta Hortic Sin 19(4):333–340 (in Chinese)

Kumashiro T, Asahi T, Komari T (1988) A new source of cytoplasmic male sterile tobacco obtained by fusion between *N. tabacum* and Xirradiated *N. afrncana* protoplasts. Plant Sci 55:247–254

Lal R, Lal S (1990) Crop improvement using biotechnology. CRC Press, Boca Raton

Laser KD, Lersten NR (1972) Anatomy and cytology of microsporogenesis in cytoplasmic male sterile angiosperms. Bot Rev 38:425–454

Le-Clercq P (1983) Etude de diverscasdesterilite male cytoplasmique chez Ietoumesol (Helianthus spp.). Agron Sci Prod Veg Environ 3:185–187

Li DR (1986) The successful development and largely spreading of the male sterile lines, maintenance lines and restored lines of *Brassica napus*. Sci Agric Sin 19(4):94 (in Chinese)

Li SL, Zhou XR, Zhou ZJ (1990) Inheritance of genetic male sterility (GMS) and its utilization in rape (*Brassica napus* L.). Crop Breed Cultivation Res Inst; Shanghai Acad Agric Sci 4(3): 321–322 (in Chinese)

Li XO, Sun RF, Wu FY, Si JG, Niu XK (1995) *In vitro* propagation of Chinese cabbage for seed production. Acta Hortic 402:306–312 (in Chinese)

Liu YM, Fang ZY, Sun PT (2002) The main way to obtain the male sterility in Cruciferous crop and its utilization. China Veg 2:52–55 (in Chinese)

Liu Z, Guan C, Zhao F, Chen S (2005) Inheritance and mapping of a restorer gene for the rapeseed cytoplasmic male sterile line 681A. Plant Breed 124:5–8

Mathias R (1985) Transfer of cytoplasmic male sterility from brown mustard (*Brassica juncea* (L.) Coss.) into rapeseed (*Brassica napus* L.). Z PXanzenzüchtg 95:371–374

Menczel L, Morgan A, Brown S, Maliga P (1987) Fusion-mediated combination of Ogura-type cytoplasmic male sterility with *Brassica napus* plastids using X-irradiated CMS protoplasts. Plant Cell Rep 6:98–101

Menczel L, Nagy F, Lazar G, Maliga P (1983) Transfer of cytoplasmic male sterility by selection of streptomycin resistance after protoplast fusion in Nicotiana. Mol Gen Genet 189:365–369

Nikova V, Zagorska N, Pundeva R (1991) Development of four tobacco cytoplasmic male sterile sources using in vitro techniques. Plant Cell Tiss Organ Cult 27:289–295

Niu XK, Wu FY, Zhong HH, Li XS (1980) The selection and utilization of Chinese cabbage (*B. pekineensis*) of male sterile AB line. Acta Hortic Sin 2:25–32

Ogura H (1968) Studies on the new male-sterility in Japanese radish, with special reference to the utilization of this sterility towards the practical raising of hybrid seeds. Mem Fac Agric Kagoshima Univ 6:39–78

Pelletier G (1993) Somatic hybridization. Plant Breed 93–106

Pelletier G, Primard C, Vedel F, Chetrit P, Remy R, Rousselle RM (1983) Intergeneric cytoplasmic hybridization in Cruciferae by protoplast fusion. Mol Gen Genet 191:244–250

Pradhan AK, Mukhopadhyay A, Pental D (1991) IdentiWcation of the putative cytoplasmic donor of a CMS system in *Brassica juncea*. Plant Breed 106:204–208

Prakash S, Ahuja I, Upreti HC, Dinesh Kumar V, Bhat SR, Kirti PB, Chopra VL (2001) Expression of male sterility in alloplasmic *Brassica juncea* with *Erucastrum canariense* cytoplasm and the development of a fertility restoration system. Plant Breed 120:479–482

Prakash S, Bhat SR, Quiros CF, Kirti PB, Chopra VL (2009) Brassica and its close allies: cytogenetics and evolution. Plant Breed Rev 31:21–187

Prakash S, Chopra VL (1990) Male sterility caused by cytoplasm of *Brassica oxyrrhina* in *B. campestris* and *B. juncea*. Theor Appl Genet 79:285–287

Prakash S, Kirti PB, Bhat SR, Gaikwad K, Dinesh Kumar V, Chopra VL (1998) A *Moricandia arvensis*-based cytoplasmic male sterility and fertility restoration system in *Brassica juncea*. Theor Appl Genet 97:488–492

Rao GV, Batra-Sarup V, Prakash S, Shivanna KR (1994) Development of a new cytoplasmic male-sterile system in *Brassica juncea* through wide hybridization. Plant Breed 112:171–174

Röbbelen G (1991) Citation at the occasion of presenting the GCIRC Superior Scientist Award to FU Tingdong. In: Proceedings of eighth international rapeseed congress, vol 1. Sasktoon, pp 2–5

Ruffio-Chable V, Bellis H, Herve Y (1993) A dominant gene for male sterility in cauliflower (*Brassica oleracea* var. *botrytis*): phenotype expression, inheritance, and use in F_1 hybrid production. Euphytica 67:9–17

Sakai T, Imamura J (1990) Intergeneric transfer of cytoplasmic male sterility between *Raphanussativus* (cms line) and *Brassica napus* through cytoplast-protoplast fusion. Theor Appl Genet 80:421–427

Shi HQ, Gong RF, Zhuang LL (1991) Studies on the utilization of heterosis in mustard (*Brassica Junces*). Acta Agron Sin 1:32–41 (in Chinese)

Shiga T, Baba S (1971) Cytoplasmic male sterility in rape plants (*Brassica napus* L.). Jpn J Breed 21:16–17

Shiga T, Baba S (1973) Cytoplasmic male sterility in oilseed rape, *Brassica napus* L., and its utilization to breeding. Jpn J Breed 23:187–197

Shull GH (1908) The Composition of Field Maize. Rep Am Breed Assoc 4:296–301

Sigareva MA, Earle ED (1997) Direct transfer of a cold-tolerant Ogura male-sterile cytoplasm into cabbage (*Brassica oleracea* ssp. *capitata*) via protoplast fusion. Theor Appl Genet 94:213–220

Sodhi YS, Chandra A, Verma JK, Arumugam N, Mukhopadhyay A, Gupta V, Pental D, Pradhan AK (2006) A new cytoplasmic male sterility system for hybrid seed production in Indian oilseed mustard *Brassica juncea*. Theor Appl Genet 114:93–99

Stiewe G, Röbbelen G (1994) Establishing cytoplasmic male sterility in *Brassica napus* by mitochondrial recombination with *B. tournefortii*. Plant Breed 113:294–304

Thompson KF (1972) Cytoplasmic male sterility in oil-seed rape plants. Heredity 29:253–257

Van Der Meer QP (1987) Chromosomal monogenic dominant male sterility in Chinese Cabbage (*Brassica rapa* subsp. *Pekinensis* (Lour.) Hanelt). Euphytica 36:927–931

Waiters TW, Mutschler MA, Earle ED (1992) Protoplast fusion-derived Ogura male sterile cauliflower with cold tolerance. Plant Cell Rep 10:624–628

Wan Z, Jing B, Tu J, Ma C, Shen J, Yi B, Wen J, Hang T, Wang X, Fu T (2008) Genetic characterization of a new cytoplasmic male sterility system (hau) in *Brassica juncea* and its transfer to *B. napus*. Theor Appl Genet 116:355–362

Wang XJ (2000) Vegetable breeding (each discussion). Chinese Agriculture, Beijing

Wei YT, Feng H, Zhang SN (1992) The inheritance of gene male sterility in Chinese cabbage (*Brassica pekinensis*). J Shenyang Agric Univ 23:260–266

Wen FY, Zhang B, Liu XH, Wang YL, Song LJ, Zhao B (2003) The research on genetic model of nuclear male sterile character in Qingmaye type of Chinese cabbage. Hua Bei J Agric Sci 18(4):42–45 (in Chinese)

Yang GS, Fu TD, Brown GG (1998) The genetic classiWcation of cytoplasmic male sterility systems in *Brassica napus* L. Sci Agric Sin 31:27–31

Yang LM, Fang ZY, Liu YM, Wang XW, Zhuang M, Zhang YY, Sun PT (2004) Zhonggan 18 a new cabbage hybrid variety with the hybridization of dominant male sterile line and inbred line. Acta Hortic Sin 34(6):837–837 (in Chinese)

Yu X, Xiao Q, Cao J, Chen Z, Hirata Y (2009) Development of two new molecular markers specific to cytoplasmic male sterility in tuber mustard (*Brassica juncea* var. tumidaTsen et Lee). Euphytica 166:367–378

Zhang SF, Song ZZ, Zhao XY (1990) Breeding of interactive genic male sterile line in Chinese cabbage and utilization model. Acta Hortic Sin 17(2):117–125 (in Chinese)

Zhu P, Wei Y (2006) Studies on interspecific cross compatipility between Ogura *Brassica campestris*sp. *Pekinensis* and *B. oleracea* var. *acephala*. Acta Hortic Sin 33:1090–1092

Chapter 12
Genetic Modifications for Pest Resistance

Hongbo Liu, Bizeng Mao, Peng Cui, Tian Tian, Changrong Huang,
Xi Xu, and Weijun Zhou

Abstract Diseases and insect pests are serious threat to the growth and yield of *Brassica* crops. As such, breeding for resistance to pests has been considered as a major objective in oilseed rape (*Brassica napus* L.) plant. The traditional method of genetic modification is utilizing the wild species which have resistance to one of diseases or insects to improve the cultivated species by distant hybridization. So far, the availability of resistant sources against pests has been greatly explored in many kinds of wild species. However, the narrow genetic resource (germplasm) also inhibits the development of pest resistant breeding program. On the other hand, with the development of biotechnology, genetic transformation has become possible to bring about quick and dramatic improvements in the tolerance to diseases and insect pests. In the past decades, more and more resistant genes were cloned and characterized, then transferred to cultivated species to obtain the resistant traits. The present chapter focuses on genetic modification of disease and insect pest resistance by conventional hybridization and transgene breeding in *Brassica* crops.

H. Liu, Ph.D.
Institute of Crop Science and Zhejiang Key Laboratory of Crop Germplasm,
Zhejiang University, Hangzhou 310058, China

College of Agricultural and Food Sciences, Zhejiang A&F University,
Lin'an 311300, Zhejiang, China

B. Mao, Ph.D.
Institute of Biotechnology, Zhejiang University, Hangzhou 310058, China

P. Cui, M.S. • T. Tian, BS • C. Huang, M.S. • X. Xu, BS
Institute of Crop Science and Zhejiang Key Laboratory of Crop Germplasm,
Zhejiang University, Hangzhou 310058, China

W. Zhou, Ph.D. (⊠)
Institute of Crop Science and Zhejiang Key Laboratory of Crop Germplasm,
Zhejiang University, Hangzhou 310058, China

Agricultural Experiment Station, Zhejiang University, Hangzhou 310058, China
e-mail: wjzhou@zju.edu.cn

S.K. Gupta (ed.), *Biotechnology of Crucifers*, DOI 10.1007/978-1-4614-7795-2_12,
© Springer Science+Business Media, LLC 2013

Keywords *Brassica* • Pest resistance • Introgression • Cultivar resource • Genetic transformation

12.1 Introduction

The Genus *Brassica* and its wild species belong to the tribe Brassiceae in the Cruciferae family. Some cultivated species can be used to produce edible oil and vegetable (Momoh et al. 2002; Zhou 2001), such as *Brassica nupas*, *Brassica rapa* and *Brassica oleracea*, etc. Meanwhile, these cultivated species were attacked by diseases and insect pests, which is a big challenge in the production for oilseed rape over the years. On the Indian subcontinent and North America, some major diseases cause considerable economic yield losses, such as blackleg, leaf blight and root rot, etc. (Rai et al. 2007). While Aphid and Diamondback moth are nefarious insect pest of Brassicaceae crops (Bhatia et al. 2011; Mohan and Gujar 2003).

Chemical control and cultivar resistance were considered to be more effective methods to the management practices of diseases and insect pests. The former is expensive and raises healthy or environmental concerns in oilseed rape production. Therefore, adoption of resistant cultivars has been considered as the most economic and environmentally friendly strategy for pest management. To date, most of wild relative species which possess pest resistant traits have been reported (Gerdemann-Knorck et al. 1995; Liu et al. 1995; Snowdon et al. 2000; Kumar et al. 2011). They used distant hybridization and backcross to integrate the important resistant traits into the cultivated species for pest resistant breeding. Meanwhile, the gradual adaptation of pests or narrow insect-resistant spectrum to transgene is another problem for agricultural application of genetically modified organism. With the development of biotechnology, a number of pest resistant genes were cloned and characterized, then transferred to cultivated species by transgene techniques (Cho et al. 2001; Chen et al. 2006; Mondal et al. 2007; Dong et al. 2008; Liu et al. 2011a). It has opened up new horizon to improve the level of pest resistance and development of *Brassica* transgene breeding.

12.2 Conventional Breeding for Pest Resistance

Successful disease and insect control requires a thorough knowledge of corresponding resistant genetic resources. There are more than 74,000 accessions of oilseed and vegetable *Brassica* germplasm lines exploited all over the world, and more than 60 % of which were held in China, India, the United Kingdom, the United States and Germany (Singh and Sharma 2007). Among these species, many resistant resources were identified and utilized to improve cultivated species by conventional breeding methods.

12.2.1 Disease

Blackleg is a serious disease of *Brassica* caused by *Leptosphaeria maculans* (Desm.) Ces. Et de Not., especially in oilseed rape (*Brassica napus* L.) cultivation in Europe, Canada and Australia. The non-aggressive isolates of *L. maculans* cause no or a little mild symptoms comparing to the aggressive isolates which result in yield losses ranged from 13 % to 50 % (Purwantara et al. 1998; Williams and Fitt 1999; Hu et al. 2009). There are few resistant cultivations which were screened to the pathogen in *B. napus*, *B. campestris* (*B. rapa*), *B. oleracea*. In contrast, wild relative species and most lines of species containing the B genome were characterized to resistant this pathogen by molecular markers (Chevre et al. 1997; Plieske et al. 1998; Dixelius 1999; Dixelius and Wahlberg 1999; Saal and Struss 2005; Dusabenyagasani and Fernando 2008). Therefore, the transferring of disease resistant traits to cultivation, both sexual and somatic, has been considered as an effective method for resistant breeding.

Chromosome introgression is an important approach to the transfer of genes between species by sexual and somatic hybridization. A number of studies demonstrated that the recombinant lines expressed high resistant level similar to that of the donor parents in interspecific hybrids (Plieske et al. 1998; Saal et al. 2004; Chevre et al. 2008). The genetic resistance to anamorph *P. lingam* was investigated in addition lines of *B. napus* and *B. nigra*. The results showed that at least 3 different *B. nigra* chromosomes had contribution to the blackleg resistance. Thus, this resistance was assumed to be polygenic (Zhu et al. 1993). In Plieske's et al. (1998) study, among homozygous recombinant lines, there are no significant differences in the level of resistance and in the phenotype of the resistance mechanisms, no matter where the B genome origins, i.e. *B. nigra*, *B. juncea* and *B. carinata*. Therefore, it was assumed that the resistance genes might be located in the conserved domain, and the mechanisms of each B-genome species are identical. For another studies, the resistance to blackleg is monogenic and highly efficient under field conditions in *B. napus-B. juncea* recombinant lines. In these addition lines, three markers totally linked to resistant *L. maculans* were also located on the B4 and B8 chromosomes in *B. napus-B. nigra* and *B. oleracea-B. nigra* addition lines. It was confirmed that there involved chromosome rearrangements between the two B genomes of *B. nigra* and *B. juncea* (Chevre et al. 1997). Thus, the majority of species containing B genome were used to create addition lines for improving the cultivation. The resistant gene, termed $r_j lm2$, was identified by screening spring type addition lines from *B. napus-B. juncea*. The introgression individuals can be resistant to all the tested virulent pathotypes at seeding stage, especially to two isolates which have shown to overcome dominant B genome-derived resistance genes (Saal et al. 2004).

When the goal of genetic modification is to transfer several genes or when the genes of interest are unidentified, asymmetric somatic hybridization such as protoplast fusion may be a more efficient method. So far, there have been shown that many successful examples in transferring resistant genes to susceptible species by asymmetric somatic hybridization (Sjodin and Glimelius 1989; Gerdemann-Knorck et al. 1995; Liu et al. 1995; Dixelius 1999). Sometimes, a specific aggressive isolate of *L. maculans* has been found to overcome resistance originating from the B genome in *Brassica* species, but the distant *Brassica* genus which has specific resistance can solve this problem. In *Brassica napus-Sinapis arvensis* addition lines, some individuals with adult plant resistance but cotyledon susceptibility were observed in possible resistant introgression lines. In addition, phytopathological analysis of selfing progenies from 3 different highly resistant BC$_3$ plants showed that seedling and adult plant resistance are probably conferred by different loci (Snowdon et al. 2000). The similar resistant mechanism between cotyledon and adult plants was observed in some other researches (Chevre et al. 1996; Chevre et al. 1997).

Apart from blackleg, some other diseases could bring about considerable economic losses to *Brassica* species, such as clubroot (*Plasmodiophora brassicae*), black spot (*Alternaria brassicae*), cottony rot (*Sclerotinia sclerotiorum*), root rot (*Rhizoctonia solani*), and so on. In addition, white rust (*Albugo candida*) obtained from *B. campestris* and *B. juncea* could attack many other different hosts which contained *B. napus*, *B. oleracea* and *Raphanus sativus* (Pidskalny and Rimmer 1985).

Clubroot, it is a root infection caused by the fungus *P. brassicae*. High resistance recourses have been found in black mustard (*B. nigra*) and radish (*R. sativus*) (Hu et al. 2009). Transfer of clubroot resistance was studied between *B. napus* and *B. nigra* by asymmetric somatic hybridization (Gerdemann-Knorck et al. 1995) and between *B. oleracea* and *R. sativus* by protoplast electrofusion (Hagimori et al. 1992). In all causes, the resistance of somatic hybrids retained were identified and progenies from some of the hybrids with high resistance inherited. But many of *B. napus* and *B. nigra* line expressed resistance to just only one isolate.

A. brassicae also known as *Alternaria* blight or *Alternaria* black spot, is the most prevalent causal agent of black spot on oilseed *Brassica* species. In serious regions, yield losses can be measured up to 70 % due to this pathogen in seed *Brassica* (Downey and Rimmer 1993). A few of successful instance of *A. brassicae* resistance transferring from *S. alba* to *B. napus* and *B. oleracea* by somatic hybridization has been reported (Chevre et al. 1991; Ryschka et al. 1996). Among the introgression plants of *B. napus*, the level of resistant to *A. brassicae* was similar to that of *S. alba* (Chevre et al. 1991). In hybrids of *B. oleracea* + *S. alba*, plants of the *S. alba* chromosome addition ranged from moderately to absolutely resistant to *A. brassicae*. For interesting, the hybrids were female fertile, and thus it is promise that resistant genetic resources can be created for *Brassica* breeding by backcross (Ryschka et al. 1996).

In some areas, *Sclerotinia sclerotiorum* (Lib.) de Bary (also be known as cottony rot) could cause an equal or even greater damage to cultivation of *Brassicas* than that of blackleg. The resistant resources to *S. sclerotiorum* were screened in the greenhouse and field among different lines of *B. oleracea* var. *capitata*. Several resistant lines PIs showed good tolerance, such as non heading cabbages, a savoy cabbage and some red cabbage. Furthermore, the backcross populations using the best non heading cabbage suggested a major recessive gene for resistance (Dickson and Petzoldt 1996).

12.2.2   Insect

The cruciferous crop is susceptible to a number of insect pests, including large white (*Pieris brassicae*), diamondback moth (*Plutella xylostella*), mustard aphid (*Lipaphis. erysimi*), mustard sawfly (*Sathalia proxima*), leaf minor (*Bagrada cruciferarum*), and so on. Chemical control was considered to be a most common method to manage these pests, but it is not a sustainable pest control strategy for continuing reliance on insecticides and environmental issue. The concept of integrated pest management (IPM) is the more effective approach for insect pest control (Panda and Khush 1995). Identifying resistant sources and understanding the mechanisms of resistance are the major components of IPM.

Pieris brassicae, also called cabbage butterfly, is one of the destructive pests mainly on cruciferous crops such as rapeseed, cauliflower and broccoli. To determine the performance of *P. brassicae* on various *Brassica* crops, assessment of different brassicaceous host plants on the fitness of *Pieris brassicae* (L.) was investigated (Hasan and Ansari 2011). Survival, development and reproduction of pest were quantified under laboratory conditions on cabbage (*B. oleracea capitata*), cauliflower (*B. oleracea botrytis*), radish (*R. sativus*), broccoli (*B. oleracea italica*), and mustard (*B. campestris*). According to the results, cabbage was recognized as the most suitable host plant for *P. brassicae* because of shorter developmental time from eggs to adult eclosion, higher percentage survival, lower doubling time (6.00), and higher number of adult emergence (29.7 %). The similar investigation was also conducted in the preferences and performance of *P. xylostella* on nine commercial cultivars of canola (*Brassica napus* L.) under greenhouse conditions. It has the potential to screen a most resistant host to be used in the integrated management of *P. xylostella* (Fathi et al. 2011). In view of these experiments conducted under the laboratory conditions, a further study should be required to give more detailed information in the field. Thus, it can help breeder to employ the most appropriate control tactics towards integrated pest management of a particular *Brassica* crop in the field.

The glucosinolate and their hydrolysis products were considered as a stimulant for larval feeding initiation and oviposition of *P. brassicae* and other *Pieridae* in *Alternaria thalinan* or other Brassicaceae (Renwick 2002; Raybold and Moyes 2001; Miles et al. 2005). The glucosinolate was used as a recognition cue to identify an appropriate host for oviposition by adult females of *P. brassicae* and other *Pieridae*. This may explain why the low level of glucosinolate in cruciferous crops could potentially reduce plant damage caused by these pests (Huang and Renwick 1994; Renwick 2002). Thus, manipulating the glucosinolate system may be an effective method to increase resistance to *P. brassicae* and other *Pieridae*.

Aphid has the preference in different species of *Brassica*. The oilseed *Brassica* is susceptible infested by green peach aphid (*Myzus persicae* S.), cabbage aphid (*Brevicoryne brassicae* L.) and turnip aphid (*L. erysimi* Kalt.). *B. juncea* is predominantly infested by *L. erysimi*, whereas *B. brassicae* and *M. persicae* also occur (Bhatia et al. 2011). In the India subcontinent and other growing regions, mustard aphid is a serious insect pest which brings about destructive damages that the mean loss in yield ranged from 35.4 % to 91.3 % under various agro-climatic conditions (Singh and Sachan 1994; Rai et al. 2007). It can cause damage at all growth stages during the crop development, especially in the flowering and seeds production stages. Over the past few decades, a number of attempts had been made to screen resistant genetic sources in crop *Brassica* species (Brar and Sandhu 1978; Amjad and Peters 1992; Bhadoria et al. 1995; Saxena et al. 1995). However, it has not been possible to breed resistant cultivars by conventional techniques owing to the non-resistance source identified in the crossable germplasms and lack of knowledge of the resistant mechanisms.

For Crucifereae species, lectins are important chemicals for its insecticidal activity against a wide range of insects (Rahbe et al. 1995; Sadasivam and Thayumanavan 2003), especially the sap sucking insects (Foissac et al. 2000; Powell 2001). A number of wild relative *Brassicas* have been reported to possess resistance against cabbage aphid, *B. brassicae* (Cole 1994a, b; Ellis and Farrell 1995; Ellis et al. 2000). Recently, *Brassica fruticulosa* and *Brassica montana* were used to screen the resistance against mustard aphid under laboratory conditions. These two wild species exhibited stronger resistance to mustard aphid than that of susceptible *B. rapa* ssp. brown sarson cv. BSH-1 (Kumar et al. 2011). The bio-chemical analysis also showed that high concentration of lectins had been detected in *B. fruticulosa*. It is attempted to transfer resistant trait from *B. fruticulosa* to *B. juncea*, and excellent resistance was observed in *B. juncea* introgression lines, indicating heritable nature of *fruticulosa* resistance.

12.3   Genetic Transformation for Pest Resistance

With the development of genetic engineering, heterologous genes have been uti-lized to ameliorate most of the economically important *Brassica* species. In this context, the *Brassica* breeding objective of disease and insect resistance can be real-ized by tissue culture and *Agrobacterium tumefaciens* mediated transformation. To date, a number of regeneration and transformation procedures were constructed in many kinds of cultivars such as *B. napus* (Moloney et al. 1989; Khan et al. 2003; Wang et al. 2005; Sakhno et al. 2008; Liu et al. 2011a), *B. juncea* (Barfield and Pua 1991; Kanrar et al. 2002; Cao et al. 2008), *B. rapa* (Cho et al. 2001; Zhang et al. 2011), *B. campestris* (Xiang et al. 2000) and *B. oleracea* (Ding et al. 1998; Chen et al. 2006; Yi et al. 2011). It was considered to be a quick and desirable approach of genetic modification for pest resistance in *Brassica* species.

Chitinase is one of the earliest resistant genes for controlling fungal and insect pests in transformation research. It can catalyze hydrolysis of chitin, which is a homopolymer of β-1,4-N-acetylglucosamine and one of the important structural components of the fungal cell wall and insect cuticle (Bartnick 1968; Kramer and Muthukrishnan 1997). These hydrolytic enzymes extracted from plants and microbes were found to retard or inhibit fungus growth in vitro (Chen et al. 2007; Mauch et al. 1988a, b; Melander et al. 2006), and a number of successful transformations have been reported that transgene crops with *chitinase* gene also have high level resistance to insects and fungi (Broglie et al. 1991; Corrado et al. 2008; Grison et al. 1996; Tohidfar et al. 2005).

However, the transgenic broccoli plants expressing a *Trichoderma harzianum* endochitinase gene just showed the resistance to *A. brassicicola*, but no affection to *S. sclerotiorum* (Mora and Earle 2001). In addition, Melander et al. (2006) found that some of the transgene doubled haploid (DH) plants with a barley *chitinase* gene just had an improvement of resistance to *L. maculans*, but none to *A. brassicicola*, *S. sclerotiorum* and *V. longisporum*. It is assumed that the reason of the differences in resistance level might be the genetic variation naturally present within the cultivars. Thus, the development of alternative approaches involving more effective resistance would be exploited.

The pathogenesis related (PR) protein is a type of toxic protein to fungal pathogens. Glucanase, as a plant defense barrier, can hydrolyze glucan which is a major cell-wall component of pathogenic fungi. The resistance to *A. brassicae* was identified in transgene plants of *B. juncea* with transferring a class I basic glucanase of tomato. Protein extracted from independent transformants, arrested 15–54% hyphal growth of *A. brassicae*. Under poly house bio-assay, the number, size and spread of lesions caused by *A. brassicae* were restricted in transgenic lines compared to that of wild type (Mondal et al. 2007).

The insect resistance *Bacillus thuringiensis* (Bt) gene has been used in controlling insect larvae infestation in transgene crops for many years. There are also many Bt genes such as *cry1Ac*, *cry1C*, *cry1Ab* and *cry1Ba3* were used in various *Brassica* transgene studies to resistant lepidopteran insects (Cao et al. 1999; Cho et al. 2001; Cao et al. 2008; Xiang et al. 2000; Yi et al. 2011). The transgenic Chinese cabbage and their progeny with *cry1C* gene were identified not only to against diamondback moths (DBM) (*P. xylostella*), cabbage loopers (*Trichoplusia ni*) and cabbage worms (*Pieris rapae*), but also to susceptible Geneva DBM and a DBM population resistant to Cry1A protein (Cho et al. 2001). In addition, pyramided two Bt genes in vegetable Indian mustard and broccoli were conducted by sequential transformation and sexual crosses respectively. The levels of two Bt proteins were stably and highly produced compared to that of wild type. Transformants pyramided cry1Ac + cry1C effectively controlled diamondback moth and the types resistant to Cry1A and Cry1C (Cao et al. 2002, 2008). These Bt-transgenic plants could be used either for controlling DBM and other lepidopteran insect pests or for integrated management as a trap crop to protect other high value non-transgenic crucifer vegetables.

Some insects have already developed resistance to Bt toxin genes due to the wide use of Bt-transgenic plants (Roush and Shelton 1997). It is imperative that requiring new insect-resistant genes or introducing more than one insect-resistant gene simultaneously into plants to avoid gradual pest adaptation. Thus, several other novel genes were also being used to enhance resistance against insects for solving this problem. It is found that protease inhibitors (PIs) exhibited the feature to be effective against leaf chewers. Oryzacystatin I (OC-I), a cysteine protease inhibitors, can inhibit the growth of *A. pisum*, *A. gossypii* and *M. persicae* when fed by protease inhibitors. It has been used to against *M. persicae* in *B. napus* transgene research.

The mean adult weight, fecundity and biomass of *M. persicae* fed on tansformants expressing OC-I showed significantly reduction, when compared to those fed wild type (Rahbe et al. 2003). The potato proteinase inhibitor II (*pinII*) is a serine peptidase inhibitor that usually possessed the function to digestive serine proteases trypsin and chymotrypsin in the insect gut (Lawrence and Koundal 2002; Mosolov and Valueva 2008). So it has great potential in crop protection for the broad anti-inset spectrum. A bioassay of DMB resistance was conducted in the second generation of individual transgene lines which integrated *pinII*. The results showed that the larvae fed on transgenic leaves had a higher mortality than those fed on the wild-type leaves (Zhang et al. 2011). Wheat germ agglutinin (WGA), a chitinbinding lectin from wheat germ, has been shown to be antimetabolic, antifeedant and insecticidal to the mustard aphid (*L. erysimi.* Kalt). The transgenic plants with WGA showed that high resistance to aphids (Kanrar et al. 2002).

Moreover, there is another option that introducing two combined genes with different mechanisms can slow down or minimize the risk of developing resistance produced by insects and fungal diseases. The novel insect-resistant gene combination, containing scorpion toxin and chitinase gene was introduced into *B. napus*. The bioassay of artificial inoculation with diamondback moth (*P. maculipenis*) larvae indicated that transformants performed high resistance against the tested pest infestation (Wang et al. 2005). Similarly, a binary vector carrying *sporamin* and *chitinase PjChi-1* genes in tandem was introduced into *B. napus* for dual resistance against disease (*S. sclerotiorum*) and insect (*P. xylostella*) attack in our transgene study. Transgenic plants exhibited high levels of resistance to *S. sclerotiorum* and *P. xylostella* compared to untransformed wild-type plants. (Liu et al. 2011a). To better understand the inheritance of resistant in progeny, artificial inoculation with *S. sclerotiorum* and *P. xylostella* in in vitro bioassays showed not only significant resistance compared to the wild type but also the resistant levels closely to the parental line (Liu et al. 2011b). In addition, we also inoculated another phytopathogenic fungus (*Botrytis cinerea*) to evaluate the resistance in progeny. For all T_1 population individuals showed a good performance to inhibit the fungal growth, especially in T_1-11#1, T_1-11#3 and T_1-11#9 of *B. napus* (Fig. 12.1). Stable and consistent expression of foreign genes in serial propagation is a crucial factor to induce effective and efficient resistance through genetic engineering in insects and fungi. It would be better if a transgenic plant sticks to the characters such as stable expression and inheritance, and low inter- or intra-transformant variability of foreign genes in serial propagation. That is a crucial factor to induce effective and efficient resistance through genetic engineering in insects and fungi.

12.4 Future Prospects

Diseases and insect pests have been a big challenge in the production of *Brassica* crops over the years. Over the past few decades, there were many attempts to improve the resistance by genetic modification. This review surmised the major

WT   T$_1$-11 # 1

T$_1$-11 # 3   T$_1$-11 # 9

Fig. 12.1 Results of positive *Brassica napus* T1-11 progeny resistance against *Botrytis cinerea*. WT: wild type, and T1-11#1, 3, 9: positive T1 progeny plants. Fully expanded leaves were wounded on the midrib with a sterilized needle and inoculated with a 3-mm mycelial segment taken from the edge of a 3-day-old colony. The cultures were then placed in Petri dish containing 3 ml of distilled water to keep the leaves fresh (at 25 ± 1 °C, keeping moisture in Petri dish). The diameter of plaque was measured after 2 days of inoculation under photoperiod of 16/8 h (light/dark). Scale bar = 1 cm

methods and developing strategy for pest resistance. In future perspective, exploitation and discussion of novel approaches and emerging possibilities will be required in the breeding program of pest resistance.

Recently, secondary metabolites of plants especially volatile substances had been explored to develop resistance to aphids (Beale et al. 2006; Schnee et al. 2006). Thus, it may be an effective method of controlling other insect infestations for *Brassica* crops by altering the volatile substance of plants to disrupt the host recognized cue of pest. In addition, signal elicitors induced by biotic stresses play an important role in plant defense against diseases and insect pests. Understanding the interactions and associated signaling pathways of plants with pests can also provide new approaches for exploiting *Brassica* defense which can be achieved through genetic modification.

Acknowledgements This work was supported by National High Technology Research and Development Program of China (2011AA10A206), the Science and Technology Department of Zhejiang Province (2012C12902-1, 2011R50026-5), Scientific Research Foundation of Zhejiang A&F University (2013FR022), China Postdoctoral Science Foundation (20110491819, 2012T50555), National Natural Science Foundation of China (31000678, 31071698, 31170405), and National Key Science and Technology Supporting Program of China (2010BAD01B01, 2010BAD01B04). Weijun Zhou (the corresponding author) is grateful to the 985-Institute of Agrobiology and Environmental Sciences of Zhejiang University for providing convenience in using the experimental equipments.

References

Amjad MD, Peters C (1992) Survival, development and reproduction of turnip aphids (Homoptera: Aphididae) on oilseeds *Brassica*. J Econ Entomol 85:2003–2007
Barfield DG, Pua EC (1991) Gene transfer in plants of *Brassica juncea* using *Agrobacterium tumefaciens* mediated transformation. Plant Cell Rep 10:308–314
Bartnick S (1968) Cell wall chemistry morphogenesis and taxonomy of fungi. Annu Rev Microbiol 22:87–108
Beale MH, Birkett MA, Bruce TJA, Chamberlain K, Field LM, Huttly AK, Martin JL, Parker R, Phillips AL, Pickett JA, Prosser IM, Shewry PR, Smart LE, Wadhams LJ, Woodcock CM, Zhang Y (2006) Aphid alarm pheromone produced by transgenic plants affects aphid and parasitoid behaviour. PNAS 103:10509–10513
Bhadoria NS, Jakhmola SS, Dhamdhere SV (1995) Relative susceptibility of mustard cultivars to *Lipaphis erysimi* in North West Madhya Pradesh (India). J Entomol Res 19:143–146
Bhatia V, Uniyal PL, Bhattacharya R (2011) Aphid resistance in *Brassica* crops: challenges, biotechnological progress and emerging possibilities. Biotechnol Adv 29:879–888
Brar KS, Sandhu GS (1978) Comparative resistance of different *Brassica* species/varieties to the mustard aphid (*Lipaphis erysimi* Kalt.) under natural and artificial conditions. Indian J Agric Res 12:198–200
Broglie K, Chet I, Holliday M, Cressman R, Biddle P, Knowlton S, Mauvais CJ, Broglie R (1991) Transgenic plants with enhanced resistance to the fungal pathogen *Rhizoctonia solani*. Science 254:1194–1197
Cao J, Tang JD, Strizhov N, Shelton AM, Earle ED (1999) Transgenic broccoli with high levels of *Bacillus thuringiensis* Cry1C protein, control diamondback moth larvae resistant to Cry1A or Cry1C. Mol Breed 5:131–141
Cao J, Zhao JZ, Tang JD, Shelton AM (2002) Broccoli plants with pyramided *cry1Ac* and *cry1C* Bt genes control diamondback moths resistant to Cry1A and Cry1C proteins. Theor Appl Genet 105:258–264
Cao J, Shelton AM, Earle ED (2008) Sequential transformation to pyramid two Bt genes in vegetable Indian mustard (*Brassica juncea* L.) and its potential for control of diamondback moth larvae. Plant Cell Rep 27:479–487
Chen HJ, Wang SJ, Chen CC, Yeh KW (2006) New gene construction strategy in T-DNA vector to enhance expression level of sweet potato sporamin and insect resistance in transgenic *Brassica oleracea*. Plant Sci 171:367–374
Chen CC, Kumar HGA, Kumar S, Tzean SS, Yeh KW (2007) Molecular cloning, characterization, and expression of a *chitinase* from the entomopathogenic fungus *Paecilomyces javanicus*. Curr Microbiol 55:8–13
Chevre AM, Eber F, Brun H, Plessis J, Primard C, Renard M (1991) Cytogemetic studies of *Brassica napus-Sinapis alba* hybrids from ovary culture and protoplast fusion. Attempts to introduce *Alternaria* resistance into rapeseed. In: Proceedings of the 8th International Rapeseed Conference, Saskatoon, Canada, 346–351
Chevre AM, Eber F, This P, Barret P, Tanguy X, Brun H, Delseny M, Renard M (1996) Characterization of *Brassica nigra* chromosomes and of blackleg resistance in *B. napus-B. nigra* addition lines. Plant Breed 115:113–118

Chevre AM, Barret P, Eber F, Dupuy P, Brun H, Tanguy X, Renard M (1997) Selection of stable *Brassica napus-B. juncea* recombinant lines resistant to blackleg (*Leptosphaeria maculans*). 1. Identification of molecular markers, chromosomal and genomic origin of the introgression. Theor Appl Genet 95:1104–1111

Chevre AM, Brun H, Eber F, Letanneur JC, Vallee P, Ermel M, Glais I, Li H, Sivasithamparam K, Barbetti MJ (2008) Stabilization of resistance to *Leptosphaeria maculans* in *Brassica napus* – *B. juncea* recombinant lines and its introgression into spring-type *Brassica napus*. Plant Dis 92:1208–1214

Cho HS, Cao J, Ren JP, Earle ED (2001) Control of Lepidopteran insect pests in transgenic Chinese cabbage (*Brassica rapa* ssp. *pekinensis*) transformed with a synthetic *Bacillus thuringiensis* cry1C gene. Plant Cell Rep 20:1–7

Cole RA (1994a) Locating a resistance mechanism to the cabbage aphid in two wild *Brassicas*. Entomol Exp Appl 71:23–31

Cole RA (1994b) Isolation of a chitin binding lectin, with insecticidal activity in chemically defined synthetic diets, from two wild *brassica* species with resistance to cabbage aphid, *Brevicoryne brassicae*. Entomol Exp Appl 72:181–187

Corrado G, Arciello S, Fanti P, Fiandra L, Garonna A, Digilio MC, Lorito M, Giordana B, Pennacchio F, Rao R (2008) The Chitinase A from the *baculovirus AcMNPV* enhances resistance to both fungi and herbivorous pests in tobacco. Transgenic Res 17:557–571

Dickson MH, Petzoldt R (1996) Breeding for resistance to *Sclerotinia sclerotiorum* in *Brassica oleracea*. Acta Hortic 407:103–108

Ding LC, Hu C, Yeh KW, Wang PJ (1998) Development of insect resistant transgenic cauliflower plants expressing the trypsin inhibitor gene isolated from local sweet potato. Plant Cell Rep 17:854–860

Dixelius C (1999) Inheritance of the resistance to *Leptosphaeria maculans* of *Brassica nigra* and *B. juncea* in near-isogenic lines of *B. napus*. Plant Breed 118:151–156

Dixelius C, Wahlberg S (1999) Resistance to *Leptosphaeria maculans* is conserved in a specific region of the *Brassica* B genome. Theor Appl Genet 99:368–372

Dong X, Ji R, Guo X, Foster SJ, Chen H, Dong C, Liu Y, Hu Q, Liu S (2008) Expressing a gene encoding wheat oxalate oxidase enhances resistance to *Sclerotinia sclerotiorum* in oilseed rape. Planta 228:331–340

Downey RK, Rimmer SR (1993) Agronomic improvement in oilseed *Brassica*. Adv Agron 50:1–50

Dusabenyagasani M, Fernando WGD (2008) Development of a SCAR marker to track canola resistance against blackleg caused by *Leptosphaeria maculans* pathogenicity group 3. Plant Dis 92:903–908

Ellis PR, Farrell JA (1995) Resistance to cabbage aphid (*Brevicoryne brassicae*) in six *Brassica* accessions in New Zealand. N Z J Crop Hortic Sci 23:25–29

Ellis PR, Kiff NB, Pink DAC, Jukes PL, Lynn J, Tatchell GM (2000) Variation in resistance to the cabbage aphid (*Brevicoryne brassicae*) between and within wild and cultivated *brassica* species. Genet Resour Crop Evol 47:395–401

Fathi SAA, Bozorg-Amirkalaee M, Sarfaraz RM (2011) Preference and performance of *Plutella xylostella* (L.) (Lepidoptera: Plutellidae) on canola cultivars. J Pest Sci 84:41–47

Foissac X, Nguyen TL, Christou P, Gatehouse AMR, Gatehouse JA (2000) Resistance to green leaf hopper (*Nephotettix virescens*) and brown plant hopper (*Nilaparvata lugens*) in transgenic rice expressing snowdrop lectin (*Galanthus nivalis* agglutinin; GNA). J Insect Physiol 46:573–583

Gerdemann-Knorck M, Nielen S, Tzscheetzsch C, Iglisch J, Schieder O (1995) Transfer of disease resistance within the genus *Brassica* through asymmetric somatic hybridization. Euphytica 85:247–253

Grison R, Grezes-Besset B, Schneider M, Lucante N, Olsen L, Leguay JJ, Toppan A (1996) Field tolerance to fungal pathogens of *Brassica napus* constitutively expressing a chimeric *chitinase* gene. Nat Biotechnol 14:643–646

Hagimori M, Nagaoka M, Kato N, Yoshikawa H (1992) Production and characterization of somatic hybrids between the Japanese radish and cauliflower. Theor Appl Genet 83:655–662

Hasan F, Ansari MS (2011) Effects of different brassicaceous host plants on the fitness of *Pieris brassicae*. Crop Prot 30:854–862

Hu Q, Li YC, Mei DS (2009) Introgression of genes from wild crucifers. In: Gupta SK (ed) Biology and breeding of crucifers. CRC Press, Boca Raton, pp 261–283

Huang X, Renwick JAA (1994) Relative activities of glucosinolates as oviposition stimulants for *Pieris rapae* and *P. napi oleracea*. J Chem Ecol 20:1025–1037

Kanrar S, Venkateswari J, Kirti PB, Chopra VL (2002) Transgenic Indian mustard (*Brassica juncea*) with resistance to the mustard aphid (*Lipaphis erysimi* Kalt.). Plant Cell Rep 20:976–981

Khan MR, Rashid H, Ansar M, Chaudry Z (2003) High frequency shoot regeneration and *Agrobacterium*-mediated DNA transfer in canola (*Brassica napus*). Plant Cell Tiss Org Cult 75:223–231

Kramer KJ, Muthukrishnan S (1997) Insect chitinases: molecular biology and potential use as biopesticides. Insect Biochem Mol Biol 27:887–900

Kumar S, Atri C, Sangha MK, Banga SS (2011) Screening of wild crucifers for resistance to mustard aphid, *Lipaphis erysimi* (Kaltenbach) and attempt at introgression of resistance gene(s) from *Brassica fruticulosa* to *Brassica juncea*. Euphytica 179:461–470

Lawrence PK, Koundal KR (2002) Plant protease inhibitors in control of phytophagous insects. Electron J Biotechnol 5:1–17

Liu JH, Dixelius C, Eriksson I, Glimelius K (1995) *Brassica napus* (+) *B. tournefortii*, a somatic hybrid containing traits of agronomic importance for rapeseed breeding. Plant Sci 109:75–86

Liu HB, Guo X, Naeem MS, Liu D, Xu L, Zhang WF, Tang GX, Zhou WJ (2011a) Transgenic *Brassica napus* L. lines carrying a two gene construct demonstrate enhanced resistance against *Plutella xylostella* and *Sclerotinia sclerotiorum*. Plant Cell Tiss Org Cult 106:143–151

Liu HB, Naeem MS, Liu D, Zhu YN, Guo X, Cui P, Zhou WJ (2011b) Analyses of inheritance patterns and consistent expression of *sporamin* and *chitinase PjChi-1* genes in *Brassica napus*. Plant Breed 130:345–351

Mauch F, Hadwiger LA, Boller T (1988a) Antifungal hydrolases in pea tissue: purification and characterization of 2 chitinases and 2 β-1, 3-glucanases differentially regulated during development and in response to fungal infection. Plant Physiol 87:325–333

Mauch F, Mauch-Mani B, Boller T (1988b) Antifungal hydrolases in pea tissue: inhibition of fungal growth by combinations of chitinase and β-1, 3-glucanases. Plant Physiol 88:936–942

Melander M, Kamnert I, Happstadius I, Liljeroth E, Bryngelsson T (2006) Stability of transgene integration and expression in subsequent generations of doubled haploid oilseed rape transformed with *chitinase* and β-*1,3-glucanase* genes in a double gene construct. Plant Cell Rep 25:942–952

Miles CI, del Campo ML, Renwick JAA (2005) Behavioral and chemosensory responses to a host recognition cue by larvae of *Pieris rapae*. J Comp Physiol A 191:147–155

Mohan M, Gujar GT (2003) Local variation in susceptibility of the diamondback moth, *Plutella xylostella* (Linnaeus) to insecticides and role of detoxification enzymes. Crop Prot 22:495–504

Moloney MM, Walker JM, Sharma KK (1989) High efficiency transformation of *Brassica napus* using *Agrobacterium* vectors. Plant Cell Rep 8:238–242

Momoh EJJ, Zhou WJ, Kristiansson B (2002) Variation in the development of secondary dormancy in oilseed rape genotypes under conditions of stress. Weed Res 42:446–455

Mondal KK, Bhattacharya RC, Koundal KR, Chatterjee SC (2007) Transgenic Indian mustard (*Brassica juncea*) expressing tomato glucanase leads to arrested growth of *Alternaria brassicae*. Plant Cell Rep 26:247–252

Mora AA, Earle ED (2001) Resistance to *Alternaria brassicicola* in transgenic broccoli expressing a *Thrichoderma harzianum* endochitinase gene. Mol Breed 8:1–9

Mosolov VV, Valueva TA (2008) Proteinase inhibitors in plant biotechnology: a review. Appl Biochem Microbiol 44:233–240

Panda N, Khush GS (1995) Host plant resistance to insects. CAB Int, Wallingford

Pidskalny RS, Rimmer SR (1985) Virulence of *Albugo candida* from turnip rape (*Brassica campestris*) and mustard (*Brassica juncea*) on various crucifers. Can J Plant Pathol 7:283–286

Plieske J, Struss D, Robbelen G (1998) Inheritance of resistance derived from the B-genome of *Brassica* against *Phoma lingam* in rapeseed and the development of molecular markers. Theor Appl Genet 97:929–936

Powell KS (2001) Antimetabolic effects of plant lectins towards nymphal stages of the plant hoppers *Tarophagous proserpina* and *Nilaparvata lugens*. Entomol Exp Appl 99:71–77

Purwantara A, Salisbury PA, Burton WA, Howlett BJ (1998) Reaction of *Brassica juncea* (Indian mustard) lines to Australian isolates of *Leptosphaeria maculans* under glasshouse and field conditions. Eur J Plant Pathol 104:895–902

Rahbe Y, Sauvion N, Febvay G, Peumans WJ, Gatehouse AMR (1995) Toxicity of lectins and processing of ingested proteins in the pea aphid *Acyrthosiphon pisum*. Entomol Exp Appl 76:143–155

Rahbe Y, Deraison C, Bonade-Bottino M, Girara C, Nardon C, Jouanin L (2003) Effects of the cysteine protease inhibitor oryzacystatin (OC-I) on different aphids and reduced performance of *Myzus persicae* on OC-I expressing transgenic oilseed rape. Plant Sci 164:441–450

Rai B, Gupta SK, Pratap A (2007) Breeding methods. In: Gupta SK (ed) Advances in botanical research-rapeseed breeding, vol 45. Academic/Elsevier, San Diego, pp 21–48

Raybold AF, Moyes CL (2001) The ecological genetics of aliphatic glucosinolates. Heredity 87:383–391

Renwick JAA (2002) The chemical world of crucivores: iures, treats and traps. Entomol Exp Appl 104:35–42

Roush RT, Shelton AM (1997) Assessing the odds: the emergence of resistance to bt transgenic plants. Nat Biotechnol 15:816–817

Ryschka U, Schumann G, Klocke E, Scholze P, Neumann M (1996) Somatic hybridization in *brassiceae*. Acta Hortic 407:201–208

Saal B, Brun H, Glais I, Struss D (2004) Identification of a *Brassica juncea*-derived recessive gene conferring resistance to *Leptosphaeria maculans* in oilseed rape. Plant Breed 123:505–511

Saal B, Struss D (2005) RGA- and RAPD-derived SCAR for a *Brassica* B-Genome introgression conferring resistance to blackleg in oilseed rape. Theor Appl Genet 101:281–290

Sadasivam S, Thayumanavan B (2003) Molecular host plant resistance to pests. Marcel Dekker, New York, pp 61–83

Sakhno LA, Gocheva EA, Komarnitskii KI, Kuchuk NV (2008) Stable expression of the promoterless *bar* gene in transformed rapeseed plants. Cytol Genet 42:16–22

Saxena AK, Bhadoria SS, Gadewadikar PN, Barteria AM, Tomar SS, Dixit SC (1995) Yield losses in some improved varieties of mustard by aphid, *Lipaphis erysimi* Kalt. Agric Sci Dig 15:235–237

Schnee C, Kollner TG, Held M, Turlings TCJ, Gershenzon J, Degenhardt J (2006) The products of a single maize sesquiterpene synthase form a volatile defense signal that attracts natural enemies of maize herbivores. PNAS 103:1129–1134

Singh CP, Sachan GC (1994) Assessment of yield losses in yellow sarson due to mustard aphid, *Lipaphis erysimi* (Kalt). J Oilseeds Res 11:179–184

Singh R, Sharma SK (2007) Evaluation, maintenance and conservation of germplasm. In: Gupta SK (ed) Advances in botanical research-rapeseed breeding, vol 45. Academic, London, pp 465–481

Sjodin C, Glimelius K (1989) Transfer of resistance against *Phoma lingam* to *Brassica napus* by asymmetric somatic hybridization combined with toxin selection. Theor Appl Genet 78:513–520

Snowdon RJ, Winter H, Diestel A, Sacristan MD (2000) Development and characterization of *Brassica napus-Sinapis arvensis* addition lines exhibiting resistance to *Leptosphaeria maculans*. Theor Appl Genet 101:1008–1014

Tohidfar M, Mohammadi M, Ghareyazie B (2005) *Agrobacterium*-mediated transformation of cotton (*Gossypium hirsutum*) using a heterologous bean *chitinase* gene. Plant Cell Tiss Org Cult 83:83–96

Wang J, Chen Z, Du J, Sun Y, Liang A (2005) Novel insect resistance in *Brassica napus* developed by transformation of *chitinase* and *scorpion toxin* genes. Plant Cell Rep 24:549–555

Williams PH, Fitt BDL (1999) Differentiating A and B groups of *Leptosphaeria maculans*, causal agent of stem canker (blackleg) of oilseed rape. Plant Pathol 48:161–175

Xiang Y, Wong WKR, Ma MC, Wong RSC (2000) *Agrobacterium*-mediated transformation of *Brassica campestris* ssp. *Parachinensis* with synthetic *Bacillus thuringiensis cry1Ab* and *cry1Ac* genes. Plant Cell Rep 19:252–256

Yi D, Cui L, Liu Y, Zhang M, Zhang Y, Fang Z, Yang L (2011) Transformation of cabbage (*Brassica oleracea* L. var. *capitata*) with Bt *cry1Ba3* gene for control of diamondback moth. Agric Sci China 10:1693–1700

Zhang J, Liu F, Yao L, Luo C, Zhao Q, Huang Y (2011) Vacuum infiltration transformation of non-heading Chinese cabbage (*Brassica rapa* L. ssp. *chinensis*) with the *pinII* gene and bioassay for diamondback moth resistance. Plant Biotechnol Rep 5:217–224

Zhou WJ (2001) Oilseed rape. In: Zhang GP, Zhou WJ (eds) Crop production. Zhejiang University Press, Hangzhou, pp 153–178

Zhu JS, Struss D, Robbelen G (1993) Studies on resistance to *Phoma lingam* in *Brassies napus*-*Brassica nigra* addition lines. Plant Breed 111:192–197

Index

S.K. Gupta (ed.), *Biotechnology of Crucifers*, DOI 10.1007/978-1-4614-7795-2,
© Springer Science+Business Media, LLC 2013

Printed by Publishers' Graphics LLC
DBT130810.15.15.30